TECHNICS AND CIVILIZATION

LEWIS MUMFORD (1895–1990) was a prolific and influential writer whose scope encompassed literary criticism, architecture, history, urban sociology, and philosophy. He served as associate editor of *Dial*, editor of the *Sociological Review* (London), and coeditor of the *American Caravan*. A member of the American Philosophical Society and a fellow of the American Academy of Arts and Sciences, he was also the architectural critic for the *New Yorker* for over thirty years. He went on to publish over thirty books—including *The Culture of Cities*, *The City in History*, *The Conduct of Life*, *The Condition of Man*, and *The Myth of the Machine*—and hundreds of periodical articles. He was eventually honored with the United States Medal of Freedom and Knight of the Order of the British Empire.

TECHNICS AND CIVILIZATION

BY LEWIS MUMFORD

With a New Foreword by Langdon Winner

The University of Chicago Press
Chicago & London

The University of Chicago Press, Chicago 60637
University of Chicago Press edition 2010

Printed in the United States of America

19   4 5

ISBN-13: 978-0-226-55027-5 (paper)
ISBN-10: 0-226-55027-3 (paper)

Library of Congress Cataloging-in-Publication Data

Mumford, Lewis, 1895–1990.
  Technics and civilization / by Lewis Mumford ; with a new foreword by Langdon Winner.
    p. cm.
  Includes bibliographical references and index.
  ISBN-13: 978-0-226-55027-5 (pbk. : alk. paper)
  ISBN-10: 0-226-55027-3 (pbk. : alk. paper) 1. Technology and civilization. 2. Industrial
arts—History. I. Title.
  CB478.M8 2010
  303.48'3—dc22
                                    2010018373

♾ The paper used in this publication meets the minimum requirements of the American
National Standard for Information Sciences—Permanence of Paper for Printed Library
Materials, ANSI Z39.48-1992.

CONTENTS

FOREWORD

Anyone who studies the human dimensions of technological change must eventually come to terms with Lewis Mumford. His pathbreaking, visionary writings on the topic offer a vast store of intellectual resources to help us ponder the basic commitments and ethical dilemmas at the heart of modern material culture. *Technics and Civilization*, the first of his books on this theme, openly challenged scholarly conventions of the early twentieth century and set the stage for decades of lively debate about the prospects for our technology-centered ways of living.

When Mumford turned his attention to the study of technology in the early 1930s, he was already a young lion on the New York literary scene, having published books on world utopian thought, nineteenth-century transcendental philosophy, and the evolution of American architecture, as well as a seminal biography of Herman Melville. Seeking to understand the formative influence of tools, instruments, and production processes in world history, he devoured the standard works on the rise of industrial society and found them remarkably thin in substance and shallow in intellectual reach. Historians and economists of the time, especially those in the United States, had settled upon a cramped view of the story, emphasizing the development of "the machine" from the eighteenth century onward and its role in shaping modern society. Mumford noticed that despite the obvious importance of technology in human affairs, there was still no systematic overview of the history of technology available in English, no book that explored human involvements with "technics" in their full, rich complexity. He asked, in effect, isn't there much more to the tale than just the powerful new engines and their contributions to productivity and economic growth? Hadn't world societies enjoyed a long involvement with technologies before coming of the Industrial Revolution?

With restless curiosity, Mumford greatly extended the period of time needed to explain key developments that form the basis of the modern scientific-technological realm. He also broadened the range of creative activities to include in his chronicle of events. Hence, the book begins with inventions from the tenth century and moves steadily forward, noting technical advances in a wide variety of traditions of arts, crafts, science, engineering, philosophy, finance, commerce, and associated domains of social practice throughout the millennium. Mumford argues that the widely heralded breakthroughs of the Industrial Revolution would not have been possible without a lengthy cultural preparation that generated the resilient foundations upon which later achievements stand. In that light, the book emphasizes a wide range of developments in areas of human inquiry that at first glance seem to have little relevance to technical devices and systems at all.

Perhaps the most memorable of the moments Mumford describes is the story of the monastery and the clock. In Benedictine monasteries of medieval Europe, spiritual and working life was divided into precise units of time, the canonical hours, as a way to magnify the strength of the monks' religious devotion. This regimen gave rise to a need for devices that could measure time: hence the development of the first simple, reliable clocks. The monasteries, in Mumford's view, "helped give human enterprise the regular collective beat and rhythm of the machine; for the clock is not merely a means of keeping track of hours, but of synchronizing the actions of men."

As he explores episodes of this kind, Mumford does not dwell on lengthy explanations about how significant breakthroughs took shape. Drawing upon the best available (mainly European) scholarship, he briefly depicts signature moments of historical change and then speculates freely, even playfully, about their meaning over the years. He describes, for example, how improvements in glassmaking from the fourteenth through the seventeenth century were crucial to key developments in modern intellectual life and economic activity we now take for granted: "Glass helped put the world in a frame: it made it possible to see certain elements of reality more clearly: and it focused attention on a sharply defined field—namely what was bounded by the frame." If there had been no improvements in glasswork, he suggests, there would have been no modern "ego" to engage in discovery and invention.

Mumford's underlying lesson is perfectly clear. Contrary to widely accepted belief, the marvels of modern production did not begin with the clanking, wheezing steam engines of the industrial age and certainly do not find their origins in the material inventions or social innovations of James Watt, Richard Arkwright, and other iconic figures of the Industrial Revolution. Among the most important elements, Mumford contends, were not only the physical tools and machines, but projects that involved a range of human motives including religious devotion and the quest for aesthetic satisfaction in science, engineering, and countless everyday pursuits. His advice is perfectly clear: we would do well to draw upon such resources in our own time as well.

Another of the book's central concerns is to explore the origins of institutions that would eventually become defining features of modern society—for example, the continuing influence of the military on a wide range of technological choices. In fact, some of the book's most dramatic prose is reserved for Mumford's description of the mine as a formative model for all industrial ideas and programs. In his view, the intensification of mining in fifteenth- and sixteenth-century Germany established the basic attitudes toward nature, work, machinery, the character of human experience, and even the agendas of scientific theory for hundreds of years to come. His imagery suggests that the mines eventually opened up, belching their dark, dangerous obsessions into society as a whole. "Now, the characteristic methods of mining do not stop at the pithead: they go on, more or less, in all the accessory occupations."

As Mumford leads us through a sequence of fascinating, paradigmatic chapters, he is positively eager to serve up heaping portions of a "grand narrative" in a style that many of today's historians and philosophers regard as methodologically suspect. After all, how can anyone presume to tell the story of modern technological culture in all its messy complexity? Mumford never bothers with questions of that sort. He simply strides into the breach, addressing his audience of specialists and general readers, promising to give a reliable, philosophical report.

Central to his understanding is a systematic argument about three historical "phases" that enable us to imagine the layers of knowledge, belief, and skill embedded within the technologies and institutions that surround us. The phases, in his view, are not wholly distinct, but instead overlap and interpenetrate with the passage of time. The earliest of these, the "eotechnic phase," comprises the diverse collection of inventions and ideas introduced from AD 1000 well into the eighteenth century. Next in sequence is the "paleotechnic phase," one characterized by the materials and power sources of the industrial era, and a period that Mumford regards as reckless to the point of barbarism. "The state of paleotechnic society may be described, ideally, as one of wardom. Its typical organs, from the mine to the battle field, were at the service of death." Unfolding from both prior stages, however, is a more mature, hopeful period of development, the "neotechnic phase," one strongly evident in the early decades of the twentieth century, that brings the promise of new alloys, electricity, and improved means of communication, along with a much needed a willingness to emphasize "the organic" in new social and technological projects.

The importance of *Technics and Civilization* is not only its pioneering method and wealth of historical illustrations, but also its articulation of a novel theory. Mumford contends that technological projects express a dynamic relationship between the inner and outer worlds of human existence. Our most impressive successes in practical, material activities are often projections of deep spiritual needs along with the most rational and irrational of passions. At the same time, living in the world of material things stimulates a creative response within human consciousness: the development of language, symbols, rituals, and fruitful insights. As he later summarized his view, "Man internalizes his external world and externalizes his internal world." Seen in that light, Mumford's work is an attempt not merely to write an accurate history of technology in its full sweep, but to explain a fundamental pattern in all of human experience, an explanation far more accurate and full of possibilities than the popular but sadly ham-fisted belief that "man is a tool-making animal."

Of course, Americans have long celebrated technological change, but in terms very different from the ones Mumford recommends. The orthodox view stresses a few key notions: that the conquest and control of nature is modern society's great mission; that efficiency is a universally applicable criterion of social choice; that modern history shows a linear accumulation of scientific progress in human well-being; that our individual and collective mission is a never-ending quest to remain competitive, racing forward along technology's cutting edge. In today's parlance, the glossy term to describe commitments and projects of this kind is "innovation." Within corporations and tech parks, innovation is enthusiastically proclaimed as our great mission.

Technics and Civilization celebrates the development of technology not only for the way it alleviates the physical burdens of life, and not only for the way it boosts production, but also because it can be a wonderful manifestation of our spirituality, sensuality, and deepest connections to nature and to each other. In that sense, Mumford holds out great hopes for technology, hopes as expansive and generous as those of any modern thinker. When he surveyed the situation in the early 1930s, however, the Machine Age had entered a severe crisis: the Great Depression. No longer was it clear that the "progress" celebrated in standard views of technology was still the inevitable direction of events. Mechanisms of production were certainly not meeting the basic human needs of much of the world's population. The promise of universal abundance was contradicted by widespread unemployment, poverty, social upheaval, and political conflict around the globe.

While Mumford took note of economic, social, and environmental collapse during this dreary period, he did not dwell on it. Instead, he sought to identify deeper patterns that would remain problematic long after the consumer economy had regained its steam and moved on to new heights of productivity: regimentation of work; militarization of social organization; waste of valuable resources; violence inflicted upon natural systems; widespread pollution of the air, land, and water; and promotion of brain-dead consumerism as life's ultimate satisfaction. The real problem, in his view, extended beyond economic depression into the very nature of modern technological civilization and our ways of imagining its humane possibilities.

While Mumford's writing in this volume remains of interest to historians, philosophers, and technical professionals, many of its key questions appeal to a much broader audience. His case studies and reflections anticipate the key concerns of science fiction novels and cinema: the excesses of technological power; the creation of artificial lifeworlds; the violence that can erupt from what seem to be exquisitely rational political structures; the enduring tension between parts and wholes, between biological entities and their synthetic replacements. To cite just one example, *Avatar* (2009), the wildly popular sci-fi thriller, echoes Mumford's concerns about the clash between a high-tech, resource-hungry civilization and a more peaceful, holistic, deeply rooted, organic culture. In director James Cameron's vision of the story, the lovely, peaceful, reasonable creatures on planet Pandora actually emerge victorious at the film's end. By the same token, Mumford's hopes for sweeping reform within the modern technological realm foreshadow both our efforts in our time to imagine "sustainable" technologies and our hopes for a "green economy." Worries about environmental degradation, the decline of petroleum reserves, and the onset of global warming cause us to confront the issues raised in "Orientation," the book's last chapter. "This reawakening of the vital and organic in every department of life undermines the authority of the purely mechanical. Life, which has always paid the fiddler, now begins to call the tune."

As he surveyed the plight of modern society during the early 1930s, Mumford was surprisingly optimistic. He believed it possible, even likely, that the excesses and injustices of the paleotechnic period would finally be overcome through a process of maturation, replaced by a thoroughly neotechnic era in which humane, well-planned, ecologically sound forms of social organization would prevail. Thoughtful, caring souls would respond to ills of a crudely mechanized society, modifying its rough edges, humanizing it from within. In his later writings, such confidence was replaced by an increasingly bitter skepticism as Mumford faced the facts of the cold war, the nuclear arms race, a polluted biosphere, deranged patterns of urban and suburban living, mass-media mystification of public opinion, and the institutionalized dominance of what he named "the Power Complex." But for that very reason, perhaps, it is best for us to return to the moment in his thinking where the shadows have not yet overcome the rays of light, and where the best possibilities are still in play. Putting aside all the reasons why the task may prove impossible, we can join the young Lewis Mumford in asking: what kind of world can we hope to build now?

LANGDON WINNER

INTRODUCTION TO THE 1963 EDITION

Technics and Civilization was first published in 1934. At that time, though scholars often characterized the present period as the "Machine Age" they still looked for its beginnings in the eighteenth century; for A. J. Toynbee, a relative of the present historian, had in the eighties applied the term "The Industrial Revolution" to the technical innovations that had taken place. And while anthropologists and archaeologists paid due attention to the technical equipment of primitive peoples, sometimes exaggerating the formative effect of tools, the broader influence of technics upon human culture was hardly touched on: the useful and the practical still stood outside the realm of the good, the true, and the beautiful.

Technics and Civilization broke with this traditional neglect of technology: it not merely summarized for the first time the technical history of the last thousand years of Western civilization, but revealed the constant interplay between the social milieu—monasticism, capitalism, science, play, luxury, and war—and the more specific achievements of the inventor, the industrialist, and the engineer. While Karl Marx had erroneously assumed that technical forces (the system of production) evolved automatically and determined the character of all other institutions, this new analysis demonstrated that the relationship was reciprocal and many-sided: a child's toy might lead to a new invention, such as the motion picture, or the ancient dream of instant communication at a distance might prompt Morse to invent the electric telegraph.

The theme for this book was first sounded in an essay called "The Drama of the Machines," published in 'Scribner's' magazine, August 1930. In that essay I said:

"If we wish to have any clear notion about the machine, we must think about its psychological as well as its practical origins; and similarly, we must appraise its esthetic and ethical results. For a century we have isolated the technical triumphs of the machine; and we have bowed before the handiwork of the inventor and the scientist; we have alternatively exalted these new instruments for their practical success and despised them for the narrowness of their achievements.

'When one examines the subject freshly, however, many of these estimates are upset. We find that there are human values in machinery we did not suspect; we also find that there are wastes, losses, perversions of energy which the ordinary economist blandly concealed. The vast material displacements the machine has made in our physical environment are perhaps in the long run less important than its spiritual contributions to our culture."

The intuitions that stirred this fresh examination had their roots in my personal experience. At the age of twelve, I built my first radio set, and soon I was writing short articles for popular technical magazines describing improvements in my apparatus. This interest led me to enroll in the Stuyvesant High School, where I received the rudiments of a sound technical and scientific education, and in particular achieved familiarity with the basic tools and mechanical processes, in cabinet making, smithing, wood and metal turning, and foundry work. A few years later, I served as laboratory helper in the cement testing laboratory of the U.S. Bureau of Standards, then at Pittsburgh, and was immersed in that classic paleotechnic environment.

My 'Drama of the Machines' drew forth an invitation from Professor R. M. MacIver to give an Extension Course on 'The Machine Age' at Columbia University: so far as I know the first course of this kind, dealing with the cultural as well as the economic and practical aspects of technology, to be offered anywhere. The preparatory work on this course furnished not merely the necessary materials, but the incentive for writing this book; and in 1932 I rounded out my earlier studies by making an exhaustive study tour of the European technical museums and libraries, particularly those in Vienna, Munich, Paris, and London. As a result, both the bibliography of *Technics and Civilization* and the list of inventions since the tenth century were more adequate than anything then available, and still are useful today.

The underlying philosophy and method of *Technics and Civilization* deliberately challenged many current conventions of scholarship, in particular the stereotyped procedures that prevented the investigator from doing justice to more than a small isolated segment of his subject, and from evaluating the social and cultural by-products of technical developments. In presenting technical development within the setting of a more general social ecology, I avoided the current bias of making it the dominant and all-important factor, as people still do today when they naively characterize our period as the Jet Age, the Nuclear Age, the Rocket Age, or the Space Age. The fact that this challenge to old ways of thinking is not yet widely accepted is perhaps the best reason for publishing this new edition in its unaltered original form.

As for my not dealing with the technical developments of the last thirty years, I make no apologies: even specialized professional historians still shrink from that formidable task. For a different reason, I have made no effort to correct the original text to conform to later knowledge and my own deeper insight. Instead, I have made revisions and additions in a series of essays and chapters, some published in the review, 'Technology and Culture,' some in the 'Proceedings of the American Philosophical Society,' and some in my books *Art and Technics* (1952), *In the Name of Sanity* (1954), and *The Transformations of Man* (1956). If fortune favors me, I propose to carry these new interpretations further in another book, *The Myth of the Machine*. In that work, I will examine certain negative aspects of current technics already visible in ancient cultures, and will amplify my chapter on Orientation, to take account of the colossal technical achievements of the last generation, and the equally colossal social dangers they have brought into existence.

Technics and Civilization heralded a change of attitude among scholars both toward the history of technics as an element in human culture and, to a lesser degree, toward evaluating its social and cultural results; and it possibly helped to generate this new interest, or at least to create the audience that made such books possible. Except for Ulrich Wendt's *Die Technik als Kulturmacht* (1906) and Stuart Chase's *Men and Machines* (1929), all the more general works on technics, like Sigfried Giedion's *Mechanization Takes Command* and R. J. Forbes' *Man the Maker* came after it. For the same reason A. Wulf's *A History of Science and Technology in the Sixteenth and Seventeenth Centuries* is absent from my bibliography. At the time I wrote this book no comprehensive history of technics was available. Fortunately that gap has now been filled by the five-volume *History of Technology* published during the fifties (Oxford University Press), and by the more compact one-volume history, based on that larger work, done by T. K. Derry and T. I. Williams (Oxford, 1961).

Since I have left the main text unaltered I have not attempted to bring the bibliography up to date to include the contributions of many new workers in the field, particularly the notable work of such French scholars as Georges Friedmann, Jean Fourastié, Roger Caillois, Pierre Francastel, Bertrand Gille, and Jacques Ellul – work that carries further the tradition of an earlier group of German scholars, including Karl Bücher, Werner Sombart, Max Weber, and even Oswald Spengler. If additional evidence were needed of the increasing interest in the relation of technics to our culture as a whole, one need only mention the appearance in 1959 of the new review 'Technology and Culture,' the organ of the American 'Society for the History of Technology,' and the magnificent Italian review, 'Civiltà delle Macchine.'

A few years ago Professor Gerald Holton, as editor of 'Daedalus,' invited me to do a review of *Technics and Civilization* from the vantage point of a quarter-century after publication. The severe – indeed sardonically oversevere – analysis of my own study I then made, published in 'Daedalus' (No. 3, 1959), absolves me from the necessity for dealing here with its weaknesses and shortcomings, while I must leave to some other hand the task or revaluing its positive qualities. In going through the text once more, to be sure of my wisdom in extending its life and influence still further through a paperback edition, I have been struck, I must immodestly confess, by its intuitive insight and fresh percipience. These often permitted me to draw sound conclusions from insufficient data and to reveal significant interrelations between areas that had hitherto been kept in strict isolation.

Though contemporary reviewers properly characterized *Technics and Civilization* as a hopeful work, I now congratulate myself rather on the fact that, even then, before the savage demoralizations and irrational projections that have attended the harnessing of nuclear energy menaced the world, I drew attention to the regressive possibilities of many of our most hopeful technical advances: I foresaw the ominous linkage, as I put it later, between 'Autmaton' and the 'Id.' The reader who, a generation ago, understood the second half of my book would not have been unprepared for the overwhelming scientific and technical advancements, nor for the perversions and paranoid compulsions, that have since taken place. So, though the technical history of the last thirty years is lacking in this survey, the basic insight necessary for interpreting these events and their consequences actually pervades the whole book. Hence my readiness to sanction this unrevised text: *Nihil Obstat!*

LEWIS MUMFORD
Amenia, New York
Spring 1963

CORRIGENDA

Except for a handful of unfortunate slips, due to inattention rather than ignorance, I have come upon few errors that call for radical revision in the light of knowledge available when the book was written. The worst mistakes are those calling Leonardo's man-powered glider an airplane, giving the selenium cell a function it no longer was used for, misdating Calthrop's invention of the streamlined locomotive (it should be circa 1865), attributing copper mines to Minnesota (iron) instead of Colorado, and making Westinghouse, rather than Western Electric, the seat of Elton Mayo's experiments.

CAPTIONS TO IMAGES
FROM THE 1934 EDITION

Note from the publisher:
The 1934 edition of *Technics and Civilization* included fifteen pages of small black-and-white illustrations of the machines or methods of production. It is not practical to reproduce those images here—nor is it necessary, in an age when readers can find the same, or similar, images on the Internet (an invention Mumford would have loved). But Mumford's captions and their thematic groupings remain an integral part of his book. Below you will find the full text of these captions, as Mumford listed them, together with image-search keywords that should lead you, on the Internet, to the image reproduced in the first edition. Image sources may well change over the years, but we trust that the reader will at least find these tips helpful in searching for similar images.

I. Anticipations of Speed

1: Rapid land locomotion: the sail-wagon (1598) used by Prince Maurice of Orange, one of the first commanders to introduce modern drill. The desire for speed, proclaimed by Roger Bacon in the thirteenth century, had become insistent by the sixteenth century. Hence skates for sport. [Image-search keywords: land-yacht, 1500s, Simon Stevin.]

2: Direct foot-driven bicycle, invented by Baron von Drais in 1817. Note that Gurney's contemporary automobile also reproduced foot motion for propulsion. The original bicycle was made of wood. After various experiments in high wheels, the machine returned to its original lines. [Image-search keywords: bicycle, German baron von Drais, 1817.]

3: Henson and Stringfellow's flying machine, built from a design patented by Henson in 1842. One of the first to follow the example of soaring birds. [Image-search keywords: Henson flying machine, 1842.]

4: Church's steam-driven passenger coach: one of many types of steam automobile driven off the roads in the 1830s by railway monopolies. The development of the automobile awaited rubber tires, heavy-surfaced roads, and liquid fuel. [Image-search keywords: Dr Church's London and Birmingham Steam Coach, 1833.]

II. Perspectives

1: Dawn of naturalism in the twelfth century. [Image-search keywords: Saint-Lazare d'Autun, France, La Tentation d'Eve.]

2: Engraving from Dürer's treatise on perspective. Scientific accuracy in representation: co-ordination of size, distance, and movement. Beginning of the Cartesian logic of science. [Image-search keywords: Dürer, perspective.]

3: Tintoretto's Susanna and the Elders. The complete picture shows a mirror at Susanna's feet: see chapter 2, section 9; also chapter 3, section 6. [Image-search keywords: Tintoretto, Susanna and the Elders.]

4: Eighteenth century automaton, or the clockwork Venus: the penultimate step from naturalism to mechanism. The next move is to remove the organic symbol entirely. [Image not available.]

III. The Dance of Death

1: Sixteenth century watch. Habits of punctuality characterized the successful bourgeoisie: hence the fashion of carrying private time-keepers from the sixteenth century onward. The fantastic shapes of many early watches show how tardily the machine found its form. [Image-search keywords: sixteenth-century watch.]

2: The printing press was a powerful agent for producing uniformity in language and so, by degrees, in thought. Standardization, mass-production, and capitalistic enterprise came in with the printing press; and not without irony, the oldest known representation of the press, shown here, appeared in a Dance of Death printed at Lyons in 1499. [Image-search keywords: Dance of Death, Lyons, 1499.]

3: Fortified camp: 1573. Sixteenth century drill was the prelude to eighteenth century industrialism. Precision and standardization appeared at an early date in the formations, the exercises, and the tactics of the army. This mechanization of men is a first step toward the mechanization of things. [Image-search keywords: Siege of Breda, Callot.]

4: Jacob Fugger II, the first of the new type of financier and investment banker. He reappears in every generation, alias Baron Rothschild, alias J. Pierpont Morgan, alias Sir Basil Zaharoff, etc., etc. Financing wars, monopolizing natural resources, promoting munitions works, creating and wrecking industries as opportunities for profit dictate, he is the very model of the pure capitalist. His dominance symbolizes the perversion of life-economy into money-economy. [Image-search keywords: Jacob (or Jakob) Fugger.]

IV. Mining, Munitions, and War

1: Sixteenth century cannon foundry, showing fortification and cannons in action in the background. The heavy demand on the mining industries that followed the introduction of cannon in the fourteenth century was likewise registered in the necessary expansion of finance. Here lies the beginning of the cycle of mining, mechanization, munitions, and finance: more dangerously evident today than ever. [Image-search keywords: sixteenth-century cannon foundry.]

2: Large-scale application of water-power to grindstones for grinding and polishing armor. These methods worked over from arms production in the sixteenth century to the cheap hardware production of Sweden at the end of the seventeenth century and the knick-knacks industry of Birmingham in the eighteenth century. [Image not available.]

3: Protection against poison gas in the mines: a necessary safety device for rescue work in the perpetually dangerous environment of the mine. Not merely the products but the tactics of the mine have been steadily introduced into modern warfare from Vauban onward, thus repaying the miner's earlier debt to gunpowder. [Image-search keywords: coal gas, gas mask, miners.]

4: Protection against the deliberate use of poison gas in warfare: both the weapon and the defence against it derive from the mine. Recent scholastic discussion of the relative humanity of butchery by gas or knife or bullet emphasizes the extreme moral refinement of our contemporary Yahoos. [Image-search keywords: gas mask, WWI.]

V. Technics of Wood

1: Wood was the main foundation of eotechnic industry; not the least important use of it was in mining. Hollow logs were used in pumps and as pipes to convey water, as well as in the troughs shown here: heavy beams were used for shoring, and planks were used in the earliest form of the railroad. The use of wood for smelting, forging, and casting—as well as in glass-making—caused a great drain on the forest. Dr. Bauer's illustrator faithfully depicts this deforestation. [Image-search keywords: Bauer, Agricola, metallica.]

2: The wheelwright was one of the principal agents of improved transport and improved power production by means of water-mills and windmills. With coopering and ship-building the work of the wheelwright was one of the fundamental eotechnic crafts. [Image-search keywords: wheelwright, sixteenth century.]

3: Old Paper Mill. Note that the wheels and shafts are almost entirely composed of wood. This material lingered on in machine-building and in mill-construction well into the nineteenth century. Metal was, up to the paleotechnic period, merely an accessory, used where a cutting edge or a resistant material was imperative, as in the runner of a skate. [Image-search keywords: paper mill, seventeenth century, shafts.]

4: The lathe, perhaps the most important machine-tool, was a direct invention of the woodman, probably in Greece. It was one of the first, and remains one of the chief, instruments of precision: Plato refers to the beauty of the geometrical shapes derived from the lathe. Note that every part was originally of wood. [Image-search keywords: primitive wooden lathe.]

VI. Eotechnic Environment

1: The old crane at Lüneburg. Originally built in the fourteenth century and since repaired. A labor-saving device common in the North Sea and Baltic ports during the eotechnic period: forerunners of the delicate bird-like monsters of steel now to be found in Hamburg and elsewhere. [Image-search keywords: Lüneburg, old crane.]

2: Typical battery of windmills near Elshout in Holland: they are often even closer in formation. The amount of horsepower developed through these windmills was in part responsible for the high state of Dutch civilization in the seventeenth century. The canal was important in land-planning and agriculture as well as transportation. [Image-search keywords: Elshout, battery of windmills.]

3: Advanced horticulture and market gardening. Not merely was the glass hot-house an eotechnic invention, but the cheapness of glass enabled bell-jars to be used outdoors for the protection and warmth of individual plants, as in this illustration. Note the banking of the earth and the use of the wall for protection. [Image not available.]

4: Naarden, Holland. Excellent example of city development and fortification at the height of the eotechnic period. The ancient bastions had only to be converted into parks, as in so many modern European towns, to create a veritable garden city. The definite pattern of the town, and its sharp contrast to the country, is still immensely superior to any of the succeeding types of urban development: above all, to the amorphous dribble of paleotechnic land-speculation. [Image-search keywords: Naarden, Holland, overview.]

VII. Early Manufacture

1: Wood-turner's shop. Typical separation of labor-power and skill: increased efficiency at the price of increased servility of labor. Note, however, the remains of the older type of motor, the bent sapling, attached to a foot-treadle. Note, too, the existence of the slide rest, usually attributed to Maudslay. [Image not available.]

2: Mass-production of bottles. The standardized glass bottle, so useful for medicines and wines, was a late eotechnic achievement. Before that the finer forms of glass, goblets, alembics, mirrors, and distillation flasks had been created. Without the use of glass for spectacles, mirrors, microscopes, telescopes, windows, and containers, our modern world, as revealed by physics and chemistry, could scarcely have been conceived. [Image not available.]

3: One of a number of multiple silk-reeling machines worked by water power illustrated in the Encyclopedia. Similar types possibly date back as far as 1272 at Bologna: they are illustrated in Zonca's treatise on machinery in 1607. Power-production, labor-saving, mass-manufacture, and mechanization date from the early eotechnic period. [Image not available.]

4: Child-labor in pin-manufacture: illustration of Adam Smith's famous example of "modern" production methods. This employment of the coolie labor of children was an essential basis of paleotechnic capitalism: it still lingers in backward areas. Once human motions had become simplified, however, they were ripe for imitation by machines. [Image not available.]

VIII. Paleotechnic Products

1: "Puffing Billy": Constructed at Wylam Colliery in 1813 by William Hedley. The oldest locomotive in existence: note the eotechnic survival in wooden boiler. [Image-search keywords: Puffing Billy.]

2: Interior of coal mine, showing primitive type of ore-car and shoring. [Image-search keywords: coal mine, early.]

3: Pittsburgh: a typical paleotechnic industrial environment. Smoke pall, air-sewage, disorder—and the human dwellings reduced to the lowest terms of decency and amenity. Crowd the houses together and the result is Philadelphia, Manchester, Preston, or Lille. Intensify the congestion and the result is New York, Glasgow, Berlin, or Bombay. [Image-search keywords: Pittsburgh, old city pics.]

4: Early London underground railway: 1860–1863. The railway building era was also a tunnel-building era. Every new element in paleotechnic transportation can be traced back directly to the mine. [Image not available.]

IX. Paleotechnic Triumphs

1: Maudslay's original screw-cutting lathe: invented about 1800. Perhaps the most original artists of the period were the toolmakers, who translated the old wooden machines into metal, who perfected and standardized the component parts, and who solved some of the other difficult mechanical problems. [Image-search keywords: Maudslay, lathe.]

2: The Brooklyn Bridge: 1869–1883. Great mass juxtaposed to great delicacy: an adroit solution for a difficult problem. The builders, John A. and Washington Roebling, deserve to be ranked in that great succession of paleotechnic engineers, beginning with Smeaton and Rennie, and including Telford, the Brunels, Samuel Bentham, and Eiffel. [Image-search keywords: Brooklyn Bridge.]

3: The Machine Hall at the Paris Exposition of 1889 was one of the finest engineering structures: technically it went beyond any of the existing train-sheds in refinement of design. Created by an architect, Dutert, and an engineer, Contamin, it had perhaps greater significance than the more daring Eiffel Tower done at the same time. Note that the American steel frame skyscraper was a product of the same period. [Image-search keywords: Machine Hall, Paris Exposition.]

4: A modern steamship: still essentially paleotechnic in design, but with all the cleanness and strength of the older type of engineering. Like so many other typical paleotechnic products, it was afflicted by giantism. In its inner arrangements, with the luxury and space of the first class contrasting with the cramped quarters and poorer fare of the third class, the big ocean steamship remains a diagrammatic picture of the paleotechnic class struggle. [Image-search keywords: ocean liner.]

X. Neotechnic Automatism

1: Modern cotton spinning. During the paleotechnic period the textile industries were the pattern for advanced production, and

the term "factory" was at first applied solely to textile factories. Today the worker has
a smaller part than ever to play in them: he
lingers on as a machine-herd. [Image-search
keywords: textile factory, spinning machine.]

2: The automatic bottle-making machine
is not merely a saver of labor but of life, for
glass-blowing lays a heavy toll upon its workers. On the other hand, cheap bottles mean
greater wastage through carelessness, and
the increased demand often tends to cancel
out some of the gains of cheap automatic production. [Image-search keywords: automatic
glass-bottle-making machine.]

3: Automatic machine for making screw-
caps in the Krausswerke in Saxony. This factory, which has remained in a single family
for a century, illustrates the change from the
handicraft methods of the old-fashioned smith
to the advanced machine methods of the modern engineer. [Image not available.]

4: Like the stream-lined railroad train,
the automatic stoker was invented more than
fifty years before it came into general use.
The type shown here has done away with a
servile form of labor and has led to increased
efficiency in fuel utilization. Note the single
attendant. [Image not available.]

XI. Airplane Shapes

1: Modern airplane, designed to decrease
wind-resistance and raise lifting power, on
lines suggested by study of birds and fish.
Since 1920 the development of scientific
knowledge and technical design have gone
on steadily here; and through the use of new
alloys like duralumin both lightness and
strength have been achieved. The airplane
has set the pace for refined and exact engineering. [Image-search keywords: airplane,
1933 or 1932.]

2: Perhaps the most radical impulse to
correct motor car design came from Glenn
Curtiss, the airplane designer, when he ran
an ordinary closed car backward and bettered
its performance. The best design so far seems
to be that of the Dymaxion Car, by Buckminster Fuller and Starling Burgess, which has
greatly improved speed and comfort without
extra horsepower. [Image-search keywords:
Dymaxion Car, Buckminster Fuller.]

3: The stream-lined train, designed but
rejected as early as 1874, now is realized in
1934, thanks to the competition and the lesson of the airplane. [Image-search keywords:
stream-lined train, 1934.]

4: So-called Rail Zeppelin. Experimental
and possibly somewhat romantic attempt to
adapt to surface transportation the advantages of airplane and dirigible. A still more
radical approach to the problem of fast land-
transportation is that now under experiment
in Soviet Russia, the "sphero-train" invented
by a young soviet engineer, M. I. Yarmalchuk.
The latter runs on large motorized ball-bearings. The airplane has freed the inventor from
the stereotypes of wheel-locomotion. [Image-
search keywords: Rail Zeppelin.]

XII. Nature and the Machine

1: Roentgen photograph of Nautilus by J.
B. Polak. Nature's use of the spiral in construction. The x-ray, like the microscope,
reveals a new esthetic world. [Image-search
keywords: seashell, photograph, nautilus.]

2: Section of modern hydro-turbine: spiral form dictated by mechanical necessity.
Geometrical forms, simple and complex, are
orchestrated in machine design. [Image
not available.]

3: Grandstand of new stadium in Florence:
Pier Luigi Nervi, architect. Engineering in
which imagination and necessity are harmo-
niously composed. [Image-search keywords:
stadium, Florence, Pier Nervi, 1932.]

4: R. Duchamp-Villon's interpretation of
the organic form of a horse in terms of the machine. [Image-search keywords: Duchamp-
Villon, horse statue.]

XIII. Esthetic Assimilation

1: Sculpture by Constantin Brancusi.

Abstraction, respect for the materials, importance of fine measurements and delicate modulations, impersonality. See plate XV, no. 2. [Image-search keywords: Constantin Brancusi, abstraction.]

2: The steel workers: mural by Thomas H. Benton. Realization of the dramatic element in modern industry, and the daily heroism which often outvies that of the battlefield. [Image-search keywords: Thomas Hart Benton, The Steel Workers, New School.]

3: Modern grain elevator. Esthetic effect derived from simplicity, essentiality, repetition of elementary forms; heightened by colossal scale. See Worringer's suggestive essay on Egypt and America. [Image-search keywords: grain elevator, collossus, 1930.]

4: Breakfast Table by Ferdinand Léger. Transposition of the organic and the living into terms of the mechanical: dismemberment of natural forms and graphic re-invention. [Image not available. See Léger, Still Life with Fruit Bowl.]

XIV. Modern Machine Art

1: Self-aligning ball-bearings. High degree of accuracy and refinement in one of the most essential departments of the machine. The beauty of elementary geometrical shapes. Perfection of finish and adjustment, though already present in fine handicraft, became common—and essential—with machine-craft. [Image-search keywords: self-aligning ball-bearings.]

2: Section of spring. Although line and mass are purely utilitarian in origin, the result when isolated has an esthetic interest. The perception of the special qualities of highly finished machine-forms was one of the prime discoveries of Brancusi and the sculptors in glass and metal, like Moholy-Nagy and Grabo. [Image not available.]

3: Glass bottles with caps: typical of modern mass-production. Contrast this simple product of the Owens-Illinois Glass Co. with the extremely complicated machine that makes such mass production possible, shown on plate XI. [Image-search keywords: square glass bottles.]

4: Kitchen ladles: another example of serial production, with all the advantages of uniform design and high refinement of finish. But while the machine cannot successfully achieve decoration, handicraft often produced forms as rational as those of the machine. In functional design the two modes overlap. Even in more primitive technics, machines like the lathe, drill, and loom conditioned handicraft, and in turn eotechnic handicraft furthered the machine. [Image-search keywords: kitchen ladles.]

XV. The New Environment

1: Interior of the giant power station at Dnieprostroy. The calmness, cleanness, and order of the neotechnic environment. The same qualities prevail in the power station or the factory as in the kitchen or the bathroom of the individual dwelling. In any one of these places one could "eat off the floor." Contrast with the paleotechnic environment. [Image not available.]

2: Waterfront at Köln. The order and plan of the neotechnic economy is apparent in what used to be the most chaotic and slatternly parts of the city. The factories and docks form a common unit, which, so far from making the scene hideous, contribute to its esthetic composure. Compare this with the competitive waste, muddle, blight, and advertisement of the older regime. [Image not available.]

3: Example of modern workers' dwellings in Sweden: typical of millions of such dwellings that came into existence in Europe after 1915, thanks to the sudden crystallization of neotechnic methods in community planning and housing. Here is a return to a handsome and well-integrated human environment, in which the efficiency of neotechnic production can be registered in a higher standard of living and a wider use of leisure. [Image not available.]

4: Modern Waterworks in Sweden. What is called the new architecture is in fact a symbol of a fresh mode of thinking and feeling and living in which the Scandinavian countries in particular have often been in the forefront. But similar examples could be culled from almost every neotechnic region. [Image not available.]

TECHNICS AND CIVILIZATION

The first draft of this book was written in 1930 and the second was completed in 1931. Up to 1932 my purpose was to deal with the machine, the city, the region, the group, and the personality within a single volume. In working out the section on technics it was necessary to increase the scale of the whole project: so the present book covers only a limited area of the first draft. While Technics and Civilization is a unit, certain aspects of the machine, such as its relation to architecture, and certain aspects of civilization that may ultimately bear upon the course of technics remain to be treated at another time.   *L. M.*

OBJECTIVES

During the last thousand years the material basis and the cultural forms of Western Civilization have been profoundly modified by the development of the machine. How did this come about? Where did it take place? What were the chief motives that encouraged this radical transformation of the environment and the routine of life: what were the ends in view: what were the means and methods: what unexpected values have arisen in the process? These are some of the questions that the present study seeks to answer.

While people often call our period the "Machine Age," very few have any perspective on modern technics or any clear notion as to its origins. Popular historians usually date the great transformation in modern industry from Watt's supposed invention of the steam engine; and in the conventional economics textbook the application of automatic machinery to spinning and weaving is often treated as an equally critical turning point. But the fact is that in Western Europe the machine had been developing steadily for at least seven centuries before the dramatic changes that accompanied the "industrial revolution" took place. Men had become mechanical before they perfected complicated machines to express their new bent and interest; and the will-to-order had appeared once more in the monastery and the army and the counting-house before it finally manifested itself in the factory. Behind all the great material inventions of the last century and a half was not merely a long internal development of technics: there was also a change of mind. Before the new industrial processes could take hold on a great scale, a reorientation of wishes, habits, ideas, goals was necessary.

To understand the dominating rôle played by technics in modern civilization, one must explore in detail the preliminary period of ideological and social preparation. Not merely must one explain the existence of the new mechanical instruments: one must explain the culture that was ready to use them and profit by them so extensively. For note this: mechanization and regimentation are not new phenomena in history: what is new is the fact that these functions have been projected and embodied in organized forms which dominate every aspect of our existence. Other civilizations reached a high degree of technical proficiency without, apparently, being profoundly influenced by the methods and aims of technics. All the critical instruments of modern technology—the clock, the printing press, the water-mill, the magnetic compass, the loom, the lathe, gunpowder, paper, to say nothing of mathematics and chemistry and mechanics— existed in other cultures. The Chinese, the Arabs, the Greeks, long before the Northern European, had taken most of the first steps toward the machine. And although the great engineering works of the Cretans, the Egyptians, and the Romans were carried out mainly on an empirical basis, these peoples plainly had an abundance of technical skill at their command. They had machines; but they did not develop "the machine." It remained for the peoples of Western Europe to carry the physical sciences and the exact arts to a point no other culture had reached, and to adapt the whole mode of life to the pace and the capacities of the machine. How did this happen? How in fact could the machine take possession of European society until that society had, by an inner accommodation, surrendered to the machine?

Plainly, what is usually called *the* industrial revolution, the series of industrial changes that began in the eighteenth century, was a transformation that took place in the course of a much longer march.

The machine has swept over our civilization in three successive waves. The first wave, which was set in motion around the tenth century, gathered strength and momentum as other institutions in civilization were weakening and dispersing: this early triumph of the machine was an effort to achieve order and power by purely external means, and its success was partly due to the fact that it evaded many of the real issues of life and turned away from the momentous moral and social difficulties that it had neither confronted nor solved. The second wave heaved upward in the eighteenth century after a long steady roll through the Middle Ages, with its improvements in mining and iron-working: accepting all the ideological premises of the first effort to create the machine, the disciples of Watt and Arkwright sought to universalize them and take advantage of the practical consequences. In the course of this effort, various moral and social and political problems which had been set to one side by the exclusive development of the machine, now returned with doubled urgency: the very efficiency of the machine was drastically curtailed by the failure to achieve in society a set of harmonious and integrated purposes. External regimentation and internal resistance and disintegration went hand in hand: those fortunate members of society who were in complete harmony with the machine achieved that state only by closing up various important avenues of life. Finally, we begin in our own day to observe the swelling energies of a third wave: behind this wave, both in technics and in civilization, are forces which were suppressed or perverted by the earlier development of the machine, forces which now manifest themselves in every department of activity, and which tend toward a new synthesis in thought and a fresh synergy in action. As the result of this third movement, the machine ceases to be a substitute for God or for an orderly society; and instead of its success being measured by the mechanization of life, its worth becomes more and more measurable in terms of its own approach to the organic and the living. The receding waves of the first two phases of the machine diminish a little the force of the third wave: but the image remains accurate to the extent that it suggests that the wave with which we are now being carried forward is moving in a direction opposite to those of the past.

By now, it is plain, a new world has come into existence; but it exists only in fragments. New forms of living have for long been in process; but so far they have likewise been divided and unfocussed: indeed, our vast gains in energy and in the production of goods have manifested themselves in part in a loss of form and an impoverishment of life. What has limited the beneficence of the machine? Under what conditions may the machine be directed toward a fuller use and accomplishment? To these questions, too, the present study seeks an answer. Technics and civilization as a whole are the result of human choices and aptitudes and strivings, deliberate as well as unconscious, often irrational when apparently they are most objective and scientific: but even when they are uncontrollable they are not external. Choice manifests itself in society in small increments and moment-to-moment decisions as well as in loud dramatic struggles; and he who does not see choice in the development of the machine merely betrays his incapacity to observe cumulative effects until they are bunched together so closely that they seem completely external and impersonal. No matter how completely technics relies upon the objective procedures of the sciences, it does not form an independent system, like the universe: it exists as an element in human culture and it promises well or ill as the social groups that exploit it promise well or ill. The machine itself makes no demands and holds out no promises: it is the human spirit that makes demands and keeps promises. In order to reconquer the machine and subdue it to human purposes, one must first understand it and assimilate it. So far, we have embraced the machine without fully understanding it, or, like the weaker romantics, we have rejected the machine without first seeing how much of it we could intelligently assimilate.

The machine itself, however, is a product of human ingenuity and effort: hence to understand the machine is not merely a first step toward re-orienting our civilization: it is also a means toward understanding society and toward knowing ourselves. The world of technics is not isolated and self-contained: it reacts to forces and impulses that come from apparently remote parts of the environment. That fact makes peculiarly hopeful the development that has been going on within the domain of technics itself since around 1870: for the organic has become visible again even within the mechanical complex: some of our most characteristic mechanical instruments— the telephone, the phonograph, the motion picture—have grown out of our interest in the human voice and the human eye and our knowledge of their physiology and anatomy. Can one detect, perhaps, the characteristic properties of this emergent order—its pattern, its planes, its angle of polarization, its color? Can one, in the process of crystallization, remove the turbid residues left behind by our earlier forms of technology? Can one distinguish and define the specific properties of a technics directed toward the service of life: properties that distinguish it morally, socially, politically, esthetically from the cruder forms that preceded it? Let us make the attempt. The study of the rise and development of modern technics is a basis for understanding and strengthening this contemporary transvaluation: and the transvaluation of the machine is the next move, perhaps, toward its mastery.

CHAPTER I.　　　CULTURAL PREPARATION

1: Machines, Utilities, and "The Machine"

During the last century the automatic or semi-automatic machine has come to occupy a large place in our daily routine; and we have tended to attribute to the physical instrument itself the whole complex of habits and methods that created it and accompanied it. Almost every discussion of technology from Marx onward has tended to overemphasize the part played by the more mobile and active parts of our industrial equipment, and has slighted other equally critical elements in our technical heritage.

What is a machine? Apart from the simple machines of classic mechanics, the inclined plane, the pulley, and so forth, the subject remains a confused one. Many of the writers who have discussed the machine age have treated the machine as if it were a very recent phenomenon, and as if the technology of handicraft had employed only tools to transform the environment. These preconceptions are baseless. For the last three thousand years, at least, machines have been an essential part of our older technical heritage. Reuleaux's definition of a machine has remained a classic: "A machine is a combination of resistant bodies so arranged that by their means the mechanical forces of nature can be compelled to do work accompanied by certain determinant motions"; but it does not take us very far. Its place is due to his importance as the first great morphologist of machines, for it leaves out the large class of machines operated by man-power.

Machines have developed out of a complex of non-organic agents for converting energy, for performing work, for enlarging the mechanical or sensory capacities of the human body, or for reducing to a mensurable order and regularity the processes of life. The automaton is the last step in a process that began with the use of one part or another of the human body as a tool. In back of the development of tools and machines lies the attempt to modify the environment in such a way as to fortify and sustain the human organism: the effort is either to extend the powers of the otherwise unarmed organism, or to manufacture outside of the body a set of conditions more favorable toward maintaining its equilibrium and ensuring its survival. Instead of a physiological adaptation to the cold, like the growth of hair or the habit of hibernation, there is an environmental adaptation, such as that made possible by the use of clothes and the erection of shelters.

The essential distinction between a machine and a tool lies in the degree of independence in the operation from the skill and motive power of the operator: the tool lends itself to manipulation, the machine to automatic action. The degree of complexity is unimportant: for, using the tool, the human hand and eye perform complicated actions which are the equivalent, in function, of a well developed machine; while, on the other hand, there are highly effective machines, like the drop hammer, which do very simple tasks, with the aid of a relatively simple mechanism. The difference between tools and machines lies primarily in the degree of automatism they have reached: the skilled tool-user becomes more accurate and more automatic, in short, more mechanical, as his originally voluntary motions settle down into reflexes, and on the other hand, even in the most completely automatic machine, there must intervene somewhere, at the beginning and the end of the process, first in the original design, and finally in the ability to overcome defects and to make repairs, the conscious participation of a human agent.

Moreover, between the tool and the machine there stands another class of objects, the machine-tool: here, in the lathe or the drill, one has the accuracy of the finest machine coupled with the skilled attendance of the workman. When one adds to this mechanical complex an external source of power, the line of division becomes even more difficult to establish. In general, the machine emphasizes specialization of function, whereas the tool indicates flexibility: a planing machine performs only one operation, whereas a knife can be used to smooth wood, to carve it, to split it, or to pry open a lock, or to drive in a screw. The automatic machine, then, is a very specialized kind of adaptation; it involves the notion of an external source of power, a more or less complicated inter-relation of parts, and a limited kind of activity. From the beginning the machine was a sort of minor organism, designed to perform a single set of functions.

Along with these dynamic elements in technology there is another set, more static in character, but equally important in function. While the growth of machines is the most patent technical fact of the last thousand years, the machine, in the form of the fire-drill or the potter's wheel, has been in existence since at least neolithic times. During the earlier period, some of the most effective adaptations of the environment came, not from the invention of machines, but from the equally admirable invention of utensils, apparatus, and utilities. The basket and the pot stand for the first, the dye vat and the brick-kiln stand for the second, and reservoirs and aqueducts and roads and buildings belong to the third class. The modern period has finally given us the power utility, like the railroad track or the electric transmission line, which functions only through the operation of power machinery. While tools and machines transform the environment by changing the shape and location of objects, utensils and apparatus have been used to effect equally necessary chemical transformations. Tanning, brewing, distilling, dyeing have been as important in man's technical development as smithing or weaving. But most of these processes remained in their traditional state till the middle of the nineteenth century, and it is only since then that they have been influenced in any large degree by the same set of scientific forces and human interests that were developing the modern power-machine.

In the series of objects from utensils to utilities there is the same relation between the workman and the process that one notes in the series between tools and automatic machines: differences in the degree of specialization, the degree of impersonality. But since people's attention is directed most easily to the noisier and more

active parts of the environment, the rôle of the utility and the apparatus has been neglected in most discussions of the machine, or,
what is almost as bad, these technical instruments have all been
clumsily grouped as machines. The point to remember is that both
have played an enormous part in the development of the modern
environment; and at no stage in history can the two means of adaptation be split apart. Every technological complex includes both: not
least our modern one.

When I use the word machines hereafter I shall refer to specific
objects like the printing press or the power loom. When I use the
term "the machine" I shall employ it as a shorthand reference to
the entire technological complex. This will embrace the knowledge
and skills and arts derived from industry or implicated in the new
technics, and will include various forms of tool, instrument, apparatus
and utility as well as machines proper.

2: The Monastery and the Clock

Where did the machine first take form in modern civilization?
There was plainly more than one point of origin. Our mechanical
civilization represents the convergence of numerous habits, ideas,
and modes of living, as well as technical instruments; and some
of these were, in the beginning, directly opposed to the civilization
they helped to create. But the first manifestation of the new order
took place in the general picture of the world: during the first seven
centuries of the machine's existence the categories of time and space
underwent an extraordinary change, and no aspect of life was left
untouched by this transformation. The application of quantitative
methods of thought to the study of nature had its first manifestation
in the regular measurement of time; and the new mechanical conception of time arose in part out of the routine of the monastery.
Alfred Whitehead has emphasized the importance of the scholastic
belief in a universe ordered by God as one of the foundations of
modern physics: but behind that belief was the presence of order in
the institutions of the Church itself.

The technics of the ancient world were still carried on from
Constantinople and Baghdad to Sicily and Cordova: hence the early

lead taken by Salerno in the scientific and medical advances of the Middle Age. It was, however, in the monasteries of the West that the desire for order and power, other than that expressed in the military domination of weaker men, first manifested itself after the long uncertainty and bloody confusion that attended the breakdown of the Roman Empire. Within the walls of the monastery was sanctuary: under the rule of the order surprise and doubt and caprice and irregularity were put at bay. Opposed to the erratic fluctuations and pulsations of the worldly life was the iron discipline of the rule. Benedict added a seventh period to the devotions of the day, and in the seventh century, by a bull of Pope Sabinianus, it was decreed that the bells of the monastery be rung seven times in the twenty-four hours. These punctuation marks in the day were known as the canonical hours, and some means of keeping count of them and ensuring their regular repetition became necessary.

According to a now discredited legend, the first modern mechanical clock, worked by falling weights, was invented by the monk named Gerbert who afterwards became Pope Sylvester II near the close of the tenth century. This clock was probably only a water clock, one of those bequests of the ancient world either left over directly from the days of the Romans, like the water-wheel itself, or coming back again into the West through the Arabs. But the legend, as so often happens, is accurate in its implications if not in its facts. The monastery was the seat of a regular life, and an instrument for striking the hours at intervals or for reminding the bell-ringer that it was time to strike the bells, was an almost inevitable product of this life. If the mechanical clock did not appear until the cities of the thirteenth century demanded an orderly routine, the habit of order itself and the earnest regulation of time-sequences had become almost second nature in the monastery. Coulton agrees with Sombart in looking upon the Benedictines, the great working order, as perhaps the original founders of modern capitalism: their rule certainly took the curse off work and their vigorous engineering enterprises may even have robbed warfare of some of its glamor. So one is not straining the facts when one suggests that the monasteries—at one time there were 40,000 under the Benedictine rule—helped to give human enterprise the regular collective beat and rhythm of the machine; for the clock is not merely a means of keeping track of the hours, but of synchronizing the actions of men.

Was it by reason of the collective Christian desire to provide for the welfare of souls in eternity by regular prayers and devotions that time-keeping and the habits of temporal order took hold of men's minds: habits that capitalist civilization presently turned to good account? One must perhaps accept the irony of this paradox. At all events, by the thirteenth century there are definite records of mechanical clocks, and by 1370 a well-designed "modern" clock had been built by Heinrich von Wyck at Paris. Meanwhile, bell towers had come into existence, and the new clocks, if they did not have, till the fourteenth century, a dial and a hand that translated the movement of time into a movement through space, at all events struck the hours. The clouds that could paralyze the sundial, the freezing that could stop the water clock on a winter night, were no longer obstacles to time-keeping: summer or winter, day or night, one was aware of the measured clank of the clock. The instrument presently spread outside the monastery; and the regular striking of the bells brought a new regularity into the life of the workman and the merchant. The bells of the clock tower almost defined urban existence. Time-keeping passed into time-serving and time-accounting and time-rationing. As this took place, Eternity ceased gradually to serve as the measure and focus of human actions.

The clock, not the steam-engine, is the key-machine of the modern industrial age. For every phase of its development the clock is both the outstanding fact and the typical symbol of the machine: even today no other machine is so ubiquitous. Here, at the very beginning of modern technics, appeared prophetically the accurate automatic machine which, only after centuries of further effort, was also to prove the final consummation of this technics in every department of industrial activity. There had been power-machines, such as the water-mill, before the clock; and there had also been various kinds of automata, to awaken the wonder of the populace in the temple, or to please the idle fancy of some Moslem caliph: machines one finds illustrated in Hero and Al-Jazari. But here was a new kind of power-machine, in which the source of power and the transmission were of such a nature as to ensure the even flow of energy throughout the works and to make possible regular production and a standardized product. In its relationship to determinable quantities of energy, to standardization, to automatic action, and finally to its own special product, accurate timing, the clock has been the foremost machine in modern technics: and at each period it has remained in the lead: it marks a perfection toward which other machines aspire. The clock, moreover, served as a model for many other kinds of mechanical works, and the analysis of motion that accompanied the perfection of the clock, with the various types of gearing and transmission that were elaborated, contributed to the success of quite different kinds of machine. Smiths could have hammered thousands of suits of armor or thousands of iron cannon, wheelwrights could have shaped thousands of great water-wheels or crude gears, without inventing any of the special types of movement developed in clockwork, and without any of the accuracy of measurement and fineness of articulation that finally produced the accurate eighteenth century chronometer.

The clock, moreover, is a piece of power-machinery whose "product" is seconds and minutes: by its essential nature it dissociated time from human events and helped create the belief in an independent world of mathematically measurable sequences: the special world of science. There is relatively little foundation for this belief in common human experience: throughout the year the days are of uneven duration, and not merely does the relation between day and night steadily change, but a slight journey from East to West alters astronomical time by a certain number of minutes. In terms of the human organism itself, mechanical time is even more foreign: while human life has regularities of its own, the beat of the pulse, the breathing of the lungs, these change from hour to hour with mood and action, and in the longer span of days, time is measured not by the calendar but by the events that occupy it. The shepherd measures from the time the ewes lambed; the farmer measures back to the day of sowing or forward to the harvest: if growth has its own duration and regularities, behind it are not simply matter and motion but the facts of development: in short, history. And while mechanical time is strung out in a succession of mathematically isolated instants, organic time—what Bergson calls duration—is cumulative in its effects. Though mechanical time can, in a sense, be speeded up or run backward, like the hands of a clock or the images of a moving picture, organic time moves in only one direction—through the cycle of birth, growth, development, decay, and death—and the past that is already dead remains present in the future that has still to be born.

Around 1345, according to Thorndike, the division of hours into sixty minutes and of minutes into sixty seconds became common: it was this abstract framework of divided time that became more and more the point of reference for both action and thought, and in the effort to arrive at accuracy in this department, the astronomical exploration of the sky focussed attention further upon the regular, implacable movements of the heavenly bodies through space. Early in the sixteenth century a young Nuremberg mechanic, Peter Henlein, is supposed to have created "many-wheeled watches out of small bits of iron" and by the end of the century the small domestic clock had been introduced in England and Holland. As with the motor car and the airplane, the richer classes first took over the new mechanism and popularized it: partly because they alone could afford it, partly because the new bourgeoisie were the first to discover that, as Franklin later put it, "time is money." To become "as regular as clockwork" was the bourgeois ideal, and to own a watch was for long a definite symbol of success. The increasing tempo of civilization led to a demand for greater power: and in turn power quickened the tempo.

Now, the orderly punctual life that first took shape in the monasteries is not native to mankind, although by now Western peoples are so thoroughly regimented by the clock that it is "second nature" and they look upon its observance as a fact of nature. Many Eastern civilizations have flourished on a loose basis in time: the Hindus have in fact been so indifferent to time that they lack even an authentic chronology of the years. Only yesterday, in the midst of the industrializations of Soviet Russia, did a society come into exist-

ence to further the carrying of watches there and to propagandize
the benefits of punctuality. The popularization of time-keeping, which
followed the production of the cheap standardized watch, first in
Geneva, then in America around the middle of the last century, was
essential to a well-articulated system of transportation and production.

To keep time was once a peculiar attribute of music: it gave industrial value to the workshop song or the tattoo or the chantey of the
sailors tugging at a rope. But the effect of the mechanical clock is
more pervasive and strict: it presides over the day from the hour of
rising to the hour of rest. When one thinks of the day as an abstract
span of time, one does not go to bed with the chickens on a winter's
night: one invents wicks, chimneys, lamps, gaslights, electric lamps,
so as to use all the hours belonging to the day. When one thinks of
time, not as a sequence of experiences, but as a collection of hours,
minutes, and seconds, the habits of adding time and saving time come
into existence. Time took on the character of an enclosed space: it
could be divided, it could be filled up, it could even be expanded
by the invention of labor-saving instruments.

Abstract time became the new medium of existence. Organic functions themselves were regulated by it: one ate, not upon feeling
hungry, but when prompted by the clock: one slept, not when one
was tired, but when the clock sanctioned it. A generalized time-
consciousness accompanied the wider use of clocks: dissociating time
from organic sequences, it became easier for the men of the
Renascence to indulge the fantasy of reviving the classic past or
of reliving the splendors of antique Roman civilization: the cult
of history, appearing first in daily ritual, finally abstracted itself as
a special discipline. In the seventeenth century journalism and periodic literature made their appearance: even in dress, following
the lead of Venice as fashion-center, people altered styles every
year rather than every generation.

The gain in mechanical efficiency through co-ordination and
through the closer articulation of the day's events cannot be over-
estimated: while this increase cannot be measured in mere horsepower, one has only to imagine its absence today to foresee the
speedy disruption and eventual collapse of our entire society. The

modern industrial régime could do without coal and iron and steam easier than it could do without the clock.

3: Space, Distance, Movement

"A child and an adult, an Australian primitive and a European, a man of the Middle Ages and a contemporary, are distinguished not only by a difference in degree, but by a difference in kind by their methods of pictorial representation."

Dagobert Frey, whose words I have just quoted, has made a penetrating study of the difference in spatial conceptions between the early Middle Ages and the Renascence: he has re-enforced by a wealth of specific detail, the generalization that no two cultures live conceptually in the same kind of time and space. Space and time, like language itself, are works of art, and like language they help condition and direct practical action. Long before Kant announced that time and space were categories of the mind, long before the mathematicians discovered that there were conceivable and rational forms of space other than the form described by Euclid, mankind at large had acted on this premise. Like the Englishman in France who thought that bread was the right name for *le pain* each culture believes that every other kind of space and time is an approximation to or a perversion of the real space and time in which *it* lives.

During the Middle Ages spatial relations tended to be organized as symbols and values. The highest object in the city was the church spire which pointed toward heaven and dominated all the lesser buildings, as the church dominated their hopes and fears. Space was divided arbitrarily to represent the seven virtues or the twelve apostles or the ten commandments or the trinity. Without constant symbolic reference to the fables and myths of Christianity the rationale of medieval space would collapse. Even the most rational minds were not exempt: Roger Bacon was a careful student of optics, but after he had described the seven coverings of the eye he added that by such means God had willed to express in our bodies an image of the seven gifts of the spirit.

Size signified importance: to represent human beings of entirely different sizes on the same plane of vision and at the same distance from the observer was entirely possible for the medieval artist. This same habit applies not only to the representation of real objects but to the organization of terrestrial experience by means of the map. In medieval cartography the water and the land masses of the earth, even when approximately known, may be represented in an arbitrary figure like a tree, with no regard for the actual relations as experienced by a traveller, and with no interest in anything except the allegorical correspondence.

One further characteristic of medieval space must be noted: space and time form two relatively independent systems. First: the medieval artist introduced other times within his own spatial world, as when he projected the events of Christ's life within a contemporary Italian city, without the slightest feeling that the passage of time has made a difference, just as in Chaucer the classical legend of Troilus and Cressida is related as if it were a contemporary story. When a medieval chronicler mentions the King, as the author of The Wandering Scholars remarks, it is sometimes a little difficult to find out whether he is talking about Caesar or Alexander the Great or his own monarch: each is equally near to him. Indeed, the word anachronism is meaningless when applied to medieval art: it is only when one related events to a co-ordinated frame of time and space that being out of time or being untrue to time became disconcerting. Similarly, in Botticelli's The Three Miracles of St. Zenobius, three different times are presented upon a single stage.

Because of this separation of time and space, things could appear and disappear suddenly, unaccountably: the dropping of a ship below the horizon no more needed an explanation than the dropping of a demon down the chimney. There was no mystery about the past from which they had emerged, no speculation as to the future toward which they were bound: objects swam into vision and sank out of it with something of the same mystery in which the coming and going of adults affects the experience of young children, whose first graphic efforts so much resemble in their organization the world of the medieval artist. In this symbolic world of space and time everything was either a mystery or a miracle. The connecting link between events was the cosmic and religious order: the true order of space was Heaven, even as the true order of time was Eternity.

Between the fourteenth and the seventeenth century a revolutionary change in the conception of space took place in Western Europe. Space as a hierarchy of values was replaced by space as a system of magnitudes. One of the indications of this new orientation was the closer study of the relations of objects in space and the discovery of the laws of perspective and the systematic organization of pictures within the new frame fixed by the foreground, the horizon and the vanishing point. Perspective turned the symbolic relation of objects into a visual relation: the visual in turn became a quantitative relation. In the new picture of the world, size meant not human or divine importance, but distance. Bodies did not exist separately as absolute magnitudes: they were co-ordinated with other bodies within the same frame of vision and must be in scale. To achieve this scale, there must be an accurate representation of the object itself, a point for point correspondence between the picture and the image: hence a fresh interest in external nature and in questions of fact. The division of the canvas into squares and the accurate observation of the world through this abstract checkerboard marked the new technique of the painter, from Paolo Ucello onward.

The new interest in perspective brought depth into the picture and distance into the mind. In the older pictures, one's eye jumped from one part to another, picking up symbolic crumbs as taste and fancy dictated: in the new pictures, one's eye followed the lines of linear perspective along streets, buildings, tessellated pavements whose parallel lines the painter purposely introduced in order to make the eye itself travel. Even the objects in the foreground were sometimes grotesquely placed and foreshortened in order to create the same illusion. Movement became a new source of value: movement for its own sake. The measured space of the picture re-enforced the measured time of the clock.

Within this new ideal network of space and time all events now took place; and the most satisfactory event within this system was uniform motion in a straight line, for such motion lent itself to accurate representation within the system of spatial and temporal co-ordinates. One further consequence of this spatial order must be noted: to place a thing and to time it became essential to one's understanding of it. In Renascence space, the existence of objects must be accounted for: their passage through time and space is a clue to their appearance at any particular moment in any particular place. The unknown is therefore no less determinate than the known: given the roundness of the globe, the position of the Indies could be assumed and the time-distance calculated. The very existence of such an order was an incentive to explore it and to fill up the parts that were unknown.

What the painters demonstrated in their application of perspective, the cartographers established in the same century in their new maps. The Hereford Map of 1314 might have been done by a child: it was practically worthless for navigation. That of Ucello's contemporary, Andrea Banco, 1436, was conceived on rational lines, and represented a gain in conception as well as in practical accuracy. By laying down the invisible lines of latitude and longitude, the cartographers paved the way for later explorers, like Columbus: as with the later scientific method, the abstract system gave rational expectations, even if on the basis of inaccurate knowledge. No longer was it necessary for the navigator to hug the shore line: he could launch out into the unknown, set his course toward an arbitrary point, and return approximately to the place of departure. Both Eden and Heaven were outside the new space; and though they lingered on as the ostensible subjects of painting, the real subjects were Time and Space and Nature and Man.

Presently, on the basis laid down by the painter and the cartographer, an interest in space as such, in movement as such, in locomotion as such, arose. In back of this interest were of course more concrete alterations: roads had become more secure, vessels were being built more soundly, above all, new inventions—the magnetic needle, the astrolabe, the rudder—had made it possible to chart and to hold a more accurate course at sea. The gold of the Indies and the fabled fountains of youth and the happy isles of endless sensual delight doubtless beckoned too: but the presence of these tangible goals does not lessen the importance of the new schemata. The categories of time and space, once practically dissociated, had become united: and the abstractions of measured time and measured space undermined the earlier conceptions of infinity and eternity, since measurement must begin with an arbitrary here and now even if space and time be empty. The itch to *use* space and time had broken out: and once they were co-ordinated with movement, they could be contracted or expanded: the conquest of space and time had begun. (It is interesting, however, to note that the very concept of acceleration, which is part of our daily mechanical experience, was not formulated till the seventeenth century.)

The signs of this conquest are many: they came forth in rapid succession. In military arts the cross-bow and the ballista were revived and extended, and on their heels came more powerful weapons for annihilating distance—the cannon and later the musket. Leonardo conceived an airplane and built one. Fantastic projects for flight were canvassed. In 1420 Fontana described a velocipede: in 1589 Gilles de Bom of Antwerp apparently built a man-propelled wagon: restless preludes to the vast efforts and initiatives of the nineteenth century. As with so many elements in our culture, the original impulse was imparted to this movement by the Arabs: as early as 880 Abû l-Qâsim had attempted flight, and in 1065 Oliver of Malmesbury had killed himself in an attempt to soar from a high place: but from the fifteenth century on the desire to conquer the air became a recurrent preoccupation of inventive minds; and it was close enough to popular thought to make the report of a flight from Portugal to Vienna serve as a news hoax in 1709.

The new attitude toward time and space infected the workshop and the counting house, the army and the city. The tempo became faster: the magnitudes became greater: conceptually, modern culture launched itself into space and gave itself over to movement. What Max Weber called the "romanticism of numbers" grew naturally out of this interest. In time-keeping, in trading, in fighting men counted numbers; and finally, as the habit grew, only numbers counted.

4: The Influence of Capitalism

The romanticism of numbers had still another aspect, important for the development of scientific habits of thought. This was the rise of capitalism, and the change from a barter economy, facilitated by small supplies of variable local coinage, to a money economy with an international credit structure and a constant reference to the abstract symbols of wealth: gold, drafts, bills of exchange, eventually merely numbers.

From the standpoint of technique, this structure had its origin in the towns of Northern Italy, particularly Florence and Venice, in the fourteenth century; two hundred years later there was in existence in Antwerp an international bourse, devoted to aiding speculation in shipments from foreign ports and in money itself. By the middle of the sixteenth century book-keeping by double entry, bills of exchange, letters of credit, and speculation in "futures" were all developed in essentially their modern form. Whereas the procedures of science were not refined and codified until after Galileo and Newton, finance had emerged in its present-day dress at the very beginning of the machine age: Jacob Fugger and J. Pierpont Morgan could understand each other's methods and point of view and temperament far better than Paracelsus and Einstein.

The development of capitalism brought the new habits of abstraction and calculation into the lives of city people: only the country folk, still existing on their more primitive local basis, were partly immune. Capitalism turned people from tangibles to intangibles: its symbol, as Sombart observes, is the account book: "its life-value lies in its profit and loss account." The "economy of acquisition," which had hitherto been practiced by rare and fabulous creatures like Midas and Croesus, became once more the everyday mode: it tended to replace the direct "economy of needs" and to substitute money-values for life-values. The whole process of business took on more and more an abstract form; it was concerned with non-commodities, imaginary futures, hypothetical gains.

Karl Marx well summed up this new process of transmutation: "Since money does not disclose what has been transformed into it, everything, whether a commodity or not, is convertible into gold. Everything becomes saleable and purchasable. Circulation is the great social retort into which everything is thrown and out of which everything is recovered as crystallized money. Not even the bones of the saints are able to withstand this alchemy; and still less able to withstand it are more delicate things, sacrosanct things which are outside the commercial traffic of men. Just as all qualitative differences between commodities are effaced in money, so money, a radical leveller, effaces all distinctions. But money itself is a commodity, an external object, capable of becoming the private property of an individual. Thus social power becomes private power in the hands of a private person."

This last fact was particularly important for life and thought: the quest of power by means of abstractions. One abstraction re-enforced the other. Time was money: money was power: power required the furtherance of trade and production: production was diverted from the channels of direct use into those of remote trade, toward the acquisition of larger profits, with a larger margin for new capital expenditures for wars, foreign conquests, mines, productive enterprises . . . more money and more power. Of all forms of wealth, money alone is without assignable limits. The prince who might desire to build five palaces would hesitate to build five thousand: but what was to prevent him from seeking by conquest and taxes to multiply by thousands the riches in his treasury? Under a money economy, to speed up the process of production was to speed up the turnover: more money. And as the emphasis upon money grew in part out of the increasing mobility of late medieval society, with its international trade, so did the resulting money economy promote more trade: landed wealth, humanized wealth, houses, paintings, sculptures, books, even gold itself were all relatively difficult to transport, whereas money could be transported after pronouncing the proper abracadabra by a simple algebraic operation on one side or another of the ledger.

In time, men were more at home with abstractions than they were with the goods they represented. The typical operations of finance were the acquisition or the exchange of magnitudes. "Even the daydreams of the pecuniary day-dreamer," as Veblen observed, "take shape as a calculus of profit and loss computed in standard units of an impersonal magnitude." Men became powerful to the extent that they neglected the real world of wheat and wool, food and clothes, and centered their attention on the purely quantitative representation of it in tokens and symbols: to think in terms of mere weight and number, to make quantity not alone an indication of value but the criterion of value—that was the contribution of capitalism to the mechanical world-picture. So the abstractions of capitalism preceded the abstractions of modern science and re-enforced at every point its typical lessons and its typical methods of procedure. The clarification and the convenience, particularly for long distance trading in space and time were great: but the social price of these economies was a high one. Mark Kepler's words, published in 1595: "As the ear is made to perceive sound and the eye to perceive color, so the mind of man has been formed to understand, not all sorts of things, but quantities. It perceives any given thing more clearly in proportion as that thing is close to bare quantities as to its origins, but the further a thing recedes from quantities, the more darkness and error inheres in it."

Was it an accident that the founders and patrons of the Royal Society in London—indeed some of the first experimenters in the physical sciences—were merchants from the City? King Charles II might laugh uncontrollably when he heard that these gentlemen had spent their time weighing air; but their instincts were justified, their procedure was correct: the method itself belonged to their tradition, and there was money in it. The power that was science and the power that was money were, in final analysis, the same kind of power: the power of abstraction, measurement, quantification.

But it was not merely in the promotion of abstract habits of thought and pragmatic interests and quantitative estimations that capitalism prepared the way for modern technics. From the beginning machines and factory production, like big guns and armaments, made direct demands for capital far above the small advances necessary to provide the old-style handicraft worker with tools or keep him alive. The freedom to operate independent workshops and factories, to use machines and profit by them, went to those who had command of capital. While the feudal families, with their command over the land, often had a monopoly over such natural resources as were found in the earth, and often retained an interest in glass-making, coal-mining, and iron-works right down to modern times, the new mechanical inventions lent themselves to exploitation by the merchant classes. The incentive to mechanization lay in the greater profits that could be extracted through the multiplied power and efficiency of the machine.

Thus, although capitalism and technics must be clearly distinguished at every stage, one conditioned the other and reacted upon it. The merchant accumulated capital by widening the scale of his operations, quickening his turnover, and discovering new territories for exploitation: the inventor carried on a parallel process by exploiting new methods of production and devising new things to be produced. Sometimes trade appeared as a rival to the machine by offering greater opportunities for profit: sometimes it curbed further developments in order to increase the profit of a particular monopoly: both motives are still operative in capitalist society. From the first, there were disparities and conflicts between these two forms of exploitation: but trade was the older partner and exercised a higher authority. It was trade that gathered up new materials from the Indies and from the Americas, new foods, new cereals, tobacco, furs: it was trade that found a new market for the trash that was turned out by eighteenth century mass-production: it was trade—abetted by war—that developed the large-scale enterprises and the administrative capacity and method that made it possible to create the industrial system as a whole and weld together its various parts.

Whether machines would have been invented so rapidly and pushed so zealously without the extra incentive of commercial profit is extremely doubtful: for all the more skilled handicraft occupations were deeply entrenched, and the introduction of printing, for example, was delayed as much as twenty years in Paris by the bitter opposition of the guild of scribes and copyists. But while technics undoubtedly owes an honest debt to capitalism, as it does likewise to war, it was nevertheless unfortunate that the machine was conditioned, at the outset, by these foreign institutions and took on characteristics that had nothing essentially to do with the technical processes or the forms of work. Capitalism utilized the machine, not to further social welfare, but to increase private profit: mechanical instruments were used for the aggrandizement of the ruling classes. It was because of capitalism that the handicraft industries in both Europe and other parts of the world were recklessly destroyed by machine products, even when the latter were inferior to the thing they replaced: for the prestige of improvement and success and power was with the machine, even when it improved nothing, even when technically speaking it was a failure. It was because of the possibilities of profit that the place of the machine was overemphasized and the degree of regimentation pushed beyond what was necessary to harmony or efficiency. It was because of certain traits in private capitalism that the machine—which was a neutral agent—has often seemed, and in fact has sometimes been, a malicious element in society, careless of human life, indifferent to human interests. The machine has suffered for the sins of capitalism; contrariwise, capitalism has often taken credit for the virtues of the machine.

By supporting the machine, capitalism quickened its pace, and gave a special incentive to preoccupation with mechanical improvements: though it often failed to reward the inventor, it succeeded by blandishments and promises in stimulating him to further effort. In many departments the pace was over-accelerated, and the stimulus was over-applied: indeed, the necessity to promote continual changes and improvements, which has been characteristic of capitalism, introduced an element of instability into technics and kept society from assimilating its mechanical improvements and integrating them in an appropriate social pattern. As capitalism itself has developed and expanded, these vices have in fact grown more enormous, and the dangers to society as a whole have likewise grown proportionately. Enough here to notice the close historical association of modern technics and modern capitalism, and to point out that, for all this historical development, there is no necessary connection between them. Capitalism has existed in other civilizations, which had a relatively low technical development; and technics made steady improvements from the tenth to the fifteenth century without the special incentive of capitalism. But the style of the machine has up to the present been powerfully influenced by capitalism: the emphasis upon bigness, for example, is a commercial trait; it appeared in guild halls and merchants' houses long before it was evident in technics, with its originally modest scale of operations.

5: From Fable to Fact

Meanwhile, with the transformation of the concepts of time and space went a change in the direction of interest from the heavenly world to the natural one. Around the twelfth century the supernatural world, in which the European mind had been enveloped as in a cloud from the decay of the classic schools of thought onward, began to lift: the beautiful culture of Provence whose language Dante himself had thought perhaps to use for his Divine Comedy, was the first bud of the new order: a bud destined to be savagely blighted by the Albigensian crusade.

Every culture lives within its dream. That of Christianity was one in which a fabulous heavenly world, filled with gods, saints, devils, demons, angels, archangels, cherubim and seraphim and dominions and powers, shot its fantastically magnified shapes and images across the actual life of earthborn man. This dream pervades the life of a culture as the fantasies of night dominate the mind of a sleeper: it is reality—while the sleep lasts. But, like the sleeper, a culture lives within an objective world that goes on through its sleeping or waking, and sometimes breaks into the dream, like a noise, to modify it or to make further sleep impossible.

By a slow natural process, the world of nature broke in upon the medieval dream of hell and paradise and eternity: in the fresh naturalistic sculpture of the thirteenth century churches one can watch the first uneasy stir of the sleeper, as the light of morning strikes his eyes. At first, the craftsman's interest in nature was a confused one: side by side with the fine carvings of oak leaves and hawthorn sprays, faithfully copied, tenderly arranged, the sculptor still created strange monsters, gargoyles, chimeras, legendary beasts. But the interest in nature steadily broadened and became more consuming. The naïve feeling of the thirteenth century artist turned into the systematic exploration of the sixteenth century botanists and physiologists.

"In the Middle Ages," as Emile Mâle said, "the idea of a thing which a man formed for himself was always more real than the actual thing itself, and we see why these mystical centuries had no conception of what men now call science. The study of things for their own sake held no meaning for the thoughtful man. . . . The task for the student of nature was to discern the eternal truth that God would have each thing express." In escaping this attitude, the vulgar had an advantage over the learned: their minds were less capable of forging their own shackles. A rational common sense interest in Nature was not a product of the new classical learning of the Renascence; rather, one must say, that a few centuries after it had flourished among the peasants and the masons, it made its way by another route into the court and the study and the university. Villard de Honnecourt's notebook, the precious bequest of a great master-mason, has drawings of a bear, a swan, a grasshopper, a fly, a dragonfly, a lobster, a lion and a pair of parroquets, all done directly from life. The book of Nature reappeared, as in a palimpsest, through the heavenly book of the Word.

During the Middle Ages the external world had had no conceptual hold upon the mind. Natural facts were insignificant compared with the divine order and intention which Christ and his Church had revealed: the visible world was merely a pledge and a symbol of that Eternal World of whose blisses and damnations it gave such a keen foretaste. People ate and drank and mated, basked in the sun and grew solemn under the stars; but there was little meaning in this immediate state: whatever significance the items of daily life had was as stage accessories and costumes and rehearsals for the drama of Man's pilgrimage through eternity. How far could the mind go in scientific mensuration and observation as long as the mystic numbers three and four and seven and nine and twelve filled every relation with an allegorical significance. Before the sequences in nature could be studied, it was necessary to discipline the imagination and sharpen the vision: mystic second sight must be converted into factual first sight. The artists had a fuller part in this discipline than they have usually been credited with. In enumerating the many parts of nature that cannot be studied without the "aid and intervening of mathematics," Francis Bacon properly includes perspective, music, architecture, and engineering along with the sciences of astronomy and cosmography.

The change in attitude toward nature manifested itself in solitary figures long before it became common. Roger Bacon's experimental precepts and his special researches in optics have long been commonplace knowledge; indeed, like the scientific vision of his Elizabethan namesake they have been somewhat overrated: their significance lies in the fact that they represented a general trend. In the thirteenth century, the pupils of Albertus Magnus were led by a new curiosity to explore their environment, while Absalon of St. Victor complained that the students wished to study "the conformation of the globe, the nature of the elements, the place of the stars, the nature of animals, the violence of the wind, the life of herbs and roots." Dante and Petrarch, unlike most medieval men, no longer avoided mountains as mere terrifying obstacles that increased the hardships of travel: they sought them and climbed them, for the exaltation that comes from the conquest of distance and the attainment of a bird's-eye view. Later, Leonardo explored the hills of Tuscany, discovered fossils, made correct interpretations of the processes of geology: Agricola, urged on by his interest in mining, did the same. The herbals and treatises on natural history that came out during the fifteenth and sixteenth centuries, though they still mingled fable and conjecture with fact, were resolute steps toward the delineation of nature: their admirable pictures still witness this. And the little books on the seasons and the routine of daily life moved in the same direction. The great painters were not far behind. The Sistine Chapel, no less than Rembrandt's famous picture, was an anatomy lesson, and Leonardo was a worthy predecessor to Vesalius, whose life overlapped his. In the sixteenth century, according to Beckmann, there were numerous private natural history collections, and in 1659 Elias Ashmole purchased the Tradescant collection, which he later presented to Oxford.

The discovery of nature as a whole was the most important part of that era of discovery which began for the Western World with the Crusades and the travels of Marco Polo and the southward ventures of the Portuguese. Nature existed to be explored, to be invaded, to be conquered, and finally, to be understood. Dissolving, the medieval dream disclosed the world of nature, as a lifting mist opens to view the rocks and trees and herds on a hillside, whose existence had been heralded only by the occasional tinkling of bells or the lowing of a cow. Unfortunately, the medieval habit of separating the soul of man from the life of the material world persisted, though the theology that supported it was weakened; for as soon as the procedure of exploration was definitely outlined in the philosophy and mechanics of the seventeenth century man himself was excluded from the picture. Technics perhaps temporarily profited by this exclusion; but in the long run the result was to prove unfortunate. In attempting to seize power man tended to reduce himself to an abstraction, or, what comes to almost the same thing, to eliminate every part of himself except that which was bent on seizing power.

6: The Obstacle of Animism

The great series of technical improvements that began to crystallize around the sixteenth century rested on a dissociation of the animate and the mechanical. Perhaps the greatest difficulty in the way of this dissociation was the persistence of inveterate habits of animistic thinking. Despite animism, such dissociations had indeed been made in the past: one of the greatest of such acts was the invention of the wheel. Even in the relatively advanced civilization of the Assyrians one sees representations of great statues being moved across bare ground on a sledge. Doubtless the notion of the wheel came originally from observing that rolling a log was easier than shoving it: but trees existed for untold years and the trimming of trees had gone on for many thousands, in all likelihood, before some neolithic inventor performed the stunning act of dissociation that made possible the cart.

So long as every object, animate or inanimate, was looked upon as the dwelling place of a spirit, so long as one expected a tree or a ship to behave like a living creature, it was next to impossible to isolate as a mechanical sequence the special function one sought to serve. Just as the Egyptian workman, when he made the leg of a chair, fashioned it to represent the leg of a bullock, so the desire naïvely to reproduce the organic, and to conjure up giants and djinns for power, instead of contriving their abstract equivalent, retarded the development of the machine. Nature often assists in such abstraction: the swan's use of its wing may have suggested the sail, even as the hornet's nest suggested paper. Conversely, the body itself is a sort of microcosm of the machine: the arms are levers, the lungs are bellows, the eyes are lenses, the heart is a pump, the fist is a hammer, the nerves are a telegraph system connected with a central station: but on the whole, the mechanical instruments were invented before the physiological functions were accurately described. The most ineffective kind of machine is the *realistic* mechanical imitation of a man or another animal: technics remembers Vaucanson for his loom, rather than for his life-like mechanical duck, which not merely ate food but went through the routine of digestion and excretion.

The original advances in modern technics became possible only when a mechanical system could be isolated from the entire tissue of relations. Not merely did the first airplane, like that of Leonardo, attempt to reproduce the motion of birds' wings: as late as 1897 Ader's batlike airplane, which now hangs in the Conservatoire des Arts et Métiers in Paris had its ribs fashioned like a bat's body, and the very propellers, as if to exhaust all the zoological possibilities, were made of thin, split wood, as much as possible like birds' feathers. Similarly, the belief that reciprocating motion, as in the movement of the arms and legs, was the "natural" form of motion was used to justify opposition to the original conception of the turbine. Branca's plan of a steam-engine at the beginning of the seventeenth century showed the boiler in the form of the head and torso of a man. Circular motion, one of the most useful and frequent attributes of a fully developed machine is, curiously, one of the least observable motions in nature: even the stars do not describe a circular course, and except for the rotifers, man himself, in occasional dances and handsprings, is the chief exponent of rotary motion.

The specific triumph of the technical imagination rested on the ability to dissociate lifting power from the arm and create a crane: to dissociate work from the action of men and animals and create the water-mill: to dissociate light from the cumbustion of wood and oil and create the electric lamp. For thousands of years animism had stood in the way of this development; for it had concealed the entire face of nature behind a scrawl of human forms: even the stars were grouped together in the living figures of Castor and Pollux or the Bull on the faintest points of resemblance. Life, not content with its own province, had flowed incontinently into stones, rivers, stars, and all the natural elements: the external environment, because it was so immediately part of man, remained capricious, mischievous, a reflection of his own disordered urges and fears.

Since the world seemed, in essence, animistic, and since these "external" powers threatened man, the only method of escape that his own will-to-power could follow was either the discipline of the self or the conquest of other men: the way of religion or the way of war. I shall discuss, in another place, the special contribution that the technique and animus of warfare made to the development of the machine; as for the discipline of the personality it was essentially, during the Middle Ages, the province of the Church, and it had gone farthest, of course, not among the peasants and nobles, still clinging to essentially pagan ways of thought, with which the Church had expediently compromised: it had gone farthest in the monasteries and the universities.

Here animism was extruded by a sense of the omnipotence of a single Spirit, refined, by the very enlargement of His duties, out of any semblance of merely human or animal capacities. God had created an orderly world, and his Law prevailed in it. His acts were perhaps inscrutable; but they were not capricious: the whole burden of the religious life was to create an attitude of humility toward the ways of God and the world he had created. If the underlying faith of the Middle Ages remained superstitious and animistic, the metaphysical doctrines of the Schoolmen were in fact anti-animistic: the gist of the matter was that God's world was not man's, and that only the church could form a bridge between man and the absolute.

The meaning of this division did not fully become apparent until the Schoolmen themselves had fallen into disrepute and their inheritors, like Descartes, had begun to take advantage of the old breach by describing on a purely mechanical basis the entire world of nature—leaving out only the Church's special province, the soul of man. It was by reason of the Church's belief in an orderly independent world, as Whitehead has shown in Science and the Modern World, that the work of science could go on so confidently. The humanists of the sixteenth century might frequently be sceptics and atheists, scandalously mocking the Church even when they remained within its fold: it is perhaps no accident that the serious scientists of the seventeenth century, like Galileo, Descartes, Leibniz, Newton, Pascal, were so uniformly devout men. The next step in development, partly made by Descartes himself, was the transfer of order from God to the Machine. For God became in the eighteenth century the Eternal Clockmaker who, having conceived and created and wound up the clock of the universe, had no further responsibility until the machine ultimately broke up—or, as the nineteenth century thought, until the works ran down.

The method of science and technology, in their developed forms, implies a sterilization of the self, an elimination, as far as possible, of the human bias and preference, including the human pleasure in man's own image and the instinctive belief in the immediate presentations of his fantasies. What better preparation could a whole culture have for such an effort than the spread of the monastic system and the multiplication of a host of separate communities, dedicated to the living of a humble and self-abnegating life, under a strict rule? Here, in the monastery, was a relatively non-animistic, non-organic world: the temptations of the body were minimized in theory and, despite strain and irregularity, often minimized in practice—more often, at all events, than in secular life. The effort to exalt the individual self was suspended in the collective routine.

Like the machine, the monastery was incapable of self-perpetuation except by renewal from without. And apart from the fact that women were similarly organized in nunneries, the monastery was like the army, a strictly masculine world. Like the army, again, it sharpened

and disciplined and focussed the masculine will-to-power: a succession of military leaders came from the religious orders, while the
leader of the order that exemplified the ideals of the Counter-Reformation began his life as a soldier. One of the first experimental
scientists, Roger Bacon, was a monk; so, again, was Michael Stifel,
who in 1544 widened the use of symbols in algebraic equations; the
monks stood high in the roll of mechanics and inventors. The spiritual
routine of the monastery, if it did not positively favor the machine,
at least nullified many of the influences that worked against it. And
unlike the similar discipline of the Buddhists, that of the Western
monks gave rise to more fertile and complex kinds of machinery than
prayer wheels.

In still another way did the institutions of the Church perhaps
prepare the way for the machine: in their contempt for the body.
Now respect for the body and its organs is deep in all the classic
cultures of the past. Sometimes, in being imaginatively projected,
the body may be displaced symbolically by the parts or organs of
another animal, as in the Egyptian Horus: but the substitution is
made for the sake of intensifying some organic quality, the power
of muscle, eye, genitals. The phalluses that were carried in a
religious procession were greater and more powerful, by representation, than the actual human organs: so, too, the images of the
gods might attain heroic size, to accentuate their vitality. The whole
ritual of life in the old cultures tended to emphasize respect for the
body and to dwell on its beauties and delights: even the monks who
painted the Ajanta caves of India were under its spell. The enthrone-
ment of the human form in sculpture, and the care of the body in
the palestra of the Greeks or the baths of the Romans, re-enforced
this inner feeling for the organic. The legend about Procrustes
typifies the horror and the resentment that classic peoples felt against
the mutilation of the body: one made beds to fit human beings, one
did not chop off legs or heads to fit beds.

This affirmative sense of the body surely never disappeared, even
during the severest triumphs of Christianity: every new pair of lovers
recovers it through their physical delight in each other. Similarly,
the prevalence of gluttony as a sin during the Middle Ages was a

witness to the importance of the belly. But the systematic teachings of the Church were directed against the body and its culture: if on one hand it was a Temple of the Holy Ghost, it was also vile and sinful by nature: the flesh tended to corruption, and to achieve the pious ends of life one must mortify it and subdue it, lessening its appetites by fasting and abstention. Such was the letter of the Church's teaching; and while one cannot suppose that the mass of humanity kept close to the letter, the feeling against the body's exposure, its uses, its celebration, was there.

While public bath houses were common in the Middle Ages, contrary to the complacent superstition that developed after the Renascence abandoned them, those who were truly holy neglected to bathe the body; they chafed their skin in hair shirts, they whipped themselves, they turned their eyes with charitable interest upon the sore and leprous and deformed. Hating the body, the orthodox minds of the Middle Ages were prepared to do it violence. Instead of resenting the machines that could counterfeit this or that action of the body, they could welcome them. The forms of the machine were no more ugly or repulsive than the bodies of crippled and battered men and women, or, if they were repulsive and ugly, they were that much further away from being a temptation to the flesh. The writer in the Nürnberg Chronicle in 1398 might say that "wheeled engines performing strange tasks and shows and follies come directly from the devil"—but in spite of itself, the Church was creating devil's disciples.

The fact is, at all events, that the machine came most slowly into agriculture, with its life-conserving, life-maintaining functions, while it prospered lustily precisely in those parts of the environment where the body was most infamously treated by custom: namely, in the monastery, in the mine, on the battlefield.

7: The Road Through Magic

Between fantasy and exact knowledge, between drama and technology, there is an intermediate station: that of magic. It was in magic that the general conquest of the external environment was decisively instituted. Without the order that the Church provided the campaign would possibly have been unthinkable; but without the wild, scrambled daring of the magicians the first positions would not have been taken. For the magicians not only believed in marvels but audaciously sought to work them: by their straining after the exceptional, the natural philosophers who followed them were first given a clue to the regular.

The dream of conquering nature is one of the oldest that has flowed and ebbed in man's mind. Each great epoch in human history in which this will has found a positive outlet marks a rise in human culture and a permanent contribution to man's security and well-being. Prometheus, the fire-bringer, stands at the beginning of man's conquest: for fire not merely made possible the easier digestion of foods, but its flames kept off predatory animals, and around the warmth of it, during the colder seasons of the year, an active social life became possible, beyond the mere huddle and vacuity of the winter's sleep. The slow advances in making tools and weapons and utensils that marked the earlier stone periods were a pedestrian conquest of the environment: gains by inches. In the neolithic period came the first great lift, with the domestication of plants and animals, the making of orderly and effective astronomical observations, and the spread of a relatively peaceful big-stone civilization in many lands separated over the planet. Fire-making, agriculture, pottery, astronomy, were marvellous collective leaps: dominations rather than adaptations. For thousands of years men must have dreamed, vainly, of further short-cuts and controls.

Beyond the great and perhaps relatively short period of neolithic invention the advances, up to the tenth century of our own era, had been relatively small except in the use of metals. But the hope of some larger conquest, some more fundamental reversal of man's dependent relation upon a merciless and indifferent external world continued to haunt his dreams and even his prayers: the myths and fairy stories are a testimony to his desire for plenitude and power, for freedom of movement and length of days.

Looking at the bird, men dreamed of flight: perhaps one of the most universal of man's envies and desires: Daedalus among the Greeks, Ayar Katsi, the flying man, among the Peruvian Indians, to say nothing of Rah and Neith, Astarte and Psyche, or the Angels of Christianity. In the thirteenth century, this dream reappeared prophetically in the mind of Roger Bacon. The flying carpet of the Arabian Nights, the seven-leagued boots, the wishing ring, were all evidences of the desire to fly, to travel fast, to diminish space, to remove the obstacle of distance. Along with this went a fairly constant desire to deliver the body from its infirmities, from its early aging, which dries up its powers, and from the diseases that threaten life even in the midst of vigor and youth. The gods may be defined as beings of somewhat more than human stature that have these powers of defying space and time and the cycle of growth and decay: even in the Christian legend the ability to make the lame walk and the blind see is one of the proofs of godhood. Imhotep and Aesculapius, by reason of their skill in the medical arts, were raised into deities by the Egyptians and the Greeks. Oppressed by want and starvation, the dream of the horn of plenty and the Earthly Paradise continued to haunt man.

It was in the North that these myths of extended powers took on an added firmness, perhaps, from the actual achievements of the miners and smiths: one remembers Thor, master of the thunder, whose magic hammer made him so potent: one remembers Loki, the cunning and mischievous god of fire: one remembers the gnomes who created the magic armor and weapons of Siegfried—Ilmarinen of the Finns, who made a steel eagle, and Wieland, the fabulous German smith, who made feather clothes for flight. Back of all these fables, these collective wishes and utopias, lay the desire to prevail over the brute nature of things.

But the very dreams that exhibited these desires were a revelation of the difficulty of achieving them. The dream gives direction to human activity and both expresses the inner urge of the organism and conjures up appropriate goals. But when the dream strides too far ahead of fact, it tends to short-circuit action: the anticipatory subjective pleasure serves as a surrogate for the thought and contrivance and action that might give it a foothold in reality. The disembodied desire, unconnected with the conditions of its fulfillment or with its means of expression, leads nowhere: at most it contributes to an inner equilibrium. How difficult was the discipline required before mechanical invention became possible one sees in the part played by magic in the fifteenth and sixteenth centuries.

Magic, like pure fantasy, was a short cut to knowledge and power. But even in the most primitive form of shamanism, magic involves a drama and an action: if one wishes to kill one's enemy by magic, one must at least mould a wax figure and stick pins into it; and similarly, if the need for gold in early capitalism promoted a grand quest for the means of transmuting base metals into noble ones, it was accompanied by fumbling and frantic attempts to manipulate the external environment. Under magic, the experimenter acknowledged that it was necessary to have a sow's ear before one could make a silk purse: this was a real advance toward matter-of-fact. "The operations," as Lynn Thorndike well says of magic, "were supposed to be efficacious here in the world of external reality": magic presupposed a public demonstration rather than a merely private gratification.

No one can put his finger on the place where magic became science, where empiricism became systematic experimentalism, where alchemy became chemistry, where astrology became astronomy, in short, where the need for immediate human results and gratifications ceased to leave its smudgy imprint. Magic was marked above all perhaps by two unscientific qualities: by secrets and mystifications, and by a certain impatience for "results." According to Agricola the transmutationists of the sixteenth century did not hesitate to conceal gold in a pellet of ore, in order to make their experiment come out successfully: similar dodges, like a concealed clock-winder, were used in the numerous perpetual motion machines that were put forward. Everywhere the dross of fraud and charlatanism mingled with the occasional grains of scientific knowledge that magic utilized or produced.

But the instruments of research were developed before a method of procedure was found; and if gold did not come out of lead in the experiments of the alchemists, they are not to be reproached for their ineptitude but congratulated on their audacity: their imaginations sniffed quarry in a cave they could not penetrate, and their

baying and pointing finally called the hunters to the spot. Something
more important than gold came out of the researches of the alchemists: the retort and the furnace and the alembic: the habit of manipulation by crushing, grinding, firing, distilling, dissolving—valuable
apparatus for real experiments, valuable methods for real science.
The source of authority for the magicians ceased to be Aristotle and
the Fathers of the Church: they relied upon what their hands could
do and their eyes could see, with the aid of mortar and pestle and
furnace. Magic rested on demonstration rather than dialectic: more
than anything else, perhaps, except painting, it released European
thought from the tyranny of the written text.

In sum, magic turned men's minds to the external world: it suggested the need of manipulating it: it helped create the tools for
successfully achieving this, and it sharpened observation as to the
results. The philosopher's stone was not found, but the science of
chemistry emerged, to enrich us far beyond the simple dreams of
the gold-seekers. The herbalist, zealous in his quest for simples and
cure-alls, led the way for the intensive explorations of the botanist
and the physician: despite our boasts of accurate coal tar drugs, one
must not forget that one of the few genuine specifics in medicine,
quinine, comes from the cinchona bark, and that chaulmoogra oil,
used with success in treating leprosy, likewise comes from an exotic
tree. As children's play anticipates crudely adult life, so did magic
anticipate modern science and technology: it was chiefly the lack of
direction that was fantastic: the difficulty was not in using the instrument but in finding a field where it could be applied and finding the
right system for applying it. Much of seventeenth century science,
though no longer tainted with charlatanism, was just as fantastic. It
needed centuries of systematic effort to develop the technique which
has given us Ehrlich's salvarsan or Bayer 207. But magic was the
bridge that united fantasy with technology: the dream of power with
the engines of fulfillment. The subjective confidence of the magicians,
seeking to inflate their private egos with boundless wealth and mysterious energies, surmounted even their practical failures: their
fiery hopes, their crazy dreams, their cracked homunculi continued

to gleam in the ashes: to have dreamed so riotously was to make the technics that followed less incredible and hence less impossible.

8: Social Regimentation

If mechanical thinking and ingenious experiment produced the machine, regimentation gave it a soil to grow in: the social process worked hand in hand with the new ideology and the new technics. Long before the peoples of the Western World turned to the machine, mechanism as an element in social life had come into existence. Before inventors created engines to take the place of men, the leaders of men had drilled and regimented multitudes of human beings: they had discovered how to reduce men to machines. The slaves and peasants who hauled the stones for the pyramids, pulling in rhythm to the crack of the whip, the slaves working in the Roman galley, each man chained to his seat and unable to perform any other motion than the limited mechanical one, the order and march and system of attack of the Macedonian phalanx—these were all machine phenomena. Whatever limits the actions and movements of human beings to their bare mechanical elements belongs to the physiology, if not to the mechanics, of the machine age.

From the fifteenth century on invention and regimentation worked reciprocally. The increase in the number and kinds of machines, mills, guns, clocks, lifelike automata, must have suggested mechanical attributes for men and extended the analogies of mechanism to more subtle and complex organic facts: by the seventeenth century this turn of interest disclosed itself in philosophy. Descartes, in analyzing the physiology of the human body, remarks that its functioning apart from the guidance of the will does not "appear at all strange to those who are acquainted with the variety of movements performed by the different automata, or moving machines fabricated by human industry, and with the help of but a few pieces compared with the great multitude of bones, nerves, arteries, veins, and other parts that are found in the body of each animal. Such persons will look upon this body as a machine made by the hand of God." But the opposite process was also true: the mechanization of human habits prepared the way for mechanical imitations.

To the degree that fear and disruption prevail in society, men tend to seek an absolute: if it does not exist, they project it. Regimentation gave the men of the period a finality they could discover nowhere else. If one of the phenomena of the breakdown of the medieval order was the turbulence that made men freebooters, discoverers, pioneers, breaking away from the tameness of the old ways and the rigor of self-imposed disciplines, the other phenomenon, related to it, but compulsively drawing society into a regimented mould, was the methodical routine of the drillmaster and the book-keeper, the soldier and the bureaucrat. These masters of regimentation gained full ascendency in the seventeenth century. The new bourgeoisie, in counting house and shop, reduced life to a careful, uninterrupted routine: so long for business: so long for dinner: so long for pleasure —all carefully measured out, as methodical as the sexual intercourse of Tristram Shandy's father, which coincided, symbolically, with the monthly winding of the clock. Timed payments: timed contracts: timed work: timed meals: from this period on nothing was quite free from the stamp of the calendar or the clock. Waste of time became for protestant religious preachers, like Richard Baxter, one of the most heinous sins. To spend time in mere sociability, or even in sleep, was reprehensible.

The ideal man of the new order was Robinson Crusoe. No wonder he indoctrinated children with his virtues for two centuries, and served as the model for a score of sage discourses on the Economic Man. Robinson Crusoe was all the more representative as a tale not only because it was the work of one of the new breed of writers, the professional journalists, but because it combines in a single setting the element of catastrophe and adventure with the necessity for invention. In the new economic system every man was for himself. The dominant virtues were thrift, foresight, skillful adaptation of means. Invention took the place of image-making and ritual; experiment took the place of contemplation; demonstration took the place of deductive logic and authority. Even alone on a desert island the sober middle class virtues would carry one through. . . .

Protestantism re-enforced these lessons of middle class sobriety and gave them God's sanction. True: the main devices of finance were a product of Catholic Europe, and Protestantism has received undeserved praise as a liberating force from medieval routine and undeserved censure as the original source and spiritual justification of modern capitalism. But the peculiar office of Protestantism was to unite finance to the concept of a godly life and to turn the asceticism countenanced by religion into a device for concentration upon worldly goods and worldly advancement. Protestantism rested firmly on the abstractions of print and money. Religion was to be found, not simply in the fellowship of religious spirits, connected historically through the Church and communicating with God through an elaborate ritual: it was to be found in the word itself: the word without its communal background. In the last analysis, the individual must fend for himself in heaven, as he did on the exchange. The expression of collective beliefs through the arts was a snare: so the Protestant stripped the images from his Cathedral and left the bare stones of engineering: he distrusted all painting, except perhaps portrait painting, which mirrored his righteousness; and he looked upon the theater and the dance as a lewdness of the devil. Life, in all its sensuous variety and warm delight, was drained out of the Protestant's world of thought: the organic disappeared. Time was real: keep it! Labor was real: exert it! Money was real: save it! Space was real: conquer it! Matter was real: measure it! These were the realities and the imperatives of the middle class philosophy. Apart from the surviving scheme of divine salvation all its impulses were already put under the rule of weight and measure and quantity: day and life were completely regimented. In the eighteenth century Benjamin Franklin, who had perhaps been anticipated by the Jesuits, capped the process by inventing a system of moral book-keeping.

How was it that the power motive became isolated and intensified toward the close of the Middle Ages?

Each element in life forms part of a cultural mesh: one part implicates, restrains, helps to express the other. During this period the mesh was broken, and a fragment escaped and launched itself on a separate career—the will to dominate the environment. To dominate, not to cultivate: to seize power, not to achieve form. One cannot, plainly, embrace a complex series of events in such simple terms

alone. Another factor in the change may have been due to an intensified sense of inferiority: this perhaps arose through the humiliating
disparity between man's ideal pretensions and his real accomplishments—between the charity and peace preached by the Church and
its eternal wars and feuds and animosities, between the holy life as
preached by the saints and the lascivious life as lived by the Renascence
Popes, between the belief in heaven and the squalid disorder and
distress of actual existence. Failing redemption by grace, harmoniza-
tion of desires, the Christian virtues, people sought, perhaps, to wipe
out their sense of inferiority and overcome their frustration by seeking power.

At all events, the old synthesis had broken down in thought and in
social action. In no little degree, it had broken down because it was
an inadequate one: a closed, perhaps fundamentally neurotic conception of human life and destiny, which originally had sprung out
of the misery and terror that had attended both the brutality of
imperialistic Rome and its ultimate putrefaction and decay. So
remote were the attitudes and concepts of Christianity from the facts
of the natural world and of human life, that once the world itself
was opened up by navigation and exploration, by the new cosmology,
by new methods of observation and experiment, there was no returning to the broken shell of the old order. The split between the
Heavenly system and the Earthly one had become too grave to be
overlooked, too wide to be bridged: human life had a destiny outside that shell. The crudest science touched closer to contemporary
truth than the most refined scholasticism: the clumsiest steam engine
or spinning jenny had more efficiency than the soundest guild regulation, and the paltriest factory and iron bridge had more promise for
architecture than the most masterly buildings of Wren and Adam;
the first yard of cloth woven by machine, the first plain iron casting,
had potentially more esthetic interest than jewelry fashioned by a
Cellini or the canvas covered by a Reynolds. In short: a live machine
was better than a dead organism; and the organism of medieval
culture was dead.

From the fifteenth century to the seventeenth men lived in an
empty world: a world that was daily growing emptier. They said

their prayers, they repeated their formulas; they even sought to retrieve the holiness they had lost by resurrecting superstitions they had long abandoned: hence the fierceness and hollow fanaticism of the Counter-Reformation, its burning of heretics, its persecution of witches, precisely in the midst of the growing "enlightenment." They threw themselves back into the medieval dream with a new intensity of feeling, if not conviction: they carved and painted and wrote—who indeed ever hewed more mightily in stone than Michelangelo, who wrote with more spectacular ecstasy and vigor than Shakespeare? But beneath the surface occupied by these works of art and thought was a dead world, an empty world, a void that no amount of dash and bravura could fill up. The arts shot up into the air in a hundred pulsing fountains, for it is just at the moment of cultural and social dissolution that the mind often works with a freedom and intensity that is not possible when the social pattern is stable and life as a whole is more satisfactory: but the idolum itself had become empty.

Men no longer believed, without practical reservations, in heaven and hell and the communion of the saints: still less did they believe in the smooth gods and goddesses and sylphs and muses whom they used, with elegant but meaningless gestures, to adorn their thoughts and embellish their environment: these supernatural figures, though they were human in origin and in consonance with certain stable human needs, had become wraiths. Observe the infant Jesus of a thirteenth century altarpiece: the infant lies on an altar, apart; the Virgin is transfixed and beatified by the presence of the Holy Ghost: the myth is real. Observe the Holy Families of the sixteenth and seventeenth century painting: fashionable young ladies are coddling their well-fed human infants: the myth has died. First only the gorgeous clothes are left: finally a doll takes the place of the living child: a mechanical puppet. Mechanics became the new religion, and it gave to the world a new Messiah: the machine.

9: The Mechanical Universe

The issues of practical life found their justification and their appropriate frame of ideas in the natural philosophy of the seventeenth century: this philosophy has remained, in effect, the working creed of technics, even though its ideology has been challenged, modified, amplified, and in part undermined by the further pursuit of science itself. A series of thinkers, Bacon, Descartes, Galileo, Newton, Pascal, defined the province of science, elaborated its special technique of research, and demonstrated its efficacy.

At the beginning of the seventeenth century there were only scattered efforts of thought, some scholastic, some Aristotelian, some mathematical and scientific, as in the astronomical observations of Copernicus, Tycho Brahe,. and Kepler: the machine had had only an incidental part to play in these intellectual advances. At the end, despite the relative sterility of invention itself during this century, there existed a fully articulated philosophy of the universe, on purely mechanical lines, which served as a starting point for all the physical sciences and for further technical improvements: the mechanical *Weltbild* had come into existence. Mechanics set the pattern of successful research and shrewd application. Up to this time the biological sciences had paralleled the physical sciences: thereafter, for at least a century and a half, they played second fiddle; and it was not until after 1860 that biological facts were recognized as an important basis for technics.

By what means was the new mechanical picture put together? And how did it come to provide such an excellent soil for the propagation of inventions and the spread of machines?

The method of the physical sciences rested fundamentally upon a few simple principles. First: the elimination of qualities, and the reduction of the complex to the simple by paying attention only to those aspects of events which could be weighed, measured, or counted, and to the particular kind of space-time sequence that could be controlled and repeated—or, as in astronomy, whose repetition could be predicted. Second: concentration upon the outer world, and the elimination or neutralization of the observer as respects the *data* with which he works. Third: isolation: limitation of the field: specialization of interest and subdivision of labor. In short, what the physical sciences call the world is not the total object of common human experience: it is just those aspects of this experience that lend themselves to accurate factual observation and to generalized statements. One may define a mechanical system as one in which any random sample of the whole will serve in place of the whole: an ounce of pure water in the laboratory is supposed to have the same properties as a hundred cubic feet of equally pure water in the cistern and the environment of the object is not supposed to affect its behavior. Our modern concepts of space and time make it seem doubtful if any pure mechanical system really exists: but the original bias of natural philosophy was to discard organic complexes and to seek isolates which could be described, *for practical purposes,* as if they completely represented the "physical world" from which they had been extracted.

This elimination of the organic had the justification not only of practical interest but of history itself. Whereas Socrates had turned his back upon the Ionian philosophers because he was more concerned to learn about man's dilemmas than to learn about trees, rivers, and stars, all that could be called positive knowledge, which had survived the rise and fall of human societies, were just such nonvital truths as the Pythagorean theorem. In contrast to the cycles of taste, doctrine, fashion, there had been a steady accretion of mathematical and physical knowledge. In this development, the study of astronomy had been a great aid: the stars could not be cajoled or perverted: their courses were visible to the naked eye and could be followed by any patient observer.

Compare the complex phenomenon of an ox moving over a winding uneven road with the movements of a planet: it is easier to trace an entire orbit than to plot the varying rate of speed and the changes of position that takes place in the nearer and more familiar object. *To fix attention upon a mechanical system was the first step toward creating system:* an important victory for rational thought. By centering effort upon the non-historic and the inorganic, the physical sciences clarified the entire procedure of analysis: for the field to which they confined their attention was one in which the method could be pushed farthest without being too palpably inadequate or encountering too many special difficulties. But the real physical world was still not simple enough for the scientific method in its first stages of development: it was necessary to reduce it to such elements as could be ordered in terms of space, time, mass, motion, quantity. The amount of elimination and rejection that accompanied this was excellently described by Galileo, who gave the process such a strong impetus. One must quote him in full:

"As soon as I form a conception of a material or corporeal substance, I simultaneously feel the necessity of conceiving that it has boundaries of some shape or other; that relatively to others it is great or small; that it is in this or that place, in this or that time; that it is in motion or at rest; that it touches, or does not touch, another body; that it is unique, rare, or common; nor can I, by any act of imagination, disjoin it from these qualities. But I do not find myself absolutely compelled to apprehend it as necessarily accompanied by such conditions as that it must be white or red, bitter or sweet, sonorous or silent, smelling sweetly or disagreeably; and if the senses had not pointed out these qualities language and imagination alone could never have arrived at them. Therefore I think that these tastes, smells, colors, etc., with regard to the object in which they appear to reside, are nothing more than mere names. They exist only in the sensitive body, for when the living creature is removed all these qualities are carried off and annihilated, although we have imposed particular names upon them, and would fain persuade ourselves that they truly and in fact exist. I do not believe that there exists anything in external bodies for exciting tastes, smells, and sounds, etc., except size, shape, quantity, and motion."

In other words, physical science confined itself to the so-called primary qualities: the secondary qualities are spurned as subjective. But a primary quality is no more ultimate or elementary than a secondary quality, and a sensitive body is no less real than an insensitive body. Biologically speaking, smell was highly important for survival: more so, perhaps, than the ability to discriminate distance or weight: for it is the chief means of determining whether food is fit to eat, and pleasure in odors not merely refined the process of eating but gave a special association to the visible symbols of erotic interest, sublimated finally in perfume. The primary qualities could be called prime only in terms of mathematical analysis, because they had, as an ultimate point of reference, an independent measuring stick for time and space, a clock, a ruler, a balance.

The value of concentrating upon primary qualities was that it neutralized in experiment and analysis the sensory and emotional reactions of the observer: apart from the process of thinking, he became an instrument of record. In this manner, scientific technique became communal, impersonal, objective, within its limited field, the purely conventional "material world." This technique resulted in a valuable moralization of thought: the standards, first worked out in realms foreign to man's personal aims and immediate interests, were equally applicable to more complex aspects of reality that stood closer to his hopes, loves, ambitions. But the first effect of this advance in clarity and in sobriety of thought was to devaluate every department of experience except that which lent itself to mathematical investigation. When the Royal Society was founded in England, the humanities were deliberately excluded.

In general, the practice of the physical sciences meant an intensification of the senses: the eye had never before been so sharp, the ear so keen, the hand so accurate. Hooke, who had seen how glasses improved seeing, doubted not that "there may be found Mechanical Inventions to improve our other senses, of hearing, smelling, tasting, touching." But with this gain in accuracy, went a deformation of experience as a whole. The instruments of science were helpless in the realm of qualities. The qualitative was reduced to the subjective: the subjective was dismissed as unreal, and the unseen and unmeasurable non-existent. Intuition and feeling did not affect mechanical process or mechanical explanations. Much could be accomplished by the new science and the new technics because much that was associated with life and work in the past—art, poetry, organic rhythm, fantasy—was deliberately eliminated. As the outer world of perception grew in importance, the inner world of feeling became more and more impotent.

The division of labor and the specialization in single parts of an operation, which already had begun to characterize the economic life of the seventeenth century, prevailed in the world of thought:

they were expressions of the same desire for mechanical accuracy and for quick results. The field of research was progressively divided up, and small parts of it were subject to intensive examination: in small measures, so to say, truth might perfect be. This restriction was a great practical device. To know the complete nature of an object does not necessarily make one fit to work with it: for complete knowledge requires a plenitude of time: moreover, it tends finally to a sort of identification which lacks precisely the cool aloofness that enables one to handle it and manipulate it for external ends. If one wishes to eat a chicken, one had better treat it as food from the beginning, and not give it too much friendly attention or human sympathy or even esthetic appreciation: if one treats the life of the chicken as an end, one may even with Brahminical thoroughness preserve the lice in its feathers as well as the bird. Selectivity is an operation necessarily adopted by the organism to keep it from being overwhelmed with irrelevant sensations and comprehensions. Science gave this inevitable selectivity a new rationale: it singled out the most negotiable set of relations, mass, weight, number, motion.

Unfortunately, isolation and abstraction, while important to orderly research and refined symbolic representation, are likewise conditions under which real organisms die, or at least cease to function effectively. The rejection of experience in its original whole, besides abolishing images and disparaging the non-instrumental aspects of thought, had another grave result: on the positive side, it was a belief in the dead; for the vital processes often escape close observation so long as the organism is alive. In short, the accuracy and simplicity of science, though they were responsible for its colossal practical achievements, were not an approach to objective reality but a departure from it. In their desire to achieve exact results the physical sciences scorned true objectivity: individually, one side of the personality was paralyzed; collectively, one side of experience was ignored. To substitute mechanical or two-way time for history, the dissected corpse for the living body, dismantled units called "individuals" for men-in-groups, or in general the mechanically measurable or reproducible for the inaccessible and the complicated and the organically whole, is to achieve a limited practical mastery at the expense of truth and of the larger efficiency that depends on truth.

By confining his operations to those aspects of reality which had, so to say, market value, and by isolating and dismembering the corpus of experience, the physical scientist created a habit of mind favorable to discrete practical inventions: at the same time it was highly unfavorable to all those forms of art for which the secondary qualities and the individualized receptors and motivators of the artist were of fundamental importance. By his consistent metaphysical principles and his factual method of research, the physical scientist denuded the world of natural and organic objects and turned his back upon real experience: he substituted for the body and blood of reality a skeleton of effective abstractions which he could manipulate with appropriate wires and pulleys.

What was left was the bare, depopulated world of matter and motion: a wasteland. In order to thrive at all, it was necessary for the inheritors of the seventeenth century idolum to fill the world up again with new organisms, devised to represent the new realities of physical science. Machines—and machines alone—completely met the requirements of the new scientific method and point of view: they fulfilled the definition of "reality" far more perfectly than living organisms. And once the mechanical world-picture was established, machines could thrive and multiply and dominate existence: their competitors had been exterminated or had been consigned to a penumbral universe in which only artists and lovers and breeders of animals dared to believe. Were machines not conceived in terms of primary qualities alone, without regard to appearance, sound, or any other sort of sensory stimulation? If science presented an ultimate reality, then the machine was, like the law in Gilbert's ballad, the true embodiment of everything that was excellent. Indeed in this empty, denuded world, the invention of machines became a duty. By renouncing a large part of his humanity, a man could achieve godhood: he dawned on this second chaos and created the machine in his own image: the image of power, but power ripped loose from his flesh and isolated from his humanity.

10: The Duty to Invent

The principles that had proved effective in the development of the scientific method were, with appropriate changes, those that served as a foundation for invention. Technics is a translation into appropriate, practical forms of the theoretic truths, implicit or formulated, anticipated or discovered, of science. Science and technics form two independent yet related worlds: sometimes converging, sometimes drawing apart. Mainly empirical inventions, like the steam-engine, may suggest Carnot's researches in thermodynamics: abstract physical investigation, like Faraday's with the magnetic field, may lead directly to the invention of the dynamo. From the geometry and astronomy of Egypt and Mesopotamia, both closely connected with the practice of agriculture to the latest researches in electro-physics, Leonardo's dictum holds true: Science is the captain and practice the soldiers. But sometimes the soldiers win the battle without leadership, and sometimes the captain, by intelligent strategy, obtains victory without actually engaging in battle.

The displacement of the living and the organic took place rapidly with the early development of the machine. For the machine was a counterfeit of nature, nature analyzed, regulated, narrowed, controlled by the mind of men. The ultimate goal of its development was however not the mere conquest of nature but her resynthesis: dismembered by thought, nature was put together again in new combinations: material syntheses in chemistry, mechanical syntheses in engineering. The unwillingness to accept the natural environment as a fixed and final condition of man's existence had always contributed both to his art and his technics: but from the seventeenth century, the attitude became compulsive, and it was to technics that he turned for fulfillment. Steam engines displaced horse power, iron and concrete displaced wood, aniline dyes replaced vegetable dyes, and so on down the line, with here and there a gap. Sometimes the new product was superior practically or esthetically to the old, as in the infinite superiority of the electric lamp over the tallow candle: sometimes the new product remained inferior in quality, as rayon is still inferior to natural silk: but in either event the gain was in the creation of an equivalent product or synthesis which was less dependent upon uncertain organic variations and irregularities in either the product itself or the labor applied to it than was the original.

Often the knowledge upon which the displacement was made was insufficient and the result was sometimes disastrous. The history of the last thousand years abounds in examples of apparent mechanical and scientific triumphs which were fundamentally unsound. One need only mention bleeding in medicine, the use of common window glass which excluded the important ultra-violet rays, the establishment of the post-Liebig dietary on the basis of mere energy replacement, the use of the elevated toilet seat, the introduction of steam heat, which dries the air excessively—but the list is a long and somewhat appalling one. The point is that invention had become a duty, and the desire to use the new marvels of technics, like a child's delighted bewilderment over new toys, was not in the main guided by critical discernment: people agreed that inventions were good, whether or not they actually provided benefits, just as they agreed that child-bearing was good, whether the offspring proved a blessing to society or a nuisance.

Mechanical invention, even more than science, was the answer to a dwindling faith and a faltering life-impulse. The meandering energies of men, which had flowed over into meadow and garden, had crept into grotto and cave, during the Renascence, were turned by invention into a confined head of water above a turbine: they could sparkle and ripple and cool and revive and delight no more: they were harnessed for a narrow and definite purpose: to move wheels and multiply society's capacity for work. To live was to work: *what other life indeed do machines know?* Faith had at last found a new object, not the moving of mountains, but the moving of engines and machines. Power: the application of power to motion, and the application of motion to production, and of production to money-making, and so the further increase of power—this was the worthiest object that a mechanical habit of mind and a mechanical mode of action put before men. As everyone recognizes, a thousand salutary instruments came out of the new technics; but in origin from the seventeenth century on the machine served as a substitute religion, and a vital religion does not need the justification of mere utility.

The religion of the machine needed such support as little as the transcendental faiths it supplanted: for the mission of religion is to provide an ultimate significance and motive-force: the necessity of invention was a dogma, and the ritual of a mechanical routine was the binding element in the faith. In the eighteenth century, Mechanical Societies sprang into existence, to propagate the creed with greater zeal: they preached the gospel of work, justification by faith in mechanical science, and salvation by the machine. Without the missionary enthusiasm of the enterprisers and industrialists and engineers and even the untutored mechanics from the eighteenth century onward, it would be impossible to explain the rush of con-verts and the accelerated tempo of mechanical improvement. The impersonal procedure of science, the hard-headed contrivances of mechanics, the rational calculus of the utilitarians—these interests captured emotion, all the more because the golden paradise of finan-cial success lay beyond.

In their compilation of inventions and discoveries, Darmstaedter and Du Bois-Reymond enumerated the following inventors: between 1700 and 1750—170: between 1750 and 1800—344: between 1800 and 1850—861: between 1850 and 1900—1150. Even allowing for the foreshortening brought about automatically by historical per-spective, one cannot doubt the increased acceleration between 1700 and 1850. Technics had seized the imagination: the engines them-selves and the goods they produced both seemed immediately desir-able. While much good came through invention, much invention came irrespective of the good. If the sanction of utility had been uppermost, invention would have proceeded most rapidly in the de-partments where human need was sharpest, in food, shelter, and clothing: but although the last department undoubtedly advanced, the farm and the common dwelling house were much slower to profit by the new mechanical technology than were the battlefield and the mine, while the conversion of gains in energy into a life abundant took place much more slowly after the seventeenth century than it had done during the previous seven hundred years.

Once in existence, the machine tended to justify itself by silently taking over departments of life neglected in its ideology. Virtuosity is an important element in the development of technics: the interest in the materials as such, the pride of mastery over tools, the skilled manipulation of form. The machine crystallized in new patterns the whole set of independent interests which Thorstein Veblen grouped loosely under "the instinct of workmanship," and enriched technics as a whole even when it temporarily depleted handicraft. The very sensual and contemplative responses, excluded from love-making and song and fantasy by the concentration upon the mechanical means of production, were not of course finally excluded from life: they re-entered it in association with the technical arts themselves, and the machine, often lovingly personified as a living creature, as with Kipling's engineers, absorbed the affection and care of both inventor and workman. Cranks, pistons, screws, valves, sinuous motions, pulsations, rhythms, murmurs, sleek surfaces, all are virtual counterparts of the organs and functions of the body, and they stimulated and absorbed some of the natural affections. But when that stage was reached, the machine was no longer a means and its operations were not merely mechanical and causal, but human and final: it contributed, like any other work of art, to an organic equilibrium. This development of value within the machine complex itself, apart from the value of the products created by it, was, as we shall see at a later stage, a profoundly important result of the new technology.

11: Practical Anticipations

From the beginning, the practical value of science was uppermost in the minds of its exponents, even in those who single-mindedly pursued abstract truth, and who were as indifferent to its popularization as Gauss and Weber, the scientists who invented the telegraph for their private communication. "If my judgment be of any weight," said Francis Bacon in The Advancement of Learning, "the use of history mechanical is of all others the most radical and fundamental towards natural philosophy: such natural philosophy as shall not vanish in the fume of subtile, sublime, or delectable speculation, but such as shall be operative to the endowment and benefit of man's life." And Descartes, in his Discourse on Method, observes: "For by them [general restrictions respecting physics] I perceived it to be possible to arrive at knowledge highly useful in life; and in lieu of the speculative philosophy usually taught in the schools to discover a practical, by means of which, knowing the force and action of fire, water, air, the stars, the heavens, and all the other bodies that surround us, as distinctly as we know the various crafts of our artisans, we might also apply them in the same way to all the uses to which they are adapted, and thus render ourselves the lords and possessors of nature. And this is a result to be desired, not only in order to the invention of an infinity of arts, by which we might be able to enjoy without any trouble the fruits of the earth, and all its comforts, but also especially for the preservation of health, which is without doubt of all blessings of this life the first and fundamental one; for the mind is so intimately dependent upon the condition and relation of the organs of the body that if any means can ever be found to render men wiser and more ingenious than hitherto, I believe that it is in medicine they must be sought for."

Who is rewarded in the perfect commonwealth devised by Bacon in The New Atlantis? In Salomon's House the philosopher and the artist and the teacher were left out of account, even though Bacon, like the prudent Descartes, clung very ceremoniously to the rites of the Christian church. For the "ordinances and rites" of Salomon's House there are two galleries. In one of these "we place patterns and samples of all manner of the more rare and excellent inventions: in the other we place the statues of all principal Inventors. There we have the statue of your Columbus, that discovered the West Indies: also the Inventor of Ships: your monk that was the Inventor of Ordnance and Gunpowder: the Inventor of Music: the Inventor of Letters: the Inventor of Printing: the Inventor of observations by astronomy: the Inventor of Works in Metal: the Inventor of Glass: the Inventor of Silk of the Worm: the Inventor of Wine: the Inventor of Corn and Bread: the Inventor of Sugars. . . . For upon every invention of value, we erect a statue to the Inventor and give him a liberal and honorable reward." This Salomon's House, as Bacon fancied it, was a combination of the Rockefeller Institute and the Deutsches Museum: there, if anywhere, was the means to-wards the relief of man's estate.

Observe this: there is little that is vague or fanciful in all these conjectures about the new rôle to be played by science and the machine. The general staff of science had worked out the strategy of the campaign long before the commanders in the field had developed a tactics capable of carrying out the attack in detail. Indeed, Usher notes that in the seventeenth century invention was relatively feeble, and the power of the technical imagination had far outstripped the actual capacities of workmen and engineers. Leonardo, Andreae, Campanella, Bacon, Hooke in his Micrographia and Glanvill in his Scepsis Scientifica, wrote down in outline the specifications for the new order: the use of science for the advancement of technics, and the direction of technics toward the conquest of nature were the burden of the whole effort. Bacon's Salomon's House, though for-mulated after the actual founding of the Accademia Lynxei in Italy, was the actual starting point of the Philosophical College that first met in 1646 at the Bullhead Tavern in Cheapside, and in 1662 was duly incorporated as the Royal Society of London for Improving Natural Knowledge. This society had eight standing committees, the first of which was to "consider and improve all mechanical inven-tions." The laboratories and technical museums of the twentieth cen-tury existed first as a thought in the mind of this philosophical cour-tier: nothing that we do or practice today would have surprised him.

So confident in the results of the new approach was Hooke that he wrote: "There is nothing that lies within the power of human wit (or which is far more effectual) of human industry which we might not compass; we might not only hope for inventions to equalize those of Copernicus, Galileo, Gilbert, Harvey, and others, whose names are almost lost, that were the inventors of Gunpowder, the Seaman's Compass, Printing, Etching, Graving, Microscopes, Etc., but multitudes that may far exceed them: for even those discovered seem to have been the product of some such methods though but imperfect; what may not be therefore expected from it if thoroughly prosecuted? Talking and contention of Arguments would soon be turned into labors; all the fine dreams and opinions and universal metaphysical nature, which the luxury of subtil brains has devised, would quickly vanish and give place to solid histories, experiments, and works."

The leading utopias of the time, Christianopolis, the City of the Sun, to say nothing of Bacon's fragment or Cyrano de Bergerac's minor works, all brood upon the possibility of utilizing the machine to make the world more perfect: the machine was the substitute for Plato's justice, temperance, and courage, even as it was likewise for the Christian ideals of grace and redemption. The machine came forth as the new demiurge that was to create a new heaven and a new earth: at the least, as a new Moses that was to lead a barbarous humanity into the promised land.

There had been premonitions of all this in the centuries before. "I will now mention," said Roger Bacon, "some of the wonderful works of art and nature in which there is nothing of magic and which magic could not perform. Instruments may be made by which the largest ships, with only one man guiding them, will be carried with greater velocity than if they were full of sailors. Chariots may be constructed that will move with incredible rapidity without the help of animals. Instruments of flying may be formed in which a man, sitting at his ease and meditating in any subject, may beat the air with his artificial wings after the manner of birds . . . as also machines which will enable men to walk at the bottom of seas or rivers without ships." And Leonardo de Vinci left behind him a list of inventions and contrivances that reads like a synopsis of the present industrial world.

But by the seventeenth century the note of confidence had increased, and the practical impulse had become more universal and urgent. The works of Porta, Cardan, Besson, Ramelli, and other ingenious inventors, engineers, and mathematicians are a witness both to increasing skill and to growing enthusiasm over technics itself. Schwenter in his Délassements Physico-Mathématiques (1636) pointed out how two individuals could communicate with each other by means of magnetic needles. "To them that come after us," said Glanvill, "it may be as ordinary to buy a pair of wings to fly to remotest regions, as now a pair of boots to ride a journey; and to confer at the distance of the Indies by sympathetic conveyances may be as usual in future times as by literary correspondence." Cyrano de Bergerac conceived the phonograph. Hooke observed that it is "not impossible to hear a whisper a furlong's distance, it having been already done; and perhaps the nature of things would not make it more impossible, although that furlong be ten times multiplied." Indeed, he even forecast the invention of artificial silk. And Glanvill said again: "I doubt not posterity will find many things that are now but rumors verified into practical realities. It may be that, some ages hence, a voyage to the Southern tracts, yea, possibly to the moon, will not be more strange than one to America. . . . The restoration of grey hairs to juvenility and the renewing the exhausted marrow may at length be effected without a miracle; and the turning of the now comparatively desert world into a paradise may not improbably be effected from late agriculture." (1661)

Whatever was lacking in the outlook of the seventeenth century it was not lack of faith in the imminent presence, the speedy development, and the profound importance of the machine. Clock-making: time-keeping: space-exploration: monastic regularity: bourgeois order: technical devices: protestant inhibitions: magical explorations: finally the magistral order, accuracy, and clarity of the physical sciences themselves—all these separate activities, inconsiderable perhaps in themselves, had at last formed a complex social and ideological network, capable of supporting the vast weight of the machine and extending its operations still further. By the middle of the eighteenth century the initial preparations were over and the key inventions had been made. An army of natural philosophers, rationalists, experimenters, mechanics, ingenious people, had assembled who were clear as to their goal and confident as to their victory. Before more than a streak of grey had appeared at the horizon's rim, they proclaimed the dawn and announced how wonderful it was: how marvelous the new day would be. Actually, they were to announce a shift in the seasons, perhaps a long cyclical change in the climate itself.

CHAPTER II.   AGENTS OF MECHANIZATION

1: The Profile of Technics

The preparation for the machine that took place between the tenth and the eighteenth century gave it a broad foundation and assured its speedy and universal conquest throughout Western Civilization. But in back of this lay the long development of technics itself: the original exploration of the raw environment, the utilization of objects shaped by nature—shells and stones and animal gut—for tools and utensils: the development of fundamental industrial processes, digging, chipping, hammering, scraping, spinning, drying: the deliberate shaping of specific tools as necessities pressed and as skill increased.

Experimental sampling, as with edibles, happy accidents, as with glass, true causal insight as with the fire-drill: all these played a part in the transformation of our material environment and steadily modified the possibilities of social life. If discovery comes first, as it apparently does in the utilization of fire, in the use of meteoric iron, in the employment of hard cutting edges such as shells, invention proper follows close at its heels: indeed, the age of invention is only another name for the age of man. If man is rarely found in the "state of nature" it is only because nature is so constantly modified by technics.

To sum up these earlier developments of technics, it may be useful to associate them with the abstract scheme of the valley section: the ideal profile of a complete mountain-and-river system. In a figurative sense, civilization marches up and down the valley-section: all the great historic cultures, with the partial exception of those secluded maritime cultures in which the seas sometimes served instead of a river, have thriven through the movement of men and institutions and inventions and goods along the natural highway of a great river: the Yellow River, the Tigris, the Nile, the Euphrates, the Rhine, the Danube, the Thames. Against the primitive backgrounds of the valley section are developed the earlier forms of technics: within the cities, the processes of invention are quickened, a multitude of new needs arises, the exigencies of close living and of a limited food supply lead to fresh adaptations and ingenuities, and in the very act of putting primitive conditions at a distance men are forced to devise substitutes for the cruder artifacts which had once ensured their survival.

Taking the purely schematic valley section in profile, one finds toward the mountain top, where on the steeper slope the rocks perhaps crop out, the quarry and the mine: almost from the dawn of history itself man engages in these occupations. It is the survival, into our own times, of the prototype of all economic activity: the stage of directly seeking and picking and collecting: berries, funguses, stones, shells, dead animals. Down to modern times, mining remained technically one of the crudest of occupations: the pick and the hammer were its principal tools. But the derivative arts of mining steadily developed in historic times: indeed the use of metals is the main element that distinguishes the later crafts of Europe up to the tenth century A.D. from the stone cultures that came before: smelting, refining, smithing, casting, all increased the speed of production, improved the forms of tools and weapons, and greatly added to their strength and effectiveness. In the forest that stretches from the crown of the mountain seaward the hunter stalks his game: his is possibly the oldest deliberate technical operation of mankind for in their origin the weapon and the tool are interchangeable. The simple hammerhead serves equally as a missile: the knife kills the game and cuts it up: the ax may cut down a tree or slay an enemy. Now the hunter survives by skill of arm and eye, now by physical strength, now by the cunning contrivance of traps and pitfalls. In the pursuit of his game he does not remain in the forest but follows wherever the chase may lead him: a habit which often leads to conflicts and hostilities in the invaded areas: perhaps in the development of war as an institutional routine.

Farther down the valley, where the little mountain torrents and brooks gather together in a stream, which facilitates transportation, is the realm of the primitive woodman: the wood chopper, the forester, the millwright, the carpenter. He cuts down trees, he hollows out wooden canoes, he contrives the bow which is perhaps the most effective type of early prime mover, and he invents the fire drill, in whose widened disc Renard sees the origin of the pulley and perhaps of the wheel, to say nothing of the windlass. The woodman's ax is the chief primitive tool of mankind: his beaverlike occupation—which perhaps accidentally resulted in the human re-invention of the bridge and the dam—is apparently the original form of modern engineering; and the most important instruments of precision in the transmission of motion and the shaping of materials came from him: above all, the lathe.

Below the ideal forest line, becoming more visible with the advance of a settled culture, as the woodman's ax opens up the clearings and the seeds that are dropped in the sunny glades are nurtured through the summer and grow with a new lushness—below the primitive woodman lies the province of the herdsman and the peasant. Goatherd, shepherd, cowherd, occupy the upland pastures or the broad grasslands of the plain-plateaus in their first or final stages of erosion. Spinning itself, the art by means of which frail filatal elements are strengthened through twisting, is one of the earliest of the great inventions, and may first have been applied to the sinews of animals: thread and string were originally used where we should now utilize them only in an emergency—as in fastening an ax-head to a handle. But the spinning and weaving of fabrics for clothing, for tents, or for rugs to serve as temporary floor in the tent, are the work of the herdsman: they came in with the domestication of animals in the neolithic period, and some of the earliest forms of the spindle and loom have remained in existence among primitive peoples.

Below the more barren pastures, the peasant takes permanent possession of the land and cultivates it. He expands into the heavier

river-bottom soils as his command over tools and domesticated animals grows, or as the struggle for existence becomes more keen: he
may even reach back into the hinterland and bring under cultivation
the potentially arable pasture. The farmer's tools and machines are
relatively few: as with the herdsman, his inventive capacities are
expended directly, for the most part, upon the plants themselves in
their selection and breeding and perfection. His tools remain without
fundamental change throughout the greater part of recorded history:
the hoe, the mattock, the plow, the spade, and the scythe. But his
utensils and his utilities are many: the irrigation ditch, the cellar,
the storage-bin, the cistern, the well, and the permanent dwelling
house occupied throughout the year, belong to the peasant: partly out
of his need for defence and cooperative action grow the village and
the town. Finally, at the oceanside, plying in and out behind the
barrier beaches and the salt marshes, lives the fisherman: a sort of
aquatic hunter. The first fisherman to construct a weir possibly invented the art of weaving: the net and the basket made out of the
reeds of the marshland certainly came out of this environment, and
the most important early mode of transport and communication, the
boat, was a direct product.

The order and security of an agricultural and pastoral civilization
was the critical improvement that came in with the neolithic period.
Out of that stability grew not merely the dwelling house and the permanent community but a cooperative economic and social life, per-
petuating its institutions by means of visible buildings and memorials
as well as by the imparted word. Into the special meeting-places that
arose more and more frequently in the areas of transition between
one phase of economic activity and another, the market grew up: in
certain kinds of goods, amber, obsidian, flint, and salt, trade over
wide areas developed at a very early period. With the exchange of
more finished kinds of goods went an exchange likewise of technological skill and knowledge: in terms of our diagrammatic valley section,
special environments, special occupational types, special techniques
shifted over from one part to another and intermingled; the result
was a steady enrichment and increasing complication of the culture
itself and the technical heritage. Lacking impersonal methods of

record, the transmission of craft-knowledge tended to create occupational castes. The conservation of skill by these means led to downright conservatism: the very refinements of traditional knowledge served, perhaps, as a brake on invention.

The various elements in a civilization are never in complete equilibrium: there is always a tug and pull of forces, and in particular, there are changes in the pressure exerted by the life-destroying functions and the life-conserving ones. In the neolithic period, the peasant and the herdsman were, it seems, uppermost: the dominant ways of life were the outcome of agriculture, and the religion and science of the day were directed towards a more perfect adjustment of man to the actual earth from which he drew his nourishment. Eventually these peasant civilizations succumbed to anti-vital forces that came from two related points of the compass: on one hand from trading, with its growth of an impersonal and abstract system of relations bound together by a cash nexus: on the other from the predatory tactics of the mobile hunters and shepherds, extending their hunting grounds and their pastures or, at a more advanced stage, their power to collect tribute and to rule. Only three great cultures have a continuous history throughout the historic period: the polite and pacific peasant cultures of India and China, and the mainly urban culture of the Jews: the last two distinguished particularly for their practical intelligence, their rational morals, their kindly manners, their cooperative and life-conserving institutions; whereas the predominantly military forms of civilization have proved self-destructive.

With the dawn of modern technics in Northern Europe one sees these primitive types once more in their original character and their typical habitats. The redifferentiation of occupations and crafts goes on under our very eyes. The rulers of Europe once more are hunters and fishers: from Norway to Naples their prowess in the chase alternates with their conquest of men: one of their prime concerns when they conquer a land is to establish their hunting rights and set aside great parks as sacred to the game they pursue. When these hardy warriors finally supplement the spear and the ax and the firebrand with the cannon as a weapon of assault, the military arts become professionalized once more, and the support of war becomes one of the principal burdens of a civil society. The primitive mining and the primitive metallurgy goes on as it had existed for long in the past: but presently the simple arts of the miner and the smith break up into a score of specialized occupations. This process proceeds at an accelerating speed as commerce expands and the demand for gold and silver increases, as war becomes more mechanized and the demand for armor, for artillery, and for the sinews of war expands. So, too, the woodman appears in the forested areas, for much of Europe had gone back into forest and grass: presently the sawyer, the carpenter, the joiner, the turner, the wheelwright have become specialized crafts. In the growing cities, from the eleventh century on, these elementary occupations appear, differentiate, react upon each other, interchange techniques and forms. Within a few hundred years almost the entire drama of technics is re-enacted once more and technics reaches a higher plane of general achievement than any other civilization had known in the past—although in special departments it was again and again surpassed by the finer arts of the East. If one takes a cross-section of technics in the Middle Ages one has at hand most of the important elements derived from the past, and the germ of most of the growth that is to take place in the future. In the rear lies handicraft and the tool, supplemented by the simple chemical processes of the farm: in the van stands the exact arts and the machine and the new achievements in metallurgy and glass-making. Some of the most characteristic instruments of medieval technics, like the cross-bow, show in their form and workmanship the imprint of both the tool and the machine. Here, then, is a central vantage point.

2: De Re Metallica

Quarrying and mining are the prime extractive occupations: without stones and metals with sharp edges and resistant surfaces neither weapons nor tools could have passed beyond a very crude shape and a limited effectiveness—however ingeniously wood, shell and bone may have been used by primitive man before he had mastered stone. The first efficient tool seems to have been a stone held in the human hand as a hammer: the German word for fist is *die Faust,* and to this day the miner's hammer is called *ein Fäustel.*

Of all stones flint, because of its commonness in Northern Europe and because of its breaking into sharp scalloped edges, was perhaps the most important in the development of tools. With the aid of other rocks, or of a pick-ax made of reindeer horn, the flint miner extracted his stone, and by patient effort shaped it to his needs: the hammer itself had reached its present refinement of shape by the late neolithic period. During a great span of primitive life the slow perfection of stone tools was one of the principal marks of its advancing civilization and its control over the environment: this reached perhaps its highest point in the Big Stone culture, with its capacity for cooperative industrial effort, as shown in the transportation of the great stones of its outdoor temples and astronomical observatories, and in its relatively high degree of exact scientific knowledge. In its latest period the use of clay for pottery made it possible to preserve and store liquids, as well as to keep dried provisions from moisture and mildew: another victory for the primitive prospector who was learning to explore the earth and adapt its nonorganic contents to his uses.

There is no sharp breach between grubbing, quarrying and mining. The same outcrop that shows quartz may equally hold gold, and the same stream that has clayey banks may disclose a gleam or two of this precious metal—precious for primitive man not only because of its rarity but because it is soft, malleable, ductile, non-oxidizing, and may be worked without the use of fire. The use of gold and amber and jade antedates the so-called age of metals: they were prized for their rareness and their magical qualities, even more than for what could be directly made of them. And the hunt for these minerals had nothing whatever to do with extending the food-supply or establishing creature comforts: man searched for precious stones, as he cultivated flowers, because long before he had invented capitalism and mass production he had acquired more energy than he needed for bare physical survival on the terms of his existing culture.

In contrast to the forethought and sober plodding of the peasant, the work of the miner is the realm of random effort: irregular in routine and uncertain in result. Neither the peasant nor the herdsman can get rich quickly: the first clears a field or plants a row of trees this year from which perhaps only his grandchildren will get the full benefits. The rewards of agriculture are limited by the known qualities of soil and seed and stock: cows do not calve more quickly one year than another, nor do they have fifteen calves instead of one; and for the seven years of abundance seven lean years, on the law of averages, are pretty sure to follow. Luck for the peasant is usually a negative fact: hail, wind, blight, rot. But the rewards of mining may be sudden, and they may bear little relation, particularly in the early stages of the industry, either to the technical ability of the miner or the amount of labor he has expended. One assiduous prospector may wear out his heart for years without finding a rich seam; a newcomer in the same district may strike luck in the first morning he goes to work. While certain mines, like the salt mines of the Salzkammergut, have been in existence for centuries, the occupation in general is an unstable one.

Until the fifteenth century A.D., mining had perhaps made less technical progress than any other art: the engineering skill that Rome showed in aqueducts and roads did not extend in any degree to the mines. Not merely had the art remained for thousands of years in a primitive stage: but the occupation itself was one of the lowest in the human scale. Apart from the lure of prospecting, no one entered the mine in civilized states until relatively modern times except as a prisoner of war, a criminal, a slave. Mining was not regarded as a humane art: it was a form of punishment: it combined the terrors of the dungeon with the physical exacerbation of the galley. The actual work of mining, precisely because it was meant to be burdensome, was not improved during the whole of antiquity, from the earliest traces of it down to the fall of the Roman Empire. In general, not merely may one say that free labor did not enter the mines until the late Middle Ages; one must also remember that serfdom remained here, in the mines of Scotland for example, a considerable time after it had been abolished in agriculture. Possibly the myth of the Golden Age was an expression of mankind's sense of what it had lost when it acquired control of the harder metals.

Was the social degradation of mining an accident, or does it lie in the nature of things? Let us examine the occupation and its environment, as it existed through the greater part of history.

Except for surface mining, the art is pursued within the bowels of the earth. The darkness is broken by the timid flare of a lamp or a candle. Until the invention of the Davy safety lamp at the beginning of the nineteenth century this fire might ignite the "mine-damp" and exterminate by a single blast all who were within range: to this day, the possibility of such an explosion remains, since sparks may occur by accident even when electricity is used. Ground-water filters through the seams and often threatens to flood the passages. Until modern tools were invented, the passage itself was a cramped one: to extract ore, children and women were employed from the earliest days to crawl along the narrow tunnel, dragging a laden cart: women indeed were so used as beasts of burden in English mines right up to the middle of the nineteenth century. When primitive tools were not sufficient to break up the ore or open a new face, it was often necessary to light great fires in the difficult seams and then douse the stone with cold water in order to make it crack: the steam was suffocating, and the cracking might be dangerous: without strong shoring, whole galleries might fall upon the workers, and frequently this happened. The deeper down the seams went the greater the danger, the greater the heat, the greater the mechanical difficulties. Among the hard and brutal occupations of mankind, the only one that compares with old-fashioned mining is modern trench warfare; and this should cause no wonder: there is a direct connection. To this day, according to Meeker, the mortality rate among miners from accidents is four times as high as any other occupation.

If the use of metals came at a relatively late date in technics, the reason is not far to seek. Metals, to begin with, usually exist as compounds in ores; and the ores themselves are often inaccessible, hard to find, and difficult to bring to the surface: even if they lie in the open they are not easy to disengage. Such a common metal as zinc was not discovered till the sixteenth century. The extraction of metals, unlike the cutting down of trees or the digging of flint, requires high temperatures over considerable periods. Even after the metals are extracted they are hard to work: the easiest is one of the most precious, gold, while the hardest is the most useful, iron. In between are tin, lead, copper, the latter of which can be worked cold only in small masses or sheets. In short: the ores and metals are recalcitrant materials: they evade discovery and they resist treatment. Only by being softened do the metals respond: where there is metal there must be fire.

Mining and refining and smithing invoke, by the nature of the material dealt with, the ruthlessness of modern warfare: they place a premium on brute force. In the technique of all these arts the pounding operations are uppermost: the pick-ax, the sledge-hammer, the ore-crusher, the stamping machine, the steam-hammer: one must either melt or break the material in order to do anything with it. The routine of the mine involves an unflinching assault upon the physical environment: every stage in it is a magnification of power. When power-machines came in on a large scale in the fourteenth-century, it was in the military and the metallurgical arts that they were, perhaps, most widely applied.

Let us now turn to the mining environment. The mine, to begin with, is the first completely inorganic environment to be created and lived in by man: far more inorganic than the giant city that Spengler has used as a symbol of the last stages of mechanical desiccation. Field and forest and stream and ocean are the environment of life: the mine is the environment alone of ores, minerals, metals. Within the subterranean rock, there is no life, not even bacteria or protozoa, except in so far as they may filter through with the ground water or be introduced by man. The face of nature above the ground is good to look upon, and the warmth of the sun stirs the blood of the hunter on the track of game or the peasant in the field. Except for the crystalline formations, the face of the mine is shapeless: no friendly trees and beasts and clouds greet the eye. In hacking and digging the contents of the earth, the miner has no eye for the forms of things: what he sees is sheer matter, and until he gets to his vein it is only an obstacle which he breaks through stubbornly and sends up to the surface. If the miner sees shapes on the walls of his cavern, as the candle flickers, they are only the monstrous distortions of his pick or his arm: shapes of fear. Day has been abolished and the rhythm of nature broken: continuous day-and-night production first came into existence here. The miner must work by artificial light even though the sun be shining outside; still further down in the seams, he must work by artificial ventilation, too: a triumph of the "manu-factured environment."

In the underground passages and galleries of the mine there is nothing to distract the miner: no pretty wench is passing in the field with a basket on her head, whose proud breasts and flanks remind him of his manhood: no rabbit scurries across his path to arouse the hunter in him: no play of light on a distant river awakens his reverie. Here is the environment of work: dogged, unremitting, con-centrated work. It is a dark, a colorless, a tasteless, a perfumeless, as well as a shapeless world: the leaden landscape of a perpetual winter. The masses and lumps of the ore itself, matter in its least organized form, complete the picture. *The mine is nothing less in fact than the concrete model of the conceptual world which was built up by the physicists of the seventeenth century.*

There is a passage in Francis Bacon that makes one believe that the alchemists had perhaps a glimpse of this fact. He says: "If then it be true that Democritus said, *That the truth of nature lieth hid in certain deep mines and caves,* and if it be true likewise that the alchemists do so much inculcate, that Vulcan is a second nature, and imitateth that dexterously and compendiously, which nature worketh by ambages and length of time, it were good to divide natural philos-ophy into the mine and the furnace: and to make two professions or occupations of natural philosophers, some to be pioneers and some smiths; some to dig, and some to refine and hammer." Did the mine acclimate us to the views of science? Did science in turn prepare us to accept the products and the environment of the mine? The matter is not susceptible to proof: but the logical relations, if not the historical facts, are plain.

The practices of the mine do not remain below the ground: they affect the miner himself, and they alter the surface of the earth. Whatever could be said in defense of the art was said with great pith and good sense by Dr. Georg Bauer (Agricola), the German physician and scientist who wrote various compendious treatises on geology and mining at the beginning of the sixteenth century. He had the honesty to sum up his opponents' arguments in detail, even if he could not successfully refute them: so that his book De Re Metallica remains to this day a classic text, like Vitruvius on Architecture.

First as to the miner himself: "The critics," says Dr. Bauer, "say further that mining is a perilous occupation to pursue because the miners are sometimes killed by the pestilential air which they breathe; sometimes their lungs rot away; sometimes the men perish by being crushed in masses of rock; sometimes falling from ladders into the shafts, they break their arms, legs, or necks. . . . But since things like this rarely happen, and only so far as workmen are careless, they do not deter miners from carrying on their trade." This last sentence has a familiar note: it recalls the defenses of potters and radium watch-dial manufacturers when the dangers of their trades were pointed out. Dr. Bauer forgot only to note that though *coal* miners are not particularly susceptible to tuberculosis, the coldness and dampness, sometimes the downright wetness, predispose the miner to rheumatism: an ill they share with rice cultivators. The physical dangers of mining remain high; some are still unavoidable.

The animus of the miner's technique is reflected in his treatment of the landscape. Let Dr. Bauer again be our witness. "Besides this the strongest argument of the detractors is that the fields are devastated by mining operations, for which reason formerly Italians were warned by law that no one should dig the earth for metals and so injure their very fertile fields, their vineyards, and their olive groves. Also they argue that the woods and groves are cut down, for there is need of endless amount of wood for timbers, machines, and the smelting of metals. And when the woods and groves are felled, there are exterminate the beasts and birds, very many of which furnish pleasant and agreeable food for man. Further, when the ores are washed, the water which has been used poisons the brooks and streams, and either destroys the fish or drives them away. Therefore the inhabitants of these regions, on account of the devastation of their fields, woods, groves, brooks, and rivers, find great difficulty in procuring the necessaries of life, and by reason of the destruction of the timber they are forced to a greater expense in erecting buildings."

There is no reason to go into Dr. Bauer's lame reply: it happens that the indictment still holds, and is an unanswerable one. One must admit the devastation of mining, even if one is prepared to justify the end. "A typical example of deforestation," says a modern writer on the subject, "is to be seen on the eastern slopes of the Sierra Nevada, overlooking the Truckee Valley, where the cutting of trees to provide timber for the deep mines of the Comstock left the hillside exposed to erosion, so that today they are bleak, barren and hideous. Most of the old mining regions tell the same tale, from Lenares to Leadville, from Potosi to Porcupine." The history of the last four hundred years has underlined the truths of this indictment; for what was only an incidental and local damage in Dr. Bauer's time became a widespread characteristic of Western Civilization just as soon as it started in the eighteenth century to rest directly upon the mine and its products, and to reflect, even in territories far from the mine itself, the practices and ideals of the miner.

One further effect of this habitual destruction and disorganization must be noted: its psychological reaction on the miner. Perhaps inevitably he has a low standard of living. Partly, this is the natural effect of capitalist monopoly, often exerted and maintained by physical compulsion: but it exists even under relatively free conditions and in "prosperous" times. The explanation is not difficult: almost any sight is brighter than the pit, almost any sound is sweeter than the clang and rap of the hammer, almost any rough cabin, so long as it keeps the water out, is a more hospitable place for an exhausted man than the dark damp gallery of a mine. The miner, like the soldier coming out of the trenches, wants a sudden relief and an immediate departure from his routine. No less notorious than the slatternly disorder of the mining town are the drinking and gambling that go on in it: a necessary compensation for the daily toil. Released from his routine, the miner takes a chance at cards or dice or whippet racing, in the hope that it will bring the swift reward denied him in the drudging efforts of the mine itself. The heroism of the miner is genuine: hence his simple animal poise: his profound personal pride and self-respect. But the brutalization is also inevitably there.

Now, the characteristic methods of mining do not stop at the pithead: they go on, more or less, in all the accessory occupations. Here is the domain, in northern mythology, of the gnomes and the brownies: the cunning little people who know how to use the bellows, the forge, the hammer and the anvil. They, too, live in the depths of the mountains, and there is something a little inhuman about them: they tend to be spiteful and tricky. Shall we set this characterization down to the fear and mistrust of neolithic peoples for those who had mastered the art of working in metals? Perhaps: at all events one notes that in Hindu and Greek mythology the same general judgment prevails as in the North. While Prometheus, who stole the fire from heaven, is a hero, Hephaestus, the blacksmith, is lame and he is the sport and butt of the other gods despite his usefulness.

Usually pocketed in the mountains, the mine, the furnace, and the forge have remained a little off the track of civilization: isolation and monotony add to the defects of the activities themselves. In an old industrial domain, like the Rhine Valley, dedicated to industry since the days of the Romans and refined by the technical and civil advances of the whole community, the direct effect of the miner's culture may be greatly ameliorated: this is true in the Essen district today, thanks to the original leadership of a Krupp and the later planning of a Schmidt. But taking mining regions as a whole, they are the very image of backwardness, isolation, raw animosities and lethal struggles. From the Rand to the Klondike, from the coal mines of South Wales to those of West Virginia, from the modern iron mines of Minnesota to the ancient silver mines of Greece, barbarism colors the entire picture.

Because of their urban situation and a more humanized rural environment, the molder and the smith have often escaped this influence: goldsmithing has always been allied with jewelry and women's ornaments, but even in the early Renascence ironwork of Italy and Germany, for example, in the locks and bands of chests as well as in the delicate traceries of railings and brackets, there is a grace and ease that point directly to a more pleasant life. In the main, however, the mining and metallurgical arts were outside the social scheme of both classic and gothic civilization. That fact proved a sinister one as soon as the methods and ideals of mining became the chief pattern for industrial effort throughout the Western World. Mine: blast: dump: crush: extract: exhaust—there was indeed something devilish and sinister about the whole business. Life flourishes finally only in an environment of the living.

3: Mining and Modern Capitalism

More closely than any other industry, mining was bound up with the first development of modern capitalism. By the sixteenth century it had definitely set the pattern for capitalist exploitation.

When mining was undertaken by free men in the fourteenth century in Germany the working of the mine was a simple partnership on a share basis. The miners themselves were often ne'er-do-wells and bankrupts who had seen better days. Partly abetted no doubt by this very application of free labor, there was a rapid advancement in technique in the German mines: by the sixteenth century those in Saxony led Europe, and German miners were imported into other countries, like England, to improve their practices.

The deepening of the mines, the extension of the operations to new fields, the application of complicated machinery for pumping water, hauling ore, and ventilating the mine, and the further application of waterpower to work the bellows in the new furnaces—all these improvements called for more capital than the original workers possessed. This led to the admission of partners who contributed money instead of work: absentee ownership: and this in turn led to a gradual expropriation of the owner-workers and the reduction of their share of the profits to the status of mere wages. This capitalistic development was further stimulated by reckless speculation in mining shares which took place as early as the fifteenth century: the local landlords and the merchants in the nearer cities eagerly followed this new gamble. If the mining industry in Dr. Bauer's day showed many of the modern improvements in industrial organization—the triple shift, the eight hour day, the existence of guilds in the various metallurgical industries for social intercourse, charitable self-help and insurance—it also showed, as the result of capitalist pressure, the characteristic features of nineteenth century industry throughout the world: the division of classes, the use of the strike as a weapon of defence, the bitter class war, and finally the extinction of the guilds' power by a combination of mine-owners and the feudal nobility during the so-called Peasants' War of 1525.

The result of that conflict was to abolish the cooperative guild basis of the mining industry, which had characterized its technical resurrection in Germany, and to place it on a free basis—that is, a basis of untrammeled acquisitiveness and class domination by the shareholders and directors, no longer bound to respect any of the humane regulations that had been developed by medieval society as measures of social protection. Even the serf had the safeguard of custom and the elementary security of the land itself: the miner and the iron-worker at the furnace was a free—that is, an unprotected—worker: the forerunner of the disinherited wage-worker of the nineteenth century. The most fundamental industry of the machine technics had known only for a moment in its history the sanctions and protections and humanities of the guild system: it stepped almost directly from the inhuman exploitation of chattel slavery to the hardly less inhuman exploitation of wage slavery. And wherever it went, the degradation of the worker followed.

But in still another way mining was an important agent of capitalism. The great need of commercial enterprise in the fifteenth century was for a sound but expansible currency, and for capital to provide the necessary capital goods—boats, mills, mine-shafts, docks, cranes—for industry. The mines of Europe began to supply this need even before the mines of Mexico and Peru. Sombart calculates that in the fifteenth and sixteenth centuries German mining earned as much in ten years as trade in the old style was able to accomplish in a hundred. As two of the greatest fortunes of modern times have been founded upon monopolies of petroleum and aluminum, so the great fortune of the Fuggers in the sixteenth century was founded upon the silver and lead mines of Styria and the Tyrol and Spain. The heaping up of such fortunes was part of a cycle we have witnessed with appropriate changes in our own time.

First: improvements in the technique of warfare, especially the rapid growth of the artillery arm, increased the consumption of iron: this led to new demands upon the mine. In order to finance the ever more costly equipment and maintenance of the new paid soldiery, the rulers of Europe had recourse to the financier. As security for the loan, the lender took over the royal mines. The development of the mines themselves then became a respectable avenue of financial enterprise, with returns that compared favorably with the usurious and generally unpayable interest. Spurred by the unpaid notes, the rulers were in turn driven to new conquests or to the exploitation of remote territories: and so the cycle began over again. War, mechanization, mining, and finance played into each other's hands. Mining was the key industry that furnished the sinews of war and increased the metallic contents of the original capital hoard, the war-chest: on the other hand, it furthered the industrialization of arms, and enriched the financier by both processes. The uncertainty of both warfare and mining increased the possibilities for speculative gains: this provided a rich broth for the bacteria of finance to thrive in.

Finally, it is possible that the animus of the miner had still another effect on the development of capitalism. This was in the notion that economic value had a relation to the quantity of brute work done and to the scarcity of the product: in the calculus of cost, these emerged as the principal elements. The rarity of gold, rubies, diamonds: the gross work that must be done to get iron out of the earth and ready for the rolling mill—these tended to be the criteria of economic value all through this civilization. But real values do not derive from either rarity or crude manpower. It is not rarity that gives the air its power to sustain life, nor is it the human work done that gives milk or bananas their nourishment. In comparison with the effects of chemical action and the sun's rays the human contribution is a small one. Genuine value lies in the power to sustain or enrich life: a glass bead may be more valuable than a diamond, a deal table more valuable esthetically than the most tortuously carved one, and the juice of a lemon may be more valuable on a long ocean voyage than a hundred pounds of meat without it. The value lies directly in the life-function: not in its origin, its rarity, or in the work done by human agents. The miner's notion of value, like the financier's, tends to be a purely abstract and quantitative one. Does the defect arise out of the fact that every other type of primitive environment contains food, something that may be immediately trans· lated into life—game, berries, mushrooms, maple-sap, nuts, sheep, corn, fish—while the miner's environment alone is—salt and sac· charin aside—not only completely inorganic but completely inedible? The miner works, not for love or for nourishment, but to "make his pile." The classic curse of Midas became perhaps the dominant characteristic of the modern machine: whatever it touched was turned to gold and iron, and the machine was permitted to exist only where gold and iron could serve as foundation.

4: The Primitive Engineer

The rational conquest of the environment by means of machines is fundamentally the work of the woodman. In part, the explanation of his success can be discovered in terms of the material he uses. For wood, beyond any other natural material, lends itself to manipulation: right down to the nineteenth century it had a place in civilization that the metals themselves were to assume only after that point.

In the forests of the temperate and sub-arctic zones, which covered the greater part of Western Europe from hilltop to riverbottom, wood was of course the most common and visible part of the environment. While the digging of stones was a laborious business, once the stone ax was shaped the cutting down of trees became a relatively easy task. What other object in nature has the length and cross-section of the tree? What other kind of material presents its characteristic properties in such a large assortment of sizes: what other kind can be split and split again with the simplest tools—the wedge and the mallet? What other common material can both be broken into definite planes and carved and shaped across those planes? The sedimentary rocks, which most nearly attain to the same qualities, are poor substitutes for wood. Unlike ores, one can cut down wood without the aid of fire. Using fire locally one can hollow out an enormous log and turn it into a canoe by charring the wood and scraping out with a primitive gouge or chisel. Down to modern times the solid trunk

of the tree was used in this primitive fashion: one of Dürer's engravings shows a man hollowing out a gigantic log; and bowls and tubs
and vats and troughs and benches were long made of single blocks
close to the natural shape.

Wood, different again from stone, has exceptional qualities for
transport: the trimmed logs may be rolled over the ground, and
because wood floats, it can be transported over long distances by
means of water even before boats are built: an unrivalled advantage.
The building of neolithic villages on wooden piles over the waters
of lakes was one of the surest witnesses to the advance of civilization:
wood delivered man from servitude to the cave and to the cold earth
itself. Thanks to the lightness and mobility of this material, as well
as to its wide distribution, one finds the products of the woodman
not merely in the uplands but down by the open sea. In the marshlands of the north coastal area in Europe, one finds the woodman
sinking his piles and building his villages—using his logs and his
mats of twigs and branches to serve as bulwark against the invading
ocean and to push it back. For thousands of years wood alone made
navigation possible.

Physically speaking, wood has the qualities of both stone and
metal: stronger in cross section than is stone, wood resembles steel
in its physical properties: its relatively high tensile and compressive
strength, together with its elasticity. Stone is a mass: but wood, by
its nature, is already a structure. The difference in toughness, tensile
strength, weight, and permeability of various species of wood, from
pine to hornbeam, from cedar to teak, give wood a natural range
of adaptability to various purposes that is matched in metals only
as a result of a long evolution of metallurgical skill: lead, tin,
copper, gold, and their alloys, the original assortment, offered a
meagre choice of possibilities, and down to the end of the nineteenth
century wood presented a greater variety. Since wood can be planed,
sawed, turned, carved, split, sliced, and even softened and bent or
cast, it is the most responsive of all materials to craftsmanship: it
lends itself to the greatest variety of techniques. But in its natural
state wood keeps the shape of the tree and retains its structure: and
the original shape of the wood suggests appropriate tools and adap-

tations of form. The curve of the branch forms the bracket, the forked stick forms the handle and the primitive type of plow.

Finally, wood is combustible; and at the beginning that fact was more important and more favorable to human development than the fire-resistance of other materials. For fire was obviously man's greatest primitive achievement in manipulating his environment as a whole: the utilization of fire raised him a whole plane above his nearest sub-human contemporaries. Wherever he could gather a few dried sticks, he could have a hearth and an altar: the germs of social life and the possibility of free thought and contemplation. Long before coal was dug or peat and dung dried, wood was man's chief source of energy, beyond the food he ate or the sun that warmed him: long after power machines were invented wood continued to be used for fuel, in the first steamboats and railroads of America and Russia.

Wood, then, was the most various, the most shapeable, the most serviceable of all the materials that man has employed in his technology: even stone was at best an accessory. Wood gave man his preparatory training in the technics of both stone and metal: small wonder that he was faithful to it when he began to translate his wooden temples into stone. And the cunning of the woodman is at the base of the most important post-neolithic achievements in the development of the machine. Take away wood, and one takes away literally the props of modern technics.

The place of the woodman in technical development has rarely been appreciated; but his work is in fact almost synonymous with power production and industrialization. He is not merely the wood-cutter who thins out the forest and provides fuel: he is not merely the charcoal burner who converts the wood into the most common and effective form of fuel, and so makes possible advances in metal-lurgy: he is, together with the miner and the smith, the primitive form of engineer; and without his skills the work of the miner and the mason would be difficult, and any great advance in their arts would have been impossible. It is the wooden shoring that makes possible the deep tunnel of the mine, even as it is the scantling and centering that make possible the lofty arch of the cathedral or the wide span of the stone bridge. It was the woodman who developed the wheel: the potter's wheel, the cart-wheel, the water-wheel, the spinning-wheel, and above all, the greatest of machine-tools, the lathe. If the boat and the cart are the woodman's supreme contribution to transport, the barrel, with its skilful use of compression and tension to achieve water-tightness is one of his most ingenious utensils: a great advance in strength and lightness over clay containers.

As for the wheel-and-axle itself, so important is it that Reuleaux and others have even said that the technical advance that characterizes specifically the modern age is that from reciprocating motions to rotary motions. Without a machine for accurately turning cylinders, screws, pistons, boring instruments, it would be impossible to create further instruments of precision: the machine-tool makes the modern machine possible. The lathe was the woodman's decisive contribution to the development of machines. First recorded among the Greeks, the primitive form of the lathe consisted of two fixed parts which hold the spindles that turn the wood. The spindle is wound up by hand and rotated by release of the bent sapling to which the wound cord is attached; the turner holds a chisel or gouge against the rotating wood which, if accurately centered, becomes a true cylinder or some modification of the cylinder. In this crude form the lathe is still used—or was fifteen years ago—in the Chiltern Hills in England: good enough to produce chair-legs shaped to pattern for the market. As an instrument of fair precision, the lathe existed long before its parts were cast in metal, before the crude form of power was converted into a foot-treadle or an electric motor, before the stock was made moveable or the adjustable slide-rest to steady the chisel was invented. The final transformation of the lathe into a metal instrument of exquisite accuracy awaited the eighteenth century: Maudslay in England is often given the credit for it. But in essentials, all the important parts had been worked out by the woodman; while the foot-treadle actually gave Watt the model he needed for translating reciprocating motion into rotary motion in his steam-engine.

The specific later contributions of the woodman to the machine will be taken up in discussing the eotechnic economy. Enough to point out here the rôle of the woodman as engineer: building dams, locks, mills, building mill-wheels, controlling the flow of water. Serving directly the needs of the peasant, the woodman often merged with him. Environmentally, however, he was caught between two movements that have always threatened and sometimes painfully narrowed the realm he has reigned over. One was the demand of the farmer for more arable land: this converted to mixed farming soils more fitted for tree culture. In France, this has gone on so far that the remaining trees may be merely a small clump or a row silhouetted against the sky: in Spain and other parts of the Mediterranean it has resulted not merely in deforestation but serious soil erosion: the same curse afflicts the seat of even more ancient civilizations, like that of China. (This evil has now been remedied in the State of New York by the purchase and reforestation of the marginal agricultural lands.)

From the other side of our typical valley section came pressure from the miner and the glassmaker. By the seventeenth century the marvellous oak forests of England had already been sacrificed to the iron-maker: so serious was the shortage that the Admiralty under Sir John Evelyn was forced to pursue a vigorous policy of reforestation in order to have enough timber for the Royal Navy. The continued attack upon the woodman's environment has led to his expulsion to remoter areas—to the birch and fir forests of North Russia and Scandinavia, to the Sierras and Rockies of America. So imperious became the commercial demand, so authoritative became the miner's methods that forest-cutting was reduced during the nineteenth century to timber-mining: today whole forests are slaughtered every week to supply the presses of the Sunday newspapers alone. But wood-culture and wood-technics, which survived through the age of metals, are likely also to endure through the age of synthetic compounds: for wood itself is nature's cheaper model for these materials.

5: From Game-Hunt to Man-Hunt

Perhaps the most positive influence in the development of the machine has been that of the soldier: in back of it lies the long development of the primitive hunter. Originally the call of the hunter for weapons was an effort to increase the food-supply: hence the invention and improvement of arrowheads, spears, slings, and knives from the earliest dawn of technics onward. The projectile and the hand-weapon were the two special lines of this development: and while the bow was perhaps the most effective weapon devised before the modern gun, since it had both range and accuracy, the sharpening of edges with the introduction of bronze and iron was scarcely less important. Shock and fire still remain among the chief tactical measures of warfare.

If the miner's task is non-organic the hunter's is anti-vital: he is a beast of prey, and the needs of his appetite as well as the excitement of the chase cause him to inhibit every other reaction—pity or esthetic pleasure—in the act of killing. The herdsman domesticates animals and in turn is domesticated by them: their protection and their nurture, in itself the outcome no doubt of man's prolongation of infancy and his more tender care of the young and helpless, bring out his most humane instincts, while the peasant learns to extend his sympathies beyond the boundaries of the animal kingdom. The daily lessons of crop and herd are lessons in co-operation and solidarity and the selective nurture of life. Even when the farmer kills, extirpating the rat or pulling out the weed, his activity is directed toward the preservation of the higher forms of life as related to human ends.

But the hunter can have no respect for life as such. He has none of the responsibilities which are preliminary to the farmer's slaughter of his cattle. Trained in the use of the weapon, killing becomes his main business. Shaken by insecurity and fear, the hunter attacks not merely the game but other hunters: living things are for him potential meat, potential skins, potential enemies, potential trophies. This predatory mode of life, deeply ingrained by man's primordial efforts to survive bare-handed in a hostile world, did not unfortunately die out with the success of agriculture: in the migrations of peoples it tended to direct their animus against other groups, particularly when animals were lacking and the food-supply was dubious, and eventually the trophies of the chase assumed symbolic forms: the treasures of the temple or the palace became the object of attack.

The advance in the "arts of peace" did not in itself lead to peace: on the contrary, the improvement of weapons and the repression of naïve hostilities under the forms of organized life, tended to make war itself more savage. Unarmed hands or feet are relatively innocent: their range is limited, their effectiveness is low. It is with the collective organization and regimentation of the army that the conflicts between men have reached heights of bestiality and terrorism that primitive peoples, with their merely post-mortem cannibalism, might well envy.

Finding the instruments of warfare more effective, men sought new occasions for their use. Robbery is perhaps the oldest of labor-saving devices, and war vies with magic in its efforts to get something for nothing—to obtain women without possessing personal charm, to achieve power without possessing intelligence, and to enjoy the rewards of consecutive and tedious labor without having lifted a finger in work or learnt a single useful skill. Lured by these possibilities, the hunter as civilization advances turns himself to systematic conquest: he seeks slaves, loot, power, and he founds the political state in order to ensure and regulate the annual tribute—enforcing, in return, a necessary modicum of order.

While pottery, basketwork, wine-making, grain-growing show only superficial improvements from neolithic times onward, the improvement of the instruments of war has been constant. The three-field system lingered in British agriculture down to the eighteenth century while the tools used in the remoter areas of England would have been laughed at by a Roman farmer: but the shambling peasant with his pruning hook or his wooden club had meanwhile been replaced by the bowman and the spearman, these had given way to the musketeer, the musketeer had been turned into a smart, mechanically responsive infantryman, and the musket itself had become more deadly in close fighting by means of the bayonet, and the bayonet in turn had become more efficient by means of drill and mass tactics, and finally, all the arms of the service had been progressively co-ordinated with the most deadly and decisive arm: the artillery. A triumph of mechanical improvement: a triumph of regimentation. If the invention of the mechanical clock heralded the new will-to-order, the use of cannon in the fourteenth century enlarged the will-to-power; and the machine as we know it represents the convergence and systematic embodiment of these two prime elements.

The regimentation of modern warfare carries much farther than the actual discipline of the army itself. From rank to rank passes the word of command: that passage would be impeded if, instead of mechanical obedience, it met with a more active and participating form of adjustment, involving a knowledge of how and why and wherefore and for whom and to what end: the commanders of the sixteenth century discovered that effectiveness in mass-fighting increased in proportion as the individual soldier was reduced to a power-unit and trained to be an automaton. The weapon, even when it is not used to inflict death, is nevertheless a means for enforcing a pattern of human behavior which would not be accepted unless the alternative were physical mutilation or death: it is, in short, a means of creating a dehumanized response in the enemy or the victim.

The general indoctrination of soldierly habits of thought in the seventeenth century was, it seems probable, a great psychological aid to the spread of machine industrialism. In terms of the barracks, the routine of the factory seemed tolerable and natural. The spread of conscription and volunteer militia forces throughout the Western World after the French Revolution made army and factory, so far as their social effects went, almost interchangeable terms. And the complacent characterizations of the First World War, namely that it was a large-scale industrial operation, has also a meaning in reverse: modern industrialism may equally well be termed a large-scale military operation.

Observe the enormous increase in the army as a power unit: the power was multiplied by the use of guns and cannon, by the increase in the size and range of cannon, by the multiplication of the number of men put in the field. The first giant cannon on record had a barrel over three and a half meters long, it weighed over 4500 kilograms: it appeared in Austria in 1404. Not merely did heavy industry develop in response to war long before it had any contributions of importance to make to the arts of peace: but the quantification of life, the concentration upon power as an end in itself, proceeded as rapidly in this department as in trade. In back of that was a growing contempt for life: for life in its variety, its individuality, its natural insurgence and exuberance. With the increase in the effectiveness of weapons, came likewise a growing sense of superiority in the soldier himself: his strength, his death-dealing properties had been heightened by technological advance. With a mere pull of the trigger, he could annihilate an enemy: that was a triumph of natural magic.

6: Warfare and Invention

Within the domain of warfare there has been no psychological hindrance to murderous invention, except that due to lethargy and routine: no limits to invention suggest themselves.

Ideals of humanity come, so to say, from other parts of the environment: the herdsman or the caravaneer brooding under the stars—a Moses, a David, a St. Paul—or the city bred man, observing closely the conditions under which men may live well together—a Confucius, a Socrates, a Jesus, bring into society the notions of peace and friendly cooperation as a higher moral expression than the subjection of other men. Often this feeling, as in St. Francis and the Hindu sages, extends to the entire world of living nature. Luther, it is true, was a miner's son; but his career proves the point rather than weakens it: he was actively on the side of the knights and soldiers when they ferociously put down the poor peasants who dared to challenge them.

Apart from the savage inroads of Tartars, Huns, Turks, it was not until the machine culture became dominant that the doctrine of untrammeled power was, practically speaking, unchallenged. Though Leonardo wasted much of his valuable time in serving warlike princes and in devising ingenious military weapons, he was still sufficiently under the restraint of humane ideals to draw the line somewhere. He suppressed the invention of the submarine boat because he felt, as he explained in his notebook, it was too satanic to be placed in the hands of unregenerate men. One by one the invention of machines and the growing belief in abstract power removed these scruples, withdrew these safeguards. Even chivalry died in the unequal contests and the triumphant slaughter of the poorly armed barbarians the European encountered in his post-Columbian spread throughout the planet.

How far shall one go back in demonstrating the fact that war has been perhaps the chief propagator of the machine? Shall it be to the poison-arrow or the poison-pellet? This was the forerunner of poison-gas: while not merely was poison gas itself a natural product of the mine, but the development of gas masks to combat it took place in the mine before they were used on the battlefield. Shall it be to the armed chariot with the scythes that revolved with its movement, mowing down the foot-soldiers? That was the forerunner of the modern tank, while the tank itself, impelled by hand power furnished by the occupants, was designed as early as 1558 by a German. Shall it be to the use of burning petroleum and Greek fire, the first of which was used considerably before the Christian era? Here was the embryo of the more mobile and effective flame-thrower of the last war. Shall it be to the earliest high-powered engine that hurled stones and javelins invented apparently under Dionysius of Syracuse and used by him in his expedition against the Carthaginians in 397 B.C.? In the hands of the Romans the catapults could throw stones weighing around 57 pounds a distance of 400 to 500 yards, while their ballistas, which were enormous wooden cross-bows for shooting stones, were accurate at even greater distances: with these instruments of precision Roman society was closer to the machine than in its aqueducts and baths. The swordsmiths of Damascus, Toledo, Milan, were noted both for their refined metallurgy and their skill in manufacturing armament: forerunners of Krupp and Creuseot. Even the utilization of the physical sciences for more effective warfare was an early development: Archimedes, the story goes, concentrated the sun's rays by means of mirrors on the sails of the enemy's fleet in Syracuse and burned the boats up. Ctesibius, one of the foremost scientists of Alexandria, invented a steam cannon: Leonardo designed another. And when the Jesuit father, Francesco Lana-Terzi, in 1670 projected a vacuum dirigible balloon, he emphasized its utility in warfare. In short the partnership between the soldier, the miner, the technician, and the scientist is an ancient one. To look upon the horrors of modern warfare as the accidental result of a fundamentally innocent and peaceful technical development is to forget the elementary facts of the machine's history.

In the development of the military arts the soldier has of course borrowed freely from other branches of technics: the more mobile fighting arms, the cavalry and the fleet, come respectively from the pastoral and the fishing occupations: static warfare, from the trenches of the Roman castra to the heavy masonry fortifications of the cities, is a product of the peasant—the Roman soldier, indeed, conquered through his spade as well as his sword—while the wooden instruments of siege, the ram, the ballista, the scaling ladder, the moving-tower, the catapult, all plainly bear the stamp of the woodman. But the most important fact about modern warfare is the steady increase of mechanization from the fourteenth century onward: here militarism forced the pace and cleared a straight path to the development of modern large-scale standardized industry.

To recapitulate: the first great advance came through the introduction of gunpowder in Western Europe: it had already been used in the East. In the early thirteen hundreds came the first cannon—or firepots—and then at a much slower pace came the hand-weapons, the pistol and the musket. Early in this development multiple firing was conceived, and the organ gun, the first machine-gun, was invented.

The effect of firearms upon technics was three-fold. To begin with, they necessitated the large scale use of iron, both for the guns themselves and for cannon-balls. While the development of armor called forth the skill of the smith, the multiplication of cannon demanded cooperative manufacture on a much larger scale: the old fashioned methods of handicraft were no longer adequate. Because of the destructions of the forest, experiments were made in the use of coal in the iron furnaces, from the seventeenth century onwards, and when, a century later, the problem was finally solved by Abraham Darby in England, coal became a key to military as well as to the new industrial power. In France, the first blast furnaces were not built till about 1550, and at the end of the century France had thirteen foundries, all devoted to the manufacture of cannon—the only other important article being scythes.

Second: the gun was the starting point of a new type of power machine: it was, mechanically speaking, a one cylinder internal combustion engine: the first form of the modern gasoline engine, and some of the early experiments in using explosive mixtures in motors sought to employ powder rather than a liquid fuel. Because of the accuracy and effectiveness of the new projectiles, these machines had still another result: they were responsible for the development of the art of heavy fortification, with elaborate outworks, moats, and salients, the latter so arranged that any one bastion could come to the aid of another by means of cross-fire. The business of defence became complicated in proportion as the tactics of offence became more deadly: road-building, canal-building, pontoon-building, bridge-building became necessary adjuncts of warfare. Leonardo, typically, offered his services to the Duke of Milan, not merely to design ordnance, but to conduct all these engineering operations. In short: war established a new type of industrial director who was not a mason or a smith or a master craftsman—the military engineer. In the prosecution of war, the military engineer combined originally all the offices of the civil, the mechanical, and the mining engineer: offices that did not begin to be fully differentiated until the eighteenth century. It was to the Italian military engineers from the fifteenth century on that the machine owed a debt quite as high as it did to the ingenious British inventors of James Watt's period.

In the seventeenth century, thanks to the skill of the great Vauban, the arts of military offence and defence had almost reached a stalemate: Vauban's forts were impregnable, against every form of attack except that which he himself finally devised. How storm these solid masses of stone? Artillery was of dubious value, since it worked in both directions: the only avenue open was to call in the miner, whose business it is to overcome stone. In accordance with Vauban's suggestion, troops of engineers, called sappers, were created in 1671, and two years later the first company of miners was raised. The stalemate was over: open warfare again became necessary and possible, and it was through the invention of the bayonet, which took place between 1680 and 1700, that the finer intimacies of personal murder were restored to this art.

If the cannon was the first of the modern space-annihilating devices, by means of which man was enabled to express himself at a distance, the semaphore telegraph (first used in war) was perhaps the second: by the end of the eighteenth century an effective system had been installed in France, and a similar one was projected for the American railroad service before Morse opportunely invented the electric telegraph. At every stage in its modern development it was war rather than industry and trade that showed in complete outline the main features that characterize the machine. The topographic survey, the use of maps, the plan of campaign—long before business men devised organization charts and sales charts—the coordination of transport, supply, and production [mutilation and destruction], the broad divisions of labor between cavalry, infantry, and artillery, and the division of the process of production between each of these branches; finally, the distinction of function between staff and field activities—all these characteristics put warfare far in advance of competitive business and handicraft with their petty, empirical and short-sighted methods of preparation and operation. *The army is in fact the ideal form toward which a purely mechanical system of industry must tend.* The utopian writers of the nineteenth century like Bellamy and Cabet, who accepted this fact, were more realistic than the business men who sneered at their "idealism." But one may doubt whether the outcome was an ideal one.

7: Military Mass-Production

By the seventeenth century, before iron had begun to be used on a great scale in any of the other industrial arts, Colbert had created arms factories in France, Gustavus Adolphus had done likewise in Sweden, and in Russia, as early as Peter the Great, there were as many as 683 workers in a single factory. There were isolated examples of large-scale mills and factories, even before that of the famous Jack of Newbury in England: but the most impressive series was the arms factories. Within these factories, the division of labor was established and the grinding and polishing machinery was worked by water-power: so that Sombart well observed that Adam Smith had done better to have taken arms, rather than pin-making, as an example of the modern productive process with all the economies of specialization and concentration.

The pressure of military demand not merely hastened factory organization at the beginning: it has remained persistent throughout its entire development. As warfare increased in scope and larger armies were brought into the field, their equipment became a much heavier task. And as their tactics became mechanized, the instruments needed to make their movements precise and well-timed were necessarily reduced to uniformity too. Hence along with factory organization there came standardization on a larger scale than was to be found in any other department of technics except perhaps printing.

The standardization and mass production of muskets came at the end of the eighteenth century: in 1785 Le Blanc in France produced muskets with interchangeable parts, a great innovation in production, and the type of all future mechanical design. (Up to this time there had been no uniformity in even the minor elements like screws and threads.) In 1800 Eli Whitney, who had obtained a contract from the United States Government to produce arms in similar fashion turned out a similar standardized weapon in his new factory at Whitneyville. "The technique of interchangeable part manufacture," as Usher observes, "was thus established in general outline before the invention of the sewing machine or the harvesting machinery. The new technique was a fundamental condition of the great achievements realized by inventors and manufacturers in those fields." Behind that improvement lay the fixed mass demand of the army. A similar step in the direction of standardized production was made in the British navy at almost the same time. Under Sir Samuel Bentham and the elder Brunel the various tackleblocks and planks of the wooden ships were cut to uniform measure: building became the assemblage of accurately measured elements, rather than old-fashioned cut-and-try handicraft production.

But there was still another place in which war forced the pace. Not merely was gun-casting the "great stimulant of improved tech-

nique in the foundry," and not merely was "the claim of Henry Cort
to the gratitude of his fellow countrymen . . . based primarily on
the contribution he had made of military security," as Ashton says,
but the demand for highgrade iron in large quantities went hand in
hand with the increase of artillery bombardment as a preparation
for assault, the effectiveness of which was presently demonstrated by
the brilliant young artilleryman who was to scourge Europe with
his technological genius whilst he liquidated the French revolution.
Indeed, the rigorous mathematical basis and the increasing precision
of artillery fire itself made it a model for the new industrial arts.
Napoleon III in the middle of the nineteenth century offered a reward
for a cheap process of making steel capable of withstanding the
explosive force of the new shells. The Bessemer process was the
direct answer to this demand.

The second department in which war anticipated the machine and
helped definitely to form it was in the social organization of the
army. Feudal warfare was usually on the basis of a forty-day service: necessarily intermitted and therefore inefficient—apart from
further delays and stoppages occasioned by rain or cold or the Truce
of God. The change from feudal service to armies on a capitalist
basis, composed of workers paid by the day—the change, that is,
from the warrior to the soldier—did not entirely overcome this
inefficiency: for if the captains of the paid bands were quick to copy
the latest improvements in arms or tactics, the actual interest of the
paid soldier was to continue in the business of being a soldier: hence
warfare at times rose to the place it so often holds among savage
tribes—an exciting ritual played under carefully established rules,
with the danger reduced almost to the proportions of an old-fashioned
football game. There was always the possibility that the mercenary
band might go on a strike or desert to the other side: money, rather
than habit or interest or delusions of grandeur [patriotism] was
the chief means of discipline. Despite the new technical weapons, the
paid soldier remained inefficient.

The conversion of loose gangs of individuals with all their incalculable variations of strength and weakness, bravery and cowardice,
zeal and indifference, into the well-exercised, disciplined, unified

soldiery of the seventeenth century was a great mechanical feat. Drill itself, after the long lapse from Roman practice in the West, was re-introduced in the sixteenth century and perfected by Prince Maurice of Orange and Nassau, and the psychology of the new industrial order appeared upon the parade ground before it came, full-fledged, into the workshop. The regimentation and mass-production of soldiers, to the end of turning out a cheap, standardized, and replaceable product, was the great contribution of the military mind to the machine process. And along with this inner regimentation went an outward one which had a further effect upon the productive system: namely, the development of the military uniform itself.

Despite sumptuary laws regulating the costumes of different social and economic groups, there was no real uniformity in the costume of the Middle Ages: no matter how common the pattern there would always, by the very nature of intermittent handicraft production, be individual variations and deviations. Such uniforms as existed were the special liveries of the great princes or municipalities: Michelangelo designed such a uniform for the Papal Guards. But with the growth in size of the army, and the daily exercise of drill, it was necessary to create an outward token of the inner unison: while small companies of men knew each other by face, larger ones could be ensured from fighting each other only by a large visible badge. The uniform was such a token and badge: first used on a large scale in the seventeenth century. Each soldier must have the same clothes, the same hat, the same equipment, as every other member of his company: drill made them act as one, discipline made them respond as one, the uniform made them look as one. The daily care of the uniform was an important element in the new *esprit de corps*.

With an army of 100,000 soldiers, such as Louis XIV had, the need for uniforms made no small demand upon industry: *it was in fact the first large-scale demand for absolutely standardized goods.* Individual taste, individual judgment, individual needs, other than the dimensions of the body, played no part in this new department of production: the conditions for complete mechanization were present. The textile industries felt this solid demand, and when the sewing machine was tardily invented by Thimonnet of Lyons in 1829, one is not surprised to find that it was the French War Department that sought first to use it. From the seventeenth century on the army became the pattern not only of production but of ideal consumption under the machine system.

Mark the effect of the large standing armies of the seventeenth century, and the even larger conscript armies whose success in France during the Revolution was to be so potent in the future development of warfare. An army is a body of pure consumers. As the army grew in size it threw a heavier and heavier burden upon productive enterprise: for the army must be fed and housed and equipped, and it does not, like the other trades, supply any service in return except that of "protection" in times of war. In war, moreover, the army is not merely a pure consumer but a negative producer: that is to say, it produces illth, to use Ruskin's excellent phrase, instead of wealth—misery, mutilation, physical destruction, terror, starvation and death characterize the process of war and form a principal part of the product.

Now, the weakness of a capitalist system of production, based upon the desire to increase the abstract tokens of power and wealth, is the fact that the consumption and turnover of goods may be retarded by human weaknesses: affectionate memory and honest workmanship. These weaknesses sometimes increase the life of a product long after the time an abstract economy would have it ticketed for replacement. Such brakes on production are automatically excluded from the army, particularly during the periods of active service: for the army is the ideal consumer, in that it tends to reduce toward zero the gap in time between profitable original production and profitable replacement. The most wanton and luxurious household cannot compete with a battlefield in rapid consumption. A thousand men mowed down by bullets are a demand more or less for a thousand more uniforms, a thousand more guns, a thousand more bayonets: and a thousand shells fired from cannon cannot be retrieved and used over again. In addition to all the mischances of battle, there is a much speedier destruction of stable equipment and supplies.

Mechanized warfare, which contributed so much to every aspect of standardized mass-production, is in fact its great justification.

Is it any wonder that it always acts as a temporary tonic on the system it has done so much to produce? Quantity production must rely for its success upon quantity consumption; and nothing ensures replacement like organized destruction. In this sense, war is not only, as it has been called, the health of the State: it is the health of the machine, too. Without the non-production of war to balance accounts algebraically, the heightened capacities of machine production can be written off only in limited ways: an increase in foreign markets, an increase in population, an increase in mass purchasing power through the drastic restriction of profits. When the first two dodges have been exhausted, war helps avert the last alternative, so terrible to the kept classes, so threatening to the whole system that supports them.

8: Drill and Deterioration

The deterioration of life under the régime of the soldier is a commonplace: but just for that reason it needs to be sharpened by explicit statement.

Physical power is a rough substitute for patience and intelligence and cooperative effort in the governance of men: if used as a normal accompaniment of action instead of a last resort it is a sign of extreme social weakness. When a child is intolerably balked by another person without precisely seeing the cause of the situation and without sufficient force to carry through his own ends, he often solves the matter by a simple wish: he wishes the other person were dead. The soldier, a slave to the child's ignorance and the child's wish, differs from him only by his ability to effect a direct passage to action. Killing is the ultimate simplification of life: a whole stage beyond the pragmatically justifiable restrictions and simplifications of the machine. And while the effort of culture is toward completer differentiation of perceptions and desires and values and ends, holding them from moment to moment in a perpetually changing but stable equilibrium, the animus of war is to enforce uniformity—to extirpate whatever the soldier can neither understand nor utilize.

In his pathetic desire for simplicity, the soldier at bottom extends the empire of irrationality, and by his effort to substitute force for emotional and intellectual grasp, for natural loyalties and cohesions, in short, for the organic processes of social life, he creates that alternating rhythm of conquest and rebellion, repressions and reprisals, which has punctuated such large periods of mankind's existence. Even when the warrior's conquests are intelligently and almost beneficently made—as in the later Inca Empire of Peru—the reactions he sets in motion undermine the ends he has in view. For terrorism and fear create a low psychic state. In the act of making himself a master, the soldier helps create a race of slaves.

As for the sense of self-esteem the soldier achieves through his willingness to face death, one cannot deny that it has a perverse life-enhancing quality, but it is common to the gunman and the bandit, as well as to the hero: and there is no ground for the soldier's belief that the battlefield is the only breeder of it. The mine, the ship, the blast furnace, the iron skeleton of bridge or skyscraper, the hospital ward, the childbed bring out the same gallant response: indeed, it is a far more common affair here than it is in the life of a soldier, who may spend his best years in empty drill, having faced no more serious threat of death than that from boredom. An imperviousness to life-values other than those clustered around the soldier's underlying death-wish, is one of the most sinister effects of the military discipline.

Fortunately for mankind, the army has usually been the refuge of third-rate minds: a soldier of distinct intellectual capacity, a Caesar or a Napoleon, stands out as a startling exception. If the soldier's mind went into action as intensely as his body, and if his intellectual discipline were as unremitting as his drill, civilization might easily have been annihilated long ago. Hence the paradox in technics: war stimulates invention, but the army resists it! The rejection of Whitworth's improved cannon and rifles in the midst of the Victorian period is but a critical instance of a common process: Alfred Krupp complained of similar resistance on the part of the army and navy to technical advance. The delay in adopting the tank by the German Army in the World War shows how torpid even "great" warriors are. So in the end, the soldier has again and again become the chief victim of his own simplification and short-cuts: in achieving machine-like precision and regularity, he has lost the capacity for intelligent response and adaptation. No wonder that in English *to soldier* means to withhold efficiency in work.

The alliance of mechanization and militarization was, in sum, an unfortunate one: for it tended to restrict the actions of social groups to a military pattern, and it encouraged the rough-and-ready tactics of the militarist in industry. It was unfortunate for society at large that a power-organization like the army, rather than the more humane and cooperative craft guild, presided over the birth of the modern forms of the machine.

9: Mars and Venus

If mechanical production was heightened and shaped by the active demands of the battlefield and the parade-ground, it was also possibly influenced by the indirect effects of war during the specious intervals of repose.

War is the chief instrument by means of which the ruling classes create the state and fix their hold upon the state. These ruling classes, whatever their military animus and origin, alternate their outbursts of prowess with periods devoted to what Veblen in his Theory of the Leisure Class called the ritual of conspicuous waste.

From the sixth century onward in Western Europe military feudalism had shared economic power with the peaceful monasteries, which formed an important pillar in the social system: from the twelfth century on the feudal lords had been curbed and kept in place by the free cities. With the rise of the absolute monarchs of the sixteenth century the old estates and corporations whose power had been localized and distributed and therefore balanced by reason of their relative autonomy, were absorbed, in effect, by the state: in the great capitals of Europe power was concentrated symbolically—and in part actually—in the absolute ruler. The culture of the great capitals, crystallized and expressed with utmost potency in the Paris of Louis XIV or the St. Petersburg of Peter the Great, became one-sided, militarist, regimented, oppressive. In that milieu, the machine could grow more lustily, for institutional life had been mechanized. So the capital cities became the focus, not merely of spending, but of capitalist production; and the lead they acquired then they have retained right down to the present.

There is a psychological ground for the wastefulness and luxury that manifested itself with such overpowering splendor in the sixteenth century, and that carried the forms of the camp and the court into every hole and corner of the modern community. At bottom, this new opulence was connected with the brutal, disorderly, irreligious mode of life which prevailed throughout society: it was not a little like the raw outbursts of drunkenness and gambling that alternated with the labor of the miner.

The military life, plainly, is a hard one. It involves, during its active pursuit, a renunciation of the comforts and securities of a normal domestic existence. The denial of the body, the deprivation of the senses, the suppression of spontaneous impulses, the forced marches, the broken sleep, the exhaustion of the marrow, the neglect of cleanliness—all these conditions of active service leave no place for the normal decencies of existence, and except for short intervals for lust or rape the soldier's sexual life is limited, too. The more arduous the campaign—and it was just at this period that the mechanization of arms and the serious discipline of drill took away the last remnants of gentlemanly ease and amateur sportsmanship— the greater the rigors and the tighter the checks, the more necessary become the ultimate compensations.

When Mars comes home, Venus waits for him in bed: the theme is a favorite one with all the Renascence painters, from Tintoretto to Rubens. And Venus serves a double purpose: she not merely gives him her body directly, but she matches the *superbia* of the soldier with her own *luxuria;* and to the degree that she has been neglected during the war she demands compensatory attention in times of peace. Venus's caresses are not by themselves enough to offset the abstentions and beastly crudeness of the battlefield: after the body has been neglected, it must be glorified. She must have jewels, silks, perfumes, rare wines, anticipating and prolonging by all possible means the erotic ritual itself. And she leaves nothing undone to gain her end: she exposes her breasts, she takes off her undergarments, reveals her limbs, even mons veneris itself to the passer-by. From the housemaid to the princess, women consciously or unconsciously adopt the habits of courtesans at the end of a great period of strain and disorder and warfare: so, extravagantly, life renews itself. The women's styles that prevailed in the Western World at the end of the last martial debauch match almost point for point those that became fashionable at the end of the Directory— down to the removal of the corset and the temporary abandonment of the petticoat.

Just because the erotic impulses seek extra compensation for their denial, they flow over and pervade every activity: the courtesan consumes the substance of the warrior's conquests. A plethora of physical goods gives special point to the triumphs of the soldier and justifies the pillage he brings home with him. Shakespeare has given us an acute study of the relationship in Antony and Cleopatra; but the economic results of it are more important here than the psychological consequences. Economically, the conquest of Mars by Venus means the heightened demand for luxuries of all sorts: for satins, laces, velvets, brocades, for precious stones and gold ornaments and finely wrought caskets to hold them: for downy couches, perfumed baths, private apartments and private gardens enclosing an Arbor of Love: in short, for the substance of an acquisitive life. If the soldier does not supply it, the merchant must: if the loot be not taken from the Court of Montezuma or a Spanish Galleon, it must be earned in the counting house. Religion itself in these courts and palaces had become an empty ceremony: is it any wonder that luxury became almost a religion?

Now observe the contrast. Private luxury was not looked upon with favor during the Middle Ages: indeed, a private life, in the modern sense, scarcely existed. It was not merely that the sins of pride, avarice and covetousness, with their possible by-products of lechery and fornication, were, if not serious offenses, at least hindrances to salvation: it was not merely that the standards of living, judged by purely financial ideals, were modest. But the Middle Ages, with their constant tendency to symbolize, used gold and jewels and artful workmanship as emblems of power. The Virgin could receive such tributes because she was Queen of Heaven: the earthly king and queen, pope and prince, representatives of the heavenly powers, might also have a certain measure of luxury to indicate their station: finally, the guilds in their mysteries and pageants might spend lavishly upon public shows. But luxury here had a collective function: even among the privileged classes it did not mean merely sensuous ease.

The breakdown of the medieval economy was marked by the emergence of the ideal of private power and private possession. The merchant, the capitalist, the freebooter, the captain of the condottieri, quite as much as the original lords and princes of the land, attempted to take over and monopolize for themselves the functions of civic life. What had been a public function became a private gesture: the morality play of the church became the masque of the court: the mural paintings that belonged to a place and an institution, became the removable easel picture that belonged to a private individual. With the medieval restriction of usury flouted under the church in the fifteenth century and abandoned even in theory by the protestant reformers of the sixteenth century, the legal mechanism for acquisition on a grand scale went hand in hand with the social and psychological demand for an acquisitive life. War was not of course the only motivating condition: but the place where the new luxury was most visible and where it was carried to a pitch of refined extravagance, was in the court.

Economically, the center of gravity shifted to the court: geograpically it shifted to the capital cities where the court—and the courtesan —were both luxuriously housed. The great art of the Baroque period is in the country houses and the town palaces: when churches and monasteries were built, they were done in the same style: abstractly, one could hardly tell the difference between the nave of one and the ballroom of the other. One acquired riches in order to consume goods according to the standards of the court: to "live like a prince" became a byword. Over it all presided the courtesan. One acquired power and riches in order to please her: one built her a palace: one gave her many servants: one brought in a Titian to paint her. And her own sense of power in turn throve on all the comforts and beauties of her life, and she counted her body flattered in proportion to her skill in extracting these luxuries. The summit of the Baroque dream was reached when Louis XIV sentimentally built the monstrous palace of Versailles on the site of the old hunting lodge in which he had first wooed Madame de la Vallière. But the dream itself was a universal one: one encounters it in every memento of the period, in the mind as well as in flesh and stone and canvas: perhaps its most splendid embodiment was in Rabelais's early conception of the Abbey of Theleme. What went on at court became the criterion of a good life; and the luxurious standards of consumption erected there spread themselves gradually throughout every walk of society.

Not merely did life as a whole become the mean handiwork of coachman, cook and groom: but the court began likewise to take a leading part in industrial production, too: the new luxury of china for the table became a monopoly of the royal porcelain factories in Prussia, Saxony, Denmark, Austria, and for woven goods the big Gobelins factory became one of the main production centers in France. In the effort to put on a front, the use of adulterations and substitutes became common. Marble was imitated in plaster, gold in gilt, handwork in moulded ornament, glass was used instead of precious stones. The reproduction for mass consumption of substitutes, as in the jewelry of Birmingham, took the place of the slow original creation of genuine handicraft: the systematic cheapening through mass production and inferior materials for the sake of achieving an effect beyond one's means, occurred in ornament long before it was applied to objects of use. With the spread of courtly ideals through society, the same change took place in the eighteenth century as happened with the introduction of the "democratic" ideal of military conscription. The standardized manufacture of cheap jewelry and domestic ornaments and textiles went along directly with the standardization of military equipment. And one notes ironically that it was out of the capital Matthew Boulton had amassed in his brummagem works at Soho that he was able to support James Watt during the period when he was perfecting the steam engine.

The concentration upon insignificant luxuries as the mark of economic well-being was in many ways an unfortunate prelude to machine production; but it was not altogether sterile. As a result of

this consumptive ritual some of the great achievements in mechanization were first conceived in terms of play: elaborate clocks whose
mannikins went through a procession of stiff and elegant movements:
dolls that moved by themselves: carriages like that Camus built for
young Louis XIV which went by clockwork: birds that twittered their
tails in time to the tinkling of a music box. Vain in origin, these toys,
these playful impulses, were not altogether fruitless. Certainly the
part played by toys and non-utilitarian instruments in fostering
important inventions cannot be lightly ignored. The first "use" of
the steam engine, as suggested by Hero, was to create magical effects
in the temple to awe the populace: and steam appears as an agent of
work in the tenth century, when used by Sylvester II to operate
an organ. The helicopter was invented as a toy in 1796. Not merely
did moving images first appear as a toy in the phenakistoscope, but
the magic lantern, which was utilized in the eventual production of
these images, was a seventeenth century toy attributed to Athanasius
Kircher. The gyroscope existed as a toy before it was used seriously
as a stabilization device; and the success of toy airplanes in the
seventies helped renew interest in the possibilities of flight. The
origin of the telephone and the phonograph is to be found in playful
automata; while the most powerful engines of the seventeenth century, the water-wheels at Marly, were constructed to pump water into
the great fountains at Versailles. Even the desire for speed in travel
first appeared in a playful form before it was embodied in the railroad and the motor car: the proménade aérienne—our present scenic
railway—appeared before either of the utilitarian devices.

The mechanical truth, in short, was sometimes first spoken in jest
—just as ether was first used in parlor games in America before it
was used in surgery. Indeed, the child's naïve interest in moving
wheels remains in only faint disguises as a large part of the adult
interest in machines: "engines are buckets and shovels dressed up
for adults." The spirit of play enfranchised the mechanical imagination. Once the organization of the machine had started, however,
the idle amusements of the aristocracy did not for long remain idle.

10: Consumptive Pull and Productive Drive

The development of the machine required both a trap and a bait, a drive and a pull, a means and a destination. Without doubt, the motive power came from technics and science: they were self-sustaining interests, and with the smiths, the wheelwrights, the founders, the clockmakers, and the growing body of experimenters and inventors, the machine established itself as the center of the productive process. But why should production itself have assumed such enormous proportions? There is nothing within the machine milieu itself that can explain this fact: for in other cultures production, though it might create vast surpluses for public works and public art, remained a bare necessity of existence, often grudgingly met—not a center of continuous and overwhelming interest. In the past, even in Western Europe, men had worked to obtain the standard of living traditional to their place and class: the notion of acquiring money in order to move out of one's class was in fact foreign to the earlier feudal and corporate ideology. When their living became easy, people did not go in for abstract acquisition: they worked less. And when Nature abetted them, they often remained in the idyllic state of the Polynesians or the Homeric Greeks, giving to art, ritual, and sex the best of their energies.

The pull, as Sombart amply demonstrated in his little study of Luxus und Kapitalismus, came mainly from the court and the courtesan: they directed the energies of society toward an ever-moving horizon of consumption. With the weakening of caste lines and the development of bourgeois individualism the ritual of conspicuous expenditure spread rapidly throughout the rest of society: it justified the abstractions of the money-makers and put to wider uses the technical progress of the inventors. The ideal of a powerful expensive life supplanted the ideal of a holy or a humane one. Heaven, which had been deferred to the Hereafter in the scheme of the Christian cosmos, was now to be enjoyed immediately: its streets paved with precious stones, its glittering walls, its marbled halls, were almost at hand—provided one had acquired money enough to buy them.

Few doubted that the Palace *was* Heaven: few doubted its sacredness. Even the poor, the overworked, the exploited were hypnotized by this new ritual, and they permitted it to go on at their expense with scarcely a murmur of protest until the French Revolution provided an interlude—after which the consumptive process was pursued again with re-doubled voracity and justified by hypocritical promises of plenty to the masses who paid the fiddler without calling the tune. The abstention from earthly joys for the sake of the hereafter, a Hereafter such as was envisioned by St. John of Patmos, had proved in fact to be one of those deceptive beatitudes, like the monastic regimen, which had worked out in earthly life as the opposite of the original aim. It was not a prelude to Heaven but a preparation for capitalist enterprise. The necessity for abstention from immediate pleasures, the postponement of present goods for future rewards, indeed the very words used by nineteenth century writers to justify the accumulation of capital and the taking of interest could have been put in the mouth of any medieval preacher, endeavoring to move men to put aside the immediate temptations of the flesh in order to earn far greater rewards for their virtue in heaven. With the acceleration of the machine, the gap in time between abstention and reward could be lessened: at least for the middle classes, the golden gates opened.

Puritanism and the counter-reformation did not seriously challenge these courtly ideals. The military spirit of the Puritans, under Cromwell, for example, fitted in well with their sober, thrifty, industrious life, concentrated upon money-making, as if by the avoidance of idleness the machinations of the devil could be eluded without avoiding devilish acts. Carlyle, the belated advocate of this militaristic puritanism, knew no other key to salvation than the gospel of work: he held that even mammonism at its lowest was in connection with the nature of things and so on the way to God. But acquisitive ideals in production necessarily go hand in hand with acquisitive modes of consumption. The puritan, who perhaps put his fortune back into trade and industrial enterprise, in the long run only made the ideals of the court spread more widely. Eventually in society, if not in the life of the individual capitalist, the day of reckoning comes: saturnalia follows the puritan's sober efforts. In a society that knows no other ideals, spending becomes the chief source of delight: finally, it amounts to a social duty.

Goods became respectable and desirable apart from the life-needs they served: they could be accumulated: they could be piled in palaces and storerooms: they could, when they resulted in surfeits and duplications, be translated temporarily into the more ethereal forms of money or bills of exchange or credit. To escape the lean restrictions of poverty became a sacred duty. Idleness was in itself a sin. A life outside the purlieus of production, without special industrial effort, without money-getting, had ceased to be respectable: the aristocracy itself, moved by its own heightened demands for luxuries and services, compromised with the merchant and manufacturing classes, married into them, adopted their vocations and interests, and welcomed new arrivals to the blessed state of riches. Philosophers speculated, now with faltering attention and a distracted eye, upon the nature of the good and the true and the beautiful. Was there any doubt about it? Their nature was essentially whatever could be embodied in material goods and profitably sold: whatever made life easier, more comfortable, more secure, physically more pleasant: in a word, better upholstered.

Finally, the theory of the new age, first formulated in terms of pecuniary success, was expressed in social terms by the utilitarians of the early nineteenth century. Happiness was the true end of man, and it consisted in achieving the greatest good for the greatest number. The essence of happiness was to avoid pain and seek pleasure: the quantity of happiness, and ultimately the perfection of human institutions, could be reckoned roughly by the amount of goods a society was capable of producing: expanding wants: expanding markets: expanding enterprises: an expanding body of consumers. The machine made this possible and guaranteed its success. To cry enough or to call a limit was treason. Happiness and expanding production were one.

That life may be most intense and significant in its moments of pain and anguish, that it may be most savorless in its moments of repletion, that once the essential means of living are provided its

intensities and ecstasies and states of equilibrium cannot be measured mathematically in any relation whatever to the quantity of
goods consumed or the quantity of power exercised—in short, the
commonplaces of experience to the lover, the adventurer, the parent,
the artist, the philosopher, the scientist, the active worker of any
sort—these commonplaces were excluded from the popular working
creed of utilitarianism. If a Bentham or a Mill tried by casuistry to
meet them, a Gradgrind and a Bounderby merely ignored them.
Mechanical production had become a categorical imperative, more
strict than any Kant discovered in his bosom.

Here, plainly, even the courtesan, even the soldier, knew better
than the merchant and the utilitarian philosopher: at a pinch one
would risk his body or the comforts of the body for honor or for
love. In furthering the quantification of life, moreover, they had at
least seized concrete loot: fabrics and foods and wines and paintings
and gardens. But by the time the nineteenth century came, these
realities had turned for the most part into paper will-o'-the-wisps:
marshlights to beguile mankind from tangible goods and immediate
fruitions. What Sombart has called the fragmentary man had come
into existence: the coarse Victorian philistine whom Ruskin ironically contrasted with the cleancut "esthete" of a Greek coin. He
boasted, this fragmentary man, on the best utilitarian principles,
that he was not in business for his health. The fact was obvious. But
for what other reason should men be in business?

The belief in the good life as the *goods* life came to fruition
before the paleotechnic complex had taken shape. This conception
gave the machine its social goal and its justification, even as it
gave form to so many of its end-products. When the machine produced other machines or other mechanical utilities, its influence was
often fresh and creative: but when the desires it gratified remained
those that had been taken over uncritically from the upper classes
during the period of dynastic absolutism, power-politics, and Baroque
emptiness, its effect was to further the disintegration of human
values.

In short, the machine came into our civilization, not to save man
from the servitude to ignoble forms of work, but to make more

widely possible the servitude to ignoble standards of consumption that had grown up among the military aristocracies. From the seventeenth century on, the machine was conditioned by the disordered social life of Western Europe. The machine gave an appearance of order to that chaos: it promised fulfillment for that emptiness: but all its promises were insidiously undermined by the very forces that gave it shape—the gambling of the miner, the power-lust of the soldier, abstract pecuniary ends of the financier, and the luxurious extensions of sexual power and surrogates for sex contrived by the court and the courtesan. All these forces, all these purposes and goals, are still visible in our machine-culture; by imitation they have spread from class to class and from town to country. Good and bad, clear and contradictory, amenable and refractory—here is the ore from which we must extract the metal of human value. Beside the few ingots of precious metal we have refined, the mountains of slag are enormous. But it is not all slag: far from it. One can even now look forward to the day when the poison gases and caked refuse, the once useless by-products of the machine, may be converted by intelligence and social cooperation to more vital uses.

CHAPTER III.   THE EOTECHNIC PHASE

1: Technical Syncretism

Civilizations are not self-contained organisms. Modern man could not have found his own particular modes of thought or invented his present technical equipment without drawing freely on the cultures that had preceded him or that continued to develop about him.

Each great differentiation in culture seems to be the outcome, in fact, of a process of syncretism. Flinders Petrie, in his discussion of Egyptian civilization, has shown that the admixture which was necessary for its development and fulfillment even had a racial basis; and in the development of Christianity it is plain that the most diverse foreign elements—a Dionysian earth myth, Greek philosophy, Jewish Messianism, Mithraism, Zoroastrianism—all played a part in giving the specific content and even the form to the ultimate collection of myths and offices that became Christianity.

Before this syncretism can take place, the cultures from which the elements are drawn must either be in a state of dissolution, or sufficiently remote in time or space so that single elements can be extracted from the tangled mass of real institutions. Unless this condition existed the elements themselves would not be free, as it were, to move over toward the new pole. Warfare acts as such an agent of dissociation, and in point of time the mechanical renascence of Western Europe was associated with the shock and stir of the Crusades. For what the new civilization picks up is not the complete forms and institutions of a solid culture, but just those fragments that can be transported and transplanted: it uses inventions, patterns, ideas, in the way that the Gothic builders in England used the occa-

107

sional stones or tiles of the Roman villa in combination with the native flint and in the entirely different forms of a later architecture. If the villa had still been standing and occupied, it could not have been conveniently quarried. It is the death of the original form, or rather, the remaining life in the ruins, that permits the free working over and integration of the elements of other cultures.

One further fact about syncretism must be noted. In the first stages of integration, before a culture has set its own definite mark upon the materials, before invention has crystallized into satisfactory habits and routine, it is free to draw upon the widest sources. The beginning and the end, the first absorption and the final spread and conquest, after the cultural integration has taken place, are over a worldwide realm.

These generalizations apply to the origin of the present-day machine civilization: a creative syncretism of inventions, gathered from the technical debris of other civilizations, made possible the new mechanical body. The waterwheel, in the form of the Noria, had been used by the Egyptians to raise water, and perhaps by the Sumerians for other purposes; certainly in the early part of the Christian era watermills had become fairly common in Rome. The windmill perhaps came from Persia in the eighth century. Paper, the magnetic needle, gunpowder, came from China, the first two by way of the Arabs: algebra came from India through the Arabs, and chemistry and physiology came via the Arabs, too, while geometry and mechanics had their origins in pre-Christian Greece. The steam engine owed its conception to the great inventor and scientist, Hero of Alexandria: it was the translations of his works in the sixteenth century that turned attention to the possibilities of this instrument of power.

In short, most of the important inventions and discoveries that served as the nucleus for further mechanical development, did not arise, as Spengler would have it, out of some mystical inner drive of the Faustian soul: they were wind-blown seeds from other cultures. After the tenth century in Western Europe the ground was, as I have shown, well plowed and harrowed and dragged, ready to receive these seeds; and while the plants themselves were growing, the cultivators of art and science were busy keeping the soil friable. Taking root in medieval culture, in a different climate and soil, these seeds of the machine sported and took on new forms: perhaps, precisely because they had *not* originated in Western Europe and had no natural enemies there, they grew as rapidly and gigantically as the Canada thistle when it made its way onto the South American pampas. But at no point—and this is the important thing to remember—did the machine represent a complete break. So far from being unprepared for in human history, the modern machine age cannot be understood except in terms of a very long and diverse preparation. The notion that a handful of British inventors suddenly made the wheels hum in the eighteenth century is too crude even to dish up as a fairy tale to children.

2: The Technological Complex

Looking back over the last thousand years, one can divide the development of the machine and the machine civilization into three successive but *over-lapping and interpenetrating phases:* eotechnic, paleotechnic, neotechnic. The demonstration that industrial civilization was not a single whole, but showed two marked, contrasting phases, was first made by Professor Patrick Geddes and published a generation ago. In defining the paleotechnic and neotechnic phases, he however neglected the important period of preparation, when all the key inventions were either invented or foreshadowed. So, following the archeological parallel he called attention to, I shall call the first period the eotechnic phase: the dawn age of modern technics.

While each of these phases roughly represents a period of human history, it is characterized even more significantly by the fact that it forms a technological complex. Each phase, that is, has its origin in certain definite regions and tends to employ certain special resources and raw materials. Each phase has its specific means of utilizing and generating energy, and its special forms of production. Finally, each phase brings into existence particular types of workers, trains them in particular ways, develops certain aptitudes and discourages others, and draws upon and further develops certain aspects of the social heritage.

Almost any part of a technical complex will point to and symbolize a whole series of relationships within that complex. Take the various types of writing pen. The goose-quill pen, sharpened by the user, is a typical eotechnic product: it indicates the handicraft basis of industry and the close connection with agriculture. Economically it is cheap; technically it is crude, but easily adapted to the style of the user. The steel pen stands equally for the paleotechnic phase: cheap and uniform, if not durable, it is a typical product of the mine, the steel mill and of mass-production. Technically, it is an improvement upon the quill-pen; but to approximate the same adaptability it must be made in half a dozen different standard points and shapes. And finally the fountain pen—though invented as early as the seventeenth century—is a typical neotechnic product. With its barrel of rubber or synthetic resin, with its gold pen, with its automatic action, it points to the finer neotechnic economy: and in its use of the durable iridium tip the fountain pen characteristically lengthens the service of the point and reduces the need for replacement. These respective characteristics are reflected at a hundred points in the typical environment of each phase; for though the various parts of a complex may be invented at various times, the complex itself will not be *in working order* until its major parts are all assembled. Even today the neotechnic complex still awaits a number of inventions necessary to its perfection: in particular an accumulator with six times the voltage and at least the present amperage of the existing types of cell.

Speaking in terms of power and characteristic materials, the eotechnic phase is a water-and-wood complex: the paleotechnic phase is a coal-and-iron complex, and the neotechnic phase is an electricity-and-alloy complex. It was Marx's great contribution as a sociological economist to see and partly to demonstrate that each period of invention and production had its own specific value for civilization, or, as he would have put it, its own historic mission. The machine cannot be divorced from its larger social pattern; for it is this pattern that gives it meaning and purpose. Every period of civilization carries within it the insignificant refuse of past technologies and the important germs of new ones: but the center of growth lies within its own complex.

The dawn-age of our modern technics stretches roughly from the year 1000 to 1750. During this period the dispersed technical advances and suggestions of other civilizations were brought together, and the process of invention and experimental adaptation went on at a slowly accelerating pace. Most of the key inventions necessary to universalize the machine were promoted during this period; there is scarcely an element in the second phase that did not exist as a germ, often as an embryo, frequently as an independent being, in the first phase. This complex reached its climax, technologically speaking, in the seventeenth century, with the foundation of experimental science, laid on a basis of mathematics, fine manipulation, accurate timing, and exact measurement.

The eotechnic phase did not of course come suddenly to an end in the middle of the eighteenth century: just as it reached its climax first of all in Italy in the sixteenth century, in the work of Leonardo and his talented contemporaries, so it came to a delayed fruition in the America of 1850. Two of its finest products, the clipper ship and the Thonet process of making bent-wood furniture, date from the eighteeen-thirties. There were parts of the world, like Holland and Denmark, which in many districts slipped directly from an eotechnic into the neotechnic economy, without feeling more than the cold shadow of the paleotechnic cloud.

With respect to human culture as a whole, the eotechnic period, though politically a chequered one, and in its later moments characterized by a deepening degradation of the industrial worker, was one of the most brilliant periods in history. For alongside its great mechanical achievements it built cities, cultivated landscapes, constructed buildings, and painted pictures, which fulfilled, in the realm of human thought and enjoyment, the advances that were being decisively made in the practical life. And if this period failed to establish a just and equitable polity in society at large, there were at least moments in the life of the monastery and the commune that were close to its dream: the afterglow of this life was recorded in More's Utopia and Andreae's Christianopolis.

Noting the underlying unity of eotechnic civilization, through all its superficial changes in costume and creed, one must look upon its successive portions as expressions of a single culture. This point is now being re-enforced by scholars who have come to disbelieve in the notion of the gigantic break supposed to have been made during the Renascence: a contemporary illusion, unduly emphasized by later historians. But one must add a qualification: namely, that with the increasing technical advances of this society there was, for reasons partly independent of the machine itself, a corresponding cultural dissolution and decay. In short, the Renascence was not, socially speaking, the dawn of a new day, but its twilight. The mechanical arts advanced as the humane arts weakened and receded, and it was at the moment when form and civilization had most completely broken up that the tempo of invention became more rapid, and the multiplication of machines and the increase of power took place.

3: New Sources of Power

At the bottom of the eotechnic economy stands one important fact: the diminished use of human beings as prime movers, and the separation of the production of energy from its application and immediate control. While the tool still dominated production energy and human skill were united within the craftsman himself: with the separation of these two elements the productive process itself tended toward a greater impersonality, and the machine-tool and the machine developed along with the new engines of power. If power machinery be a criterion, the modern industrial revolution began in the twelfth century and was in full swing by the fifteenth.

The eotechnic period was marked first of all by a steady increase in actual horsepower. This came directly from two pieces of apparatus: first, the introduction of the iron horseshoe, probably in the ninth century, a device that increased the range of the horse, by adapting him to other regions besides the grasslands, and added to his effective pulling power by giving his hoofs a grip. Second: by the tenth century the modern form of harness, in which the pull is met at the shoulder instead of at the neck, was re-invented in Western Europe—it had existed in China as early as 200 B.C.—and by the twelfth century, it had supplanted the inefficient harness the Romans had known. The gain was a considerable one, for the horse was now not merely a useful aid in agriculture or a means of transport: he became likewise an improved agent of mechanical production: mills utilizing horsepower directly for grinding corn or for pumping water came into existence all over Europe, sometimes supplementing other forms of non-human power, sometimes serving as the principal source. The increase in the number of horses was made possible, again, by improvements in agriculture and by the opening up of the hitherto sparsely cultivated or primeval forest areas in northern Europe. This created a condition somewhat similar to that which was repeated in America during the pioneering period: the new colonists, with plenty of land at their disposal, were lacking above all in labor power, and were compelled to resort to ingenious labor-saving devices that the better settled regions in the south with their surplus of labor and their easier conditions of living were never forced to invent. This fact perhaps was partly responsible for the high degree of technical initiative that marks the period.

But while horse power ensured the utilization of mechanical methods in regions not otherwise favored by nature, the greatest technical progress came about in regions that had abundant supplies of wind and water. It was along the fast flowing streams, the Rhône and the Danube and the small rapid rivers of Italy, and in the North Sea and Baltic areas, with their strong winds, that this new civilization had its firmest foundations and some of its most splendid cultural expressions.

Water-wheels for raising water in a chain of pots and for working automatic figures were described by Philo of Byzantium in the third century B.C.; and water-mills were definitely recorded in Rome in the first century B.C. Antipater of Thessalonica, a contemporary of Cicero, sang his praise of the new mills in the following poem: "Cease from grinding, ye women who toil at the mill; sleep late even if the crowing cocks announce the dawn. For

Demeter has ordered the Nymphs to perform the work of your hands, and they, leaping down on the top of the wheel, turn its axle which, with its revolving spokes, turns the heavy concave Nisyrian millstones. We taste again the joys of the primitive life, learning to feast on the products of Demeter without labor." The allusion is significant; it shows, as Marx pointed out, how much more humanely classic civilizations regarded labor-saving devices than did the enterprisers of the nineteenth century; and it proves, furthermore, that though the more primitive horizontal wheel was probably earlier, and because of its simple construction was used widely, the more complicated vertical type had come into use—and apparently likewise with the more efficient overshot wheel. Vitruvius, in his treatise on architecture, describes the design of gearing to regulate the speed.

Unlike the elaborate sanitary facilities of Rome, the water-mill never fell into complete disuse. There are allusions to such mills, as Usher points out, in a collection of Irish laws in the fifth century; and they crop out at intervals in other laws and chronicles. Though first used to grind corn, the water-mill was used to saw wood as early as the fourth century; and while, with the breakdown of the Empire and the decrease of the population, the number of mills may have decreased for a time, they came back again in the land-redemption and the land-colonization that took place under the monastic orders around the tenth century: by the time the Domesday Book survey was made there were five thousand water-mills in England alone—about one to every four hundred people—and England was then a backward country on the fringe of European civilization. By the fourteenth century, the water-mill had become common for manufacturing in all the great industrial centers: Bologna, Augsburg, Ulm. Their use possibly worked down the rivers toward the estuaries; for in the sixteenth century the low countries used water-mills to take advantage of the power of the tides.

Grinding grain and pumping water were not the only operations for which the water-mill was used: it furnished power for pulping rags for paper (Ravensburg: 1290): it ran the hammering and cutting machines of an ironworks (near Dobrilugk, Lausitz, 1320): it sawed wood (Augsburg: 1322): it beat hides in the tannery, it furnished power for spinning silk, it was used in fulling-mills to work up the felts, and it turned the grinding machines of the armorers. The wire-pulling machine invented by Rudolph of Nürnberg in 1400 was worked by water-power. In the mining and metal working operations Dr. Georg Bauer described the great convenience of water-power for pumping purposes in the mine, and suggested that if it could be utilized conveniently, it should be used instead of horses or man-power to turn the underground machinery. As early as the fifteenth century, water-mills were used for crushing ore. The importance of water-power in relation to the iron industries cannot be over-estimated: for by utilizing this power it was possible to make more powerful bellows, attain higher heats, use larger furnaces, and therefore increase the production of iron.

The extent of all these operations, compared with those undertaken today in Essen or Gary, was naturally small: but so was the society. The diffusion of power was an aid to the diffusion of population: as long as industrial power was represented directly by the utilization of energy, rather than by financial investment, the balance between the various regions of Europe and between town and country within a region was pretty evenly maintained. It was only with the swift concentration of financial and political power in the sixteenth and seventeenth centuries, that the excessive growth of Antwerp, London, Amsterdam, Paris, Rome, Lyons, Naples, took place.

Only second to waterpower in importance was windpower. Whatever the route it entered, the windmill spread rapidly in Europe, and it was widely diffused by the end of the twelfth century. The first definite knowledge of the windmill comes from a charter in 1105 authorizing the Abbot of Savigny to establish windmills in the diocese of Evreux, Bayeux, and Coutances; in England, the first date is 1143, and in Venice 1332: in 1341 the Bishop of Utrecht sought to establish authority over the winds that blew in his province: that in itself is almost enough to establish the industrial value of the windmill in the Low Countries by this time.

Apart from the wind-turbine, described as early as 1438, there were three types. In the most primitive type the entire structure faced the prevailing wind: in another, the entire structure turned to face it, sometimes being mounted on a boat to facilitate this; and in the most developed type the turret alone turned. The mill reached its greatest size and its most efficient form in the hands of the Dutch engineers toward the end of the sixteenth century, although the Italian engineers, including Leonardo himself, who is usually given credit for the turret windmill, contributed their share to the machine. In this development the Low Countries were almost as much the center of power production as England was during the later coal and iron régime. The Dutch provinces in particular, a mere film of sand, drenched with wind and water, plowed from one end to the other by the Rhine, the Amstel, the Maas, developed the windmill to the fullest possible degree: it ground the grain produced on the rich meadows, it sawed the wood brought down from the Baltic coast to make the great merchant marine, and it ground the spices—some five hundred thousand pounds per annum by the seventeenth century—that were brought from the Orient. A similar civilization spread up and down the peaty marshlands and barrier beaches from Flanders to the Elbe, for the Saxon and East Frisian shores of the Baltic had been repeopled by Dutch colonists in the twelfth century.

Above all, the windmill was the chief agent in land reclamation. The threat of inundation by the sea led these North Sea fishermen and farmers to attempt not only to control the water itself, but by keeping it back, to add to the land. The game was worth the effort, for this heavy soil provided rich pasture, after it was drained and sweetened. First carried on by the monastic orders, this reclamation had become, by the sixteenth century, one of the major industries of the Dutch. Once the dykes were built, however, the problem was how keep the area under the level of the sea clear of water: the windmill, which operates most steadily and strongly precisely when the storms are most fierce, was the means of raising the water of the rising streams and canals: it maintained the balance between the water and the land that made life possible in this precarious situation. Under the stimulus of self-imposed necessity, the Dutch became the leading engineers of Europe: their only rivals were in Italy. When the English, in the early seventeenth century, wished to drain their fens, they invited Cornelius Vermeyden, a celebrated Dutch engineer, to undertake the job.

The gain in energy through using wind and water power was not merely direct. By making possible the cultivation of the rich soil of the polder, these mechanical instruments reversed that steady degradation of the soil which had resulted from the cutting down of the forest cover and from the improvident system of agriculture that had succeeded the best Roman practice. Land building and irrigation are the signs of a planned, regenerative agriculture: the windmill added absolutely to the amount of energy avilable by helping to throw open these rich lands, as well as by protecting them and helping to work up their ultimate products.

This development of wind and water power did not reach its height in most parts of Europe until the seventeenth century: in England, not till the eighteenth century. How great was the increase of non-organic energy during this period? What was the sum total of non-human energy applied to production? It is difficult, perhaps impossible, to make even a rough guess as to the total amount of energy available: all one can say is that it kept increasing steadily from the eleventh century on. Marx observed that in Holland as late as 1836 there were 12,000 windmills in existence, giving as much as six thousand horse-power: but the estimate is too low, for one authority rates the average efficiency of the Dutch windmills as high as ten horse-power each; while Vowles notes that the ordinary old-fashioned type of Dutch windmill with four sails each twenty-four feet long and six feet wide generates about 4.5 brake horse-power in a twenty mile wind. Of course this estimate does not include the water power that was being used. Potentially, the amount of energy available for production was high as compared with any previous civilization. In the seventeenth century the most powerful prime mover in existence was the waterworks for Versailles: it developed a hundred horse-power and could raise a million gallons a day 502 feet. But as early as 1582 Peter Morice's tide-mill pumps, erected in London, raised four million gallons of water a day through a 12 inch pipe into a tank 128 feet high.

While the supply of both wind and water was subject to the

vagaries of local weather and the annual rainfall, there was probably compared to the present day less stoppage through variations
in the human labor requirement, owing to strikes, lockouts, and over-
production. In addition to this, since neither wind nor water-power
could be effectually monopolized—despite many efforts from the
thirteenth century on to prohibit small mills and querns, and to establish the custom of grinding at the lord's mill—the source of
energy itself was free: once built, the mill added nothing to the cost
of production. Unlike the later primitive steam engine, both a large
and a costly device, very small and primitive water mills could be
built, and were built; and since most of the moveable parts were of
wood and stone, the original cost was low and the deterioration
through seasonal disuse was not as great as would have been the
case had iron been used. The mill was good for a long life; the
upkeep was nominal; the supply of power was inexhaustible. And
so far from robbing the land and leaving behind debris and depopulated villages, as mining did, the mills helped enrich the land and
facilitated a conservative stable agriculture.

Thanks to the menial services of wind and water, a large
intelligentsia could come into existence, and great works of art and
scholarship and science and engineering could be created without
recourse to slavery: a release of energy, a victory for the human
spirit. Measuring the gains not in horsepower originally used but
in work finally accomplished, the eotechnic period compares favorably both with the epochs that preceded it and with the phases of
mechanical civilization that followed it. When the textile industries
attained an unheard of volume of production in the eighteenth century it was by means of water-power, not the steam engine, that this
was first achieved; and the first prime mover to exceed the poor five
or ten per cent efficiency of the early steam engines was Fourneyron's
water-turbine, a further development of the Baroque spoonwheel,
perfected in 1832. By the middle of the nineteenth century water-
turbines of 500 H.P. had been built. Plainly, the modern industrial
revolution would have come into existence and gone on steadily had
not a ton of coal been dug in England, and had not a new iron mine
been opened.

4: Trunk, Plank, and Spar

The mystic identification with the life of the old forests, which one feels in the ballads and folk-tales of the period, expressed a fact about the civilization which was emerging: wood was the universal material of the eotechnic economy.

First of all, wood was the foundation of its building. All the elaborate masonry forms were dependent upon the work of the carpenter: it was not merely that the piers themselves, in the later gothic construction, resembled tree trunks laced together or that the filtered light within the church had the dimness of the forest, while the effect of the bright glass was like that of the blue sky or a sunset seen through the tracery of branches: the fact is that none of this construction was possible without an elaborate falsework of wood: nor without wooden cranes and windlasses could the stones have been conveniently raised the necessary heights. Moreover, wood alternated with stone as a building material; and when in the sixteenth century the windows of the dwelling began to imitate in breadth and openness those of the public buildings, wooden beams carried the load across a space impossible for ordinary stone or brick construction to span: in Hamburg the burgher houses of the sixteenth century have windows across the whole front.

As for the common tools and utensils of the time, they were more often of wood than of any other material. The carpenter's tools were of wood, but for the last cutting edge: the rake, the oxyoke, the cart, the wagon, were of wood: so was the washtub in the bathhouse: so was the bucket and so was the broom: so in certain parts of Europe was the poor man's shoe. Wood served the farmer and the textile worker: the loom and the spinning-wheel, the oil presses and the wine presses were of wood, and even a hundred years after the printing press was invented, it was still made of wood. The very pipes that carried water in the cities were often tree-trunks: so were the cylinders of pumps. One rocked a wooden cradle; one slept on a wooden bed; and when one dined one "boarded." One brewed beer in a wooden vat and put the liquor in a wooden barrel. Stoppers of cork, introduced after the invention of the glass bottle, begin to be mentioned in the fifteenth century. The ships of course were made of wood and pegged together with wood: but to say that is only to say that the principal machines of industry were likewise made of wood: the lathe, the most important machine-tool of the period, was made entirely of wood—not merely the base but the moveable parts. Every part of the windmill and the water-mill except for the grinding and cutting elements was made of wood, even the gearing: the pumps were chiefly of wood, and even the steam engine, down to the nineteenth century, had a large number of wooden parts: the boiler itself might be of barrel construction, the metal being confined to the part exposed to the fire.

In all the operations of industry, wood played a part out of all proportion to that played by metals: had it not, indeed, been for the demand for metal coins, armor, cannons, and cannon-balls during this period, the need for metals would have been relatively insignificant: it was not merely the direct use of wood, but its part in mining and smelting and forging, that was responsible, as I pointed out before, for the destruction of the forests. The operations of mining demanded wooden beams to serve as shoring: wooden carts transported the ore, and wooden planks carried the load over the uneven surface of the mine.

Most of the key machines and inventions of the later industrial age were first developed in wood before they were translated into metal: wood provided the finger-exercises of the new industrialism. The debt of iron to wood was a heavy one: as late as 1820 Ithiel Town, a New Haven architect, patented a new type of lattice truss bridge, free from arch action and horizontal thrust, which became the prototype of many later iron bridges. As raw material, as tool, as machine-tool, as machine, as utensil and utility, as fuel, and as final product wood was the dominant industrial resource of the eotechnic phase.

Wind, water, and wood combined to form the basis for still another important technical development: the manufacture and operation of boats and ships.

If the twelfth century witnessed the introduction of the mariner's compass, the thirteenth brought the installation of the permanent rudder, used instead of the oar for steering, and the sixteenth introduced the use of the clock to determine longitude and the use of the quadrant to determine latitude—while the paddle-wheel, which was not to become important until the nineteenth century, was invented possibly as early as the sixth century, and was designed definitely in 1410, if not put into use until later. Out of the needs of navigation came that enormous labor-saving device, the logarithmic table, worked out by Briggs on Napier's foundation, and a little more than a century later the ship's chronometer was finally perfected by Harrison.

At the beginning of this period sails, which had hitherto been used chiefly with oars, began to supplant them and wind took the place of human muscle for working ships. In the fifteenth century the two-masted ship had come into existence: but it was dependent upon a fair wind. By 1500 the three-masted ship had appeared, and it was so far improved that it could beat against the wind: long ocean voyages were at last possible, without a Viking's daring and a Job's patience. As shipping increased and the art of navigation improved, harbors were developed, lighthouses were placed on treacherous parts of the coast, and at the beginning of the eighteenth century the first lightships were put to anchor on the Nore Sands off the English coast. With growing confidence in his ability to steer, to make headway, to find his position, and to reach port, the sailor replaced the slow land routes with his water routes. The economic gain due to water transport has been calculated for us by Adam Smith: "A broad-wheeled wagon," he observes in The Wealth of Nations, "attended by two men and drawn by eight horses, in about six weeks' time carries and brings back between London and Edinburgh near four ton weight of goods. In about the same time a ship navigated by six or eight men, and sailing between the ports of London and Leith, frequently carries and brings back two hundred ton weight of goods. Six or eight men, therefore, by the help of water carriage, can carry and bring back in the same time the same quantity of goods between London and Edinburgh, as 50 broad-wheeled waggons, attended by a hundred men, and drawn by 400 horses."

But ships served not only for facilitating international transport and trade over the ocean and along the continental rivers: boats also served for regional and local transportation. The two dominant cities, one at the beginning and the other at the end of the eotechnic period were Venice and Amsterdam: both of them built upon piles, both of them served by a network of canals. The canal itself was an ancient utility; but the widespread use of it in Western Europe definitely characterized this new economy. From the sixteenth century on canals supplemented the natural waterways: useful for the purposes of irrigation and drainage, and in both departments a boon to agriculture, canals also became the new highways in the more progressive regions of Europe. It was on the canals of Holland that the first regular and reliable transportation service came into existence: almost two centuries before the railroad. "Except in the case of ice," as Dr. H. W. Van Loon observes, "the canal boat ran as regularly as a train. It did not depend upon the wind and the condition of the roads." And the service was frequent: there were sixteen boats between Delft and Rotterdam every day.

The first big navigation canal was that between the Baltic and the Elbe; but by the seventeenth century Holland had a network of local and trans-regional canals that served to coordinate industry, agriculture, and transport. Incidentally, the contained and quiet waters of the canal, with its graded bank and its tow-path, was a great labor-saving device: the effectiveness of a man and a single horse, or a man with a pole, is incomparably higher on a water highway than on a land highway.

The order of development here is significant. Apart from beginnings in Italy—including Leonardo's plan for improving the navigation of rivers by canalization and locks—the first great system of canals was in the Low Countries, where they had been instituted by the Romans: then in France in the seventeenth century, with the Briare, Centre, and Languedoc canals, then in England in the eighteenth century, and finally in America—except for the minor city canals of New Amsterdam—in the nineteenth century. The progressive countries of the paleotechnic era were in this respect the backward ones of the eotechnic phase. And just as the windmills and water-mills served to distribute power, so the canal distributed population and goods and effected a closer union between town and country. Even in America one could see the typical eotechnic pattern of population and industry in the State of New York around 1850, when, on the basis of local saw mills, local gristmills, and an interlacing system of canals and dirt roads, the entire state was populated with remarkable evenness, and industrial opportunities were available at almost every point in the entire region. This balance between agriculture and industry, this diffusion of civilization, was one of the great social achievements of the eotechnic period: to this day, it gives to the Dutch village an outward touch of fine urbanity; and it offers a marked contrast with the atrocious lopsidedness of the period that followed.

The development of ships, harbors, lighthouses, and canals went on steadily: indeed, the eotechnic complex held together longer in maritime matters than it did in any other department of activity. The fastest type of sailing ship, the clipper, was not designed until the eighteen forties, and it was not until the twentieth century that the triangular type of mainsail replaced the topheavy polygon on the smaller craft and improved their speed. The sailing ship, like the windmill and the water-mill, was at the mercy of wind and water: but the gains in labor-saving and in horse-power, though again incalculable, were tremendously important. To speak of power as a recent acquisition of industry is to forget the kinetic energy of falling water and moving air; while to forget the part of the sailing ship in power-utilization is to betray a landlubber's ignorance of the realities of economic life from the twelfth century down to the third quarter of the nineteenth. Apart from this, the ship was indirectly a factor in rationalizing production and standardizing goods. Thus large factories for manufacturing ship's biscuits were built in Holland in the seventeenth century; and the manufacture of ready-to-wear clothing for civilians was first begun in New Bedford in the eighteen-forties because of the need for quickly outfitting sailors when they reached port.

5: Through a Glass, Brightly

But most important of all was the part played by glass in the eotechnic economy. Through glass new worlds were conceived and brought within reach and unveiled. Far more significant for civilization and culture than progress in the metallurgical arts up to the eighteenth century was the great advance in glass-making.

Glass itself was a very ancient discovery of the Egyptians, or possibly even of some more primitive people. Beads of glass have been found as far back as 1800 B.C. and openings for glass windows were found in the excavation of Pompeian houses. In the early Middle Ages, glass furnaces began to come back, first in the wooded districts near the monasteries, then near the cities: glass was used for holding liquids and for making the windows of public buildings. The early glass was of indifferent texture and finish: but by the twelfth century glass of intense color was made, and the use of these glasses in the windows of the new churches, admitting light, modifying it, transforming it, gave them a sombre brilliance that the most ornate carving and gold of the baroque churches only feebly rival.

By the thirteenth century the famous glass works at Murano, near Venice, had been founded; and glass was already used there for windows, for ship-lanterns, and for goblets. Despite the most zealous efforts to keep secret the technical methods of the Venetian glass workers, the knowledge of the art spread to other parts of Europe: by 1373 there was a guild of glassmakers in Nürnberg, and the development of glass-making went on steadily in other parts of Europe. In France it was one of the few trades that could be carried on by a noble family—thus taking on the characteristics of porcelain manufacture—and as early as 1635 Sir Robert Mansell obtained a monopoly for making flint glass in consideration of his being the first person who employed pit-coal instead of wood in his furnaces in England.

The development of glass changed the aspect of indoor life, particularly in regions with long winters and cloudy days. At first it was such a precious commodity that the glass panes were removable and were put in a safe place when the occupants left the house for any time. This high cost restricted glass to public buildings, but step by step it made its way into the private dwelling: Aeneas Sylvius de Piccolomini found in 1448 that half the houses in Wien had glass windows, and toward the end of the sixteenth century glass assumed in the design and construction of the dwelling house a place it had never had in any previous architecture. A parallel development went on in agriculture. An unedited letter, dated 1385, written in Latin and signed John, relates that "at Bois-le-Duc there are marvellous machines, even for drawing water, beating hides, and scraping cloth. There, too, they grow flowers in glass pavilions turned to the south." Hothouses, which used lapis specularis, a species of mica, instead of glass, were used by the Emperor Tiberius: but the glass hothouse was probably an eotechnic invention. It lengthened the growing period of Northern Europe, increased, so to say, the climatic range of a region, and utilized solar energy which would otherwise have been wasted: another clean gain. Even more important for industry, glass lengthened the span of the working day in cold or in inclement weather, particularly in the northern regions.

To have light in the dwelling house or the hothouse without being subject to cold or rain or snow, was the great contribution to the regularity of domestic living and business routine. This substitution of the window for the wooden shutter, or for oiled paper and muslin, was not fairly complete until the end of the seventeenth century: that is, until the processes of glass-making had been improved and cheapened, and the number of furnaces multiplied. Meanwhile, the product itself had been undergoing a change toward clarification and purification. As early as 1300 pure colorless glass was made in Murano: a fact that is established by a law imposing a heavy punishment upon the utilization of ordinary glass for eye glasses. In losing color and ceasing to serve as picture—the function it had occupied in medieval church decoration—and in letting in, instead, the forms and colors of the outside world, glass served also as a symbol of the double process of naturalism and abstraction which had begun to characterize the thought of Europe. More than that: it furthered this process. Glass helped put the world in a frame: it made it possible to see certain elements of reality more clearly: and it focussed attention on a sharply defined field—namely, that which was bounded by the frame.

The medieval symbolism dissolved and the world became a strangely different place as soon as one looked at it through glasses. The first change was effected by the use of the convex lens in spectacles: this corrected the flattening of the human lens due to age, and the defect of farsightedness: Singer has suggested that the revival of learning might in part be attributed to the number of additional years of eyesight for reading that the spectacles gave to human life. Spectacles were in wide use by the fifteenth century, when, with the invention of printing, a great need for them declared itself; and at the end of that century the concave lens was introduced to correct nearsightedness. Nature had provided lenses in every dew-drop and in the gum of every balsam tree: but it remained for the eotechnic glassmakers to utilize that fact. Roger Bacon is often given the credit for the invention of spectacles: the fact is at all events that apart from guesses and anticipations his major scientific work was in the realm of optics.

Long before the sixteenth century, the Arabs had discovered the use of a long tube for isolating and concentrating the field of stars under observation: but it was a Dutch optician, Johann Lippersheim, who in 1605 invented the telescope and thus suggested to Galileo the efficient means he needed for making astronomical observations. In 1590 another Hollander, the optician Zacharias Jansen invented the compound microscope: possibly also the telescope. One invention increased the scope of the macrocosm; the other revealed the microcosm: between them, the naïve conceptions of space that the ordinary man carried around were completely upset: one might say that these two inventions, in terms of the new perspective, extended the vanishing point toward infinity and increased almost infinitely the plane of the foreground from which those lines had their point of origin.

In the middle of the seventeenth century Leeuwenhoek, the methodical merchant and experimenter, through employing a distinguished technique, became the world's first bacteriologist. He discovered monsters in the scrapings of his teeth more mysterious and awful than any that had been encountered in the search for the Indies. If the glass did not actually add a new dimension to space, it extended its area, and it filled that space with new bodies, fixed stars at unimaginably vast distances, microcellular organisms whose existence was so incredible that, but for the researches of Spallanzani, they remained outside the sphere of serious investigation for over a century, after which their existence, their partnership, their enmity, almost became the source of a new demonology.

Glasses not merely opened people's eyes but their minds: seeing was believing. In the more primitive stages of thought the intuitions and ratiocinations of authority were sacrosanct, and the person who insisted on seeing proof of imagined events was reviled as the famous disciple had been: he was a doubting Thomas. Now the eye became the most respected organ. Roger Bacon refuted the superstition that diamonds could not be broken except by using goat's blood by resorting to experiment: he fractured the stones without using blood and reported: *"I have seen this work with my own eyes."* The use of glasses in the following centuries magnified the authority of the eye.

The development of glass had another important function. If the new astronomy were inconceivable without it, and if bacteriology would have been impossible, it is almost as true that chemistry would have been severely handicapped but for this development. Professor J. L. Myres, the classic archaeologist, has even suggested that the backwardness of the Greeks in chemistry was due to the lack of good glass. For glass has unique properties: not merely can it be made transparent, but it is, for most elements and chemical compounds, resistant to chemical change: it has the great advantage of remaining neutral to the experiment itself, while it permits the observer to see what is going on in the vessel. Easy to clean, easy to seal, easy to transform in shape, strong enough so that fairly thin globes can withstand the pressure of the atmosphere when exhausted, glass has a combination of properties that no wood or metal or clay container can rival. In addition it can be subjected to relatively high heats and—what became important during the nineteenth century—it is an insulator. The retort, the distilling flask, the test-tube: the barometer, the thermometer, the lenses and the slide of the microscope, the electric light, the x-ray tube, the audion—all these are products of glass technics, and where would the sciences be without them? A methodical analysis of temperature and pressure and the physical constitution of matter all awaited the development of glass: the accomplishments of Boyle, Torricelli, Pascal, Galileo, were specifically eotechnic works. Even in medicine glass has its triumph: the first instrument of precision to be used in diagnosis was the modification of Galileo's thermometer that Sanctorius introduced.

There is one further property of glass that had its first full effect in the seventeenth century. One sees it perhaps most clearly in the homes of the Dutch, with their enormous windows, for it was in the Netherlands that the use of glass and its manifold applications went farthest. Transparent glass lets in the light: it brings out, with merciless sincerity, moats dancing in the sunbeams and dirt lurking in the corner: for its fullest use, again, the glass itself must be clean, and no surface can be subject to a greater degree of verifiable cleanliness than the slick hard surface of glass. So, both by what it is and by what it does, glass is favorable to hygiene: the clean window, the scoured floor, the shiny utensils, are characteristic of the eotechnic household; and the plentiful supply of water, through the introduction of canals and pumping works with water pipes for circulation throughout the city, only made the process easier and more universal. Sharper eyesight: a sharper interest in the external world: a sharper response to the clarified image—these characteristics went hand in hand with the widespread introduction of glass.

6: Glass and the Ego

If the outward world was changed by glass, the inner world was likewise modified. Glass had a profound effect upon the development of the personality: indeed, it helped to alter the very concept of the self.

In a small way, glass had been used for mirrors by the Romans; but the background was a dark one, and the image was no more plain than it had been on the polished metal surface. By the sixteenth century, even before the invention of plate glass that followed a hundred years later, the mechanical surface of the glass had been improved to such an extent that, by coating it with a silver amalgam, an excellent mirror could be created. Technically this was, according to Schulz, perhaps the highest point in Venetian glass-making. Large mirrors, accordingly, became relatively cheap and the hand-mirror became a common possession.

For perhaps the first time, except for reflections in the water and in the dull surfaces of metal mirrors, it was possible to find an image that corresponded accurately to what others saw. Not merely in the privacy of the boudoir: in another's home, in a public gathering, the image of the ego in new and unexpected attitudes accompanied one. The most powerful prince of the seventeenth century created a vast hall of mirrors, and the mirror spread from one room to another in the bourgeois household. Self-consciousness, introspection, mirror-conversation developed with the new object itself: this preoccupation with one's image comes at the threshold of the mature personality when young Narcissus gazes long and deep into the face of the pool— and the sense of the separate personality, a perception of the objective attributes of one's identity, grows out of this communion.

The use of the mirror signalled the beginning of introspective biography in the modern style: that is, not as a means of edification but as a picture of the self, its depths, its mysteries, its inner dimensions. The self in the mirror corresponds to the physical world that was brought to light by natural science in the same epoch: it was the self *in abstracto,* only part of the real self, the part that one can divorce from the background of nature and the influential presence of other men. But there is a value in this mirror personality that more naïve cultures did not possess. If the image one sees in the mirror is abstract, it is not ideal or mythical: the more accurate the physical instrument, the more sufficient the light on it, the more relentlessly does it show the effects of age, disease, disappointment, frustration, slyness, covetousness, weakness—these come out quite as clearly as health and joy and confidence. Indeed, when one is completely whole and at one with the world one does not need the mirror: it is in the period of psychic disintegration that the individual personality turns to the lonely image to see what in fact is there and what he can hold on to; and it was in the period of cultural disintegration that men began to hold the mirror up to outer nature.

Who is the greatest of the introspective biographers? Where does one find him? It is none other than Rembrandt, and it is no accident that he was a Hollander. Rembrandt had a robust interest in the doctors and burghers about him: as a young man he was still enough of a guildsman and still had enough of the corporate personality to make a pass at painting those collective portraits which the members of the Nightwatch or the College of Physicians might commission—although already he was playing tricks with their conventions. But he came to the core of his art in the series of self-portraits he painted: for it was partly from the face he found in the mirror, from the knowledge of himself he developed and expressed in this communion, that he achieved the insight he applied to other men. A little later than Rembrandt, the Venice of the Alps, Annecy, harbored another portrait painter and introspectionist, Jean Jacques Rousseau who, more than Montaigne, was the father of the modern literary biography and the psychological novel.

The exploration of the solitary soul, the abstract personality, lingered on in the work of the poets and painters even after the eotechnic complex had broken up and the artists who had once dominated it were driven, by a more hostile world that was indifferent to visual images and antipathetic to the uniqueness of the individual soul—were driven to the point of complete frustration and madness. Enough here to remark that the isolation of the world from the self —the method of the physical sciences—and the isolation of the self from the world—the method of introspective biography and romantic poetry—were complementary phases of a single process. Much was learnt through that dissociation: for in the act of disintegrating the wholeness of human experience, the various atomic fragments that composed it were more clearly seen and more readily grasped. If the process itself was ultimately mad, the method that was derived from it was valuable.

The world as conceived and observed by science, the world as revealed by the painter, were both worlds that were seen through and with the aid of glasses: spectacles, microscopes, telescopes, mirrors, windows. What was the new easel picture, in fact, but a removable window opening upon an imaginary world? That acute scientific mind, Descartes, in describing the book on natural history that he failed to write, mentions how he wished finally to describe "how from these ashes, by the mere intensity of its [heat's] action, it formed glass: for as this transmutation of ashes into glass appeared to me as wonderful as any other in nature, I took a special pleasure in describing it." One can well understand his delight. Glass was in fact the peep-hole through which one beheld a new world. Through glass some of the mysteries of nature themselves became transparent. Is it any wonder then that perhaps the most comprehensive philosopher of the seventeenth century, at home alike in ethics and politics and science and religion, was Benedict Spinoza: not merely a Hollander, but a polisher of lenses.

7: The Primary Inventions

Between 1000 and 1750 in Western Europe the new technics fostered and adapted a series of fundamental inventions and discoveries: they were the foundation of the rapid advances that followed. And the speed of the ultimate movement, like the rapidity of an army's attack, was in proportion to the thoroughness of the preparation. Once the breach had been made in the line, it was easy for the rest of the army to follow through: but until that first act had been accomplished the army, however strong and eager and clamorous, could not move an inch. The primary inventions brought into being something that had not existed before: mechanical clocks, the telescope, cheap paper, print, the printing-press, the magnetic compass, the scientific method, inventions which were the means to fresh inventions, knowledge that was the center of expanding knowledge. Some of these necessary inventions, like the lathe and the loom, were far older than the eotechnic period: others, like the mechanical clock, were born with the renewed impulse toward regularity and regimentation. Only after these steps had been taken could the secondary inventions flourish: the regulation of the movement, which made the clock more accurate, the invention of the flying shuttle, which made the work of weaving swifter, the rotary press, which increased the output of printed matter.

Now an important point must be noted: the inventions of the eotechnic phase were only in a minor degree the direct product of craft skill and knowledge, proceeding out of the regular routine of industry. The tendency of organization by crafts, regulated in the interests of standardized and efficient work, guaranteed by local monopolies, was on the whole conservative, although in the building crafts, between the tenth and the fifteenth centuries, there were undoubtedly many daring innovators. In the beginning, it was knowledge, skill, experience, that had been the subjects of guild monopoly. With the growth of capitalism came the bestowing of special monopolies, first to the chartered companies, and then to the owners of special patents granted for specific original inventions. This was proposed by Bacon in 1601 and happened first in England in 1624. From this time on it *was not the past heritage that was effectively monopolized but the new departure from it.*

A special inducement was offered to those whose mechanical ingenuity supplanted the social and economic regulations of the guilds. In this situation, it was natural that invention should occupy the attention of those outside the industrial system itself—the military engineer, and even the amateur in every walk of life. Invention was a means of escaping one's class or achieving private riches within it: if the absolute monarch could say "L'Etat, c'est moi," the successful inventor could in effect say: "The Guild—that's me." While the detailed perfection of inventions was, more often than not, the work of skilled workers in the trade, the decisive idea was frequently the work of amateurs. Mechanical inventions broke the caste-lines of industry, even as they were later to threaten the caste-lines of society itself.

But the most important invention of all had no direct industrial connection whatever: namely, the invention of the experimental method in science. This was without doubt the greatest achievement of the eotechnic phase: its full effect upon technics did not begin to be felt until the middle of the nineteenth century. The experimental method, as I have already pointed out, owed a great debt to the transformation of technics: for the relative impersonality of the new instruments and machines, particularly the automata, must have helped to build up the belief in an equally impersonal world of irreducible and brute facts, operating as independently as clockwork and removed from the wishes of the observer: the reorganization of experience in terms of mechanical causality and the development of cooperative, controlled, repeatable, verifiable experiments, utilizing just such segments of reality as lent themselves to this method— *this was a gigantic labor-saving device.* It cut a short straight path through jungles of confused empiricism and laid down a rough corduroy road over swamps of superstitious and wishful thinking: to have found such a swift means of intellectual locomotion was perhaps sufficient excuse at the beginning for indifference to the scenery and for contempt for everything that did not speed the journey. None of the inventions that followed the development of the scientific method were so important in remolding the thought and activity of mankind as those that made experimental science possible. Eventually the scientific method was to repay its debt to technics a hundredfold: two centuries later, as we shall see, it was to suggest new combinations of means and turn into the realm of possibility the wildest dreams and the most irresponsible wishes of the race.

For out of the hitherto almost impenetrable chaos of existence there emerged finally, by the seventeenth century, an orderly world: the factual, impersonal order of science, articulated in every part and everywhere under the dominion of "natural law." Order, even when it was accepted as a basis for human designs, once rested on a pure act of faith: only the stars and the planets manifested it to the naked intelligence. Now order was supported by a method. Nature ceased to be inscrutable, subject to demonic incursions from another world: the very essence of Nature, as freshly conceived by the new scientists, was that its sequences were orderly and therefore predictable: even the path of a comet could be charted through the sky. It was on the model of this external physical order that men began systematically to reorganize their minds and their practical activities: this carried further, and into every department, the precepts and the practices empirically fostered by bourgeois finance. Like Emerson, men felt that the universe itself was fulfilled and justified, when ships came and went with the regularity of heavenly bodies. And they were right: there was something cosmic about it. To have made so much order visible was no little triumph.

In mechanical invention proper, the chief eotechnic innovation was of course the mechanical clock. By the end of the eotechnic phase, the domestic clock had become a common part of the household equipment, except among the poorer industrial workers and the peasants; and the watch was one of the chief articles of ornament carried by the well-to-do. The application of the pendulum to the clock, by Galileo and Huyghens, increased the accuracy of the instrument for common use.

But the indirect influence of clock-making was also important: as the first real instrument of precision, it set the pattern in accuracy and finish for all further instruments, all the more because it was regulated by the ultimate precision of the planetary movements themselves. In solving the problems of transmitting and regulating motion, the makers of clockwork helped the general development of fine mechanisms. To quote Usher once more: "The primary development of the fundamental principles of applied mechanics was . . . largely based upon the problems of the clock." Clockmakers, along with blacksmiths and locksmiths, were among the first machinists: Nicholas Forq, the Frenchman who invented the planer in 1751, was a clockmaker: Arkwright, in 1768, had the help of a Warrington clockmaker; it was Huntsman, another clockmaker, desirous of a more finely tempered steel for the watchspring, who invented the process of producing crucible steel: these are only a few of the more outstanding names. In sum, the clock was the most influential of machines, mechanically as well as socially; and by the middle of the eighteenth century it had become the most perfect: indeed, its inception and its perfection pretty well delimit the eotechnic phase. To this day, it is the pattern of fine automatism.

Second to the clock in order if not perhaps in importance was the printing press. Its development was admirably summed up by Carter, who did so much to clarify the historic facts. "Of all the world's great inventions that of printing is the most cosmopolitan and international. China invented paper and first experimented with block printing and moveable type. Japan produced the earliest block prints that are now extant. Korea first printed with type of metal, cast from a mould. India furnished the language and religion of the earliest block prints. People of Turkish race were among the most important agents in carrying block printing across Asia, and the earliest extant type are in a Turkish tongue. Persia and Egypt are the two lands of the Near East where block printing is known to have been done before it began in Europe. The Arabs were the agents who prepared the way by carrying the making of paper from China to Europe. . . . Florence and Italy were the first countries in Christendom to manufacture paper. As for block printing, and its advent into Europe, Russia's claim to have been the channel rests on the oldest authority, though Italy's claim is equally strong. Germany, Italy, and the Netherlands were the earliest centers of the block printing art. Holland and France, as well as Germany, claim to have experimented with type. Germany perfected the invention, and from Germany it spread to all the world."

The printing press and movable type were perfected by Gutenberg and his assistants at Mainz in the fourteen-forties. An astronomical calendar done in 1447 is the earliest datable example of Gutenberg's printing; but perhaps an inferior mode of printing may have been practiced earlier by Coster in Haarlem. The decisive improvement came with the invention of a hand-mold to cast uniform metal types.

Printing was from the beginning a completely mechanical achievement. Not merely that: it was the type for all future instruments of reproduction: for the printed sheet, even before the military uniform, was the first completely standardized product, manufactured in series, and the movable types themselves were the first example of completely standardized and interchangeable parts. Truly a revolutionary invention in every department.

By the end of fifty years there were over a thousand public printing presses in Germany alone, to say nothing of those in monasteries and castles; and the art had spread rapidly, despite all attempts at secrecy and monopoly, to Venice, Florence, Paris, London, Lyons, Leipzig, and Frankfort-am-Main. While there was strong competition from the well-established hand-copyists the art was encouraged by emancipation from taxes and guild regulations. Printing

lent itself to large-scale production: at the end of the fifteenth century there was in Nürnberg a large printing business with twenty-
four presses and a hundred employees—typesetters, printers, cor-
rectors, binders.

Compared with oral communication any sort of writing is a great
labor saving device, since it frees communication from the restrictions of time and space and makes discourse wait on the convenience
of the reader—who can interrupt the flow of thought or repeat it
or concentrate upon isolated parts of it. The printed page increased
the safety and permanence of the written record by manifolding it,
extended the range of communication, and economized on time and
effort. So print speedily became the new medium of intercourse:
abstracted from gesture and physical presence, the printed word
furthered that process of analysis and isolation which became the
leading achievement of eotechnic thought and which tempted Auguste
Comte to dub the whole epoch "metaphysical." By the end of the
seventeenth century time-keeping had merged with record-keeping
in the art of communication: the news-letter, the market report, the
newspaper, the periodical followed.

More than any other device, the printed book released people
from the domination of the immediate and the local. Doing so, it
contributed further to the dissociation of medieval society: print
made a greater impression than actual events, and by centering attention on the printed word, people lost that balance between the sensuous and the intellectual, between image and sound, between the
concrete and the abstract, which was to be achieved momentarily by
the best minds of the fifteenth century—Michelangelo, Leonardo,
Alberti—before it passed out, and was replaced by printed letters
alone. To exist was to exist in print: the rest of the world tended
gradually to become more shadowy. Learning became book-learning
and the authority of books was more widely diffused by printing, so
that if knowledge had an ampler province so, too, did error. The
divorce between print and firsthand experience was so extreme that
one of the first great modern educators, John Amos Komensky,
advocated the picture book for children as a means of restoring the
balance and providing the necessary visual associations.

But the printing press by itself did not perform the revolution: paper played a scarcely less important part: for its uses went far beyond the printed page: The application of power-driven machinery to paper production was one of the important developments of this economy. Paper removed the necessity for face to face contact: debts, deeds, contracts, news, were all committed to paper, so that, while feudal society existed by virtue of customs that were rigorously maintained from generation to generation, the last elements of feudal society were abolished in England by the simple device of asking peasants who had always had a customary share in the common lands for some documentary proof that they had ever owned it. Custom and memory now played second fiddle to the written word: reality meant "established on paper." Was it written in the bond? If so, it must be fulfilled. If not, it could be flouted. Capitalism, by committing its transactions to paper, could at last make and preserve a strict accountancy of time and money; and the new education for the merchant classes and their helpers consisted essentially in a mastery of the three R's. A paper world came into existence, and putting a thing on paper became the first stage in thought and action: unfortunately also often the last.

As a space-saver, a time-saver, a labor-saver—and so ultimately a life-saver—paper had a unique part to play in the development of industrialism. Through the habit of using print and paper thought lost some of its flowing, four-dimensional, organic character, and became abstract, categorical, stereotyped, content with purely verbal formulations and verbal solutions to problems that had never been presented or faced in their concrete inter-relationships.

The primary mechanical inventions of the clock and the printing press were accompanied by social inventions that were almost equally important: the university, beginning with Bologna in 1100, Paris in 1150, Cambridge in 1229 and Salamanca in 1243: a co-operative organization of knowledge on an international basis. The medical school, from Salerno and Montpellier onward, was not alone the first technical school in the modern sense; but the physicians, trained in the natural sciences at these schools and schooled by practice in the observation of nature, were among the pioneers in every department of technics and science: Paracelsus, Ambroise Paré, Cardan, Gilbert the author of *De Magnete*, Harvey, Erasmus Darwin, down to Thomas Young and Robert von Mayer were all physicians. In the sixteenth century two further social inventions were added: the scientific academy, first founded in the Accademia Secretorum Naturae in Naples in 1560, and the industrial exhibition, the first of which was held at the Rathaus in Nürnberg in 1569, the second in Paris in 1683.

By means of the university, the scientific academy and the industrial exhibition the exact arts and sciences were systematically explored, the new achievements were cooperatively exploited, and the new lines of investigation were given a common basis. One further important institution must be added: the laboratory. Here a new type of environment was created, combining the resources of the cell, the study, the library, and the workshop. Discovery and invention, like every other form of activity, consists in the interaction of an organism with its environment. New functions demand new environments, which tend to stimulate, concentrate, and perpetuate the special activity. By the seventeenth century these new environments had been created.

More direct in its effect upon technics was the creation of the factory. Down to the nineteenth century factories were always called mills, for what we call the factory grew out of the application of water-power to industrial processes; and it was the existence of a central building, divorced from the home and the craftsman's shop, in which large bodies of men could be gathered to perform the various necessary industrial operations with the benefit of large-scale co-operation that differentiated the factory in the modern sense from the largest of workshops. In this critical development the Italians again led the way, as they did in canal-building and fortification: but by the eighteenth century factories had reached the stage of large-scale operation in Sweden, in the manufacture of hardware, and this was true of Bolton's later works in Birmingham.

The factory simplified the collection of raw materials and the distribution of the finished product: it also facilitated the specialization of skill and the division of the processes of production: finally, by providing a common meeting place for the workers it partly overcame the isolation and helplessness that afflicted the handicraft worker after the structure of the town guilds had become dilapidated. The factory had finally a double rôle: it was an agent of mechanical regimentation, like the new army, and it was an example of genuine social order, appropriate to the new processes in industry. In either light, it was a significant invention. On one hand it gave a new motive for capitalistic investment in the form of the joint stock company operated for profit and it furnished the ruling classes with a powerful weapon: on the other, it served as a center for a new kind of social integration and made possible an efficient coordination of production which would be valuable under any social order.

The unison and cooperation produced by these various institutions, from the university to the factory, vastly increased the amount of effective energy in society: for energy is not merely a question of bare physical resources but of their harmonious social application. Habits of politeness, such as the Chinese have cultivated, may be quite as important in increasing efficiency, even measured in crude terms of footpounds of work performed, as economic methods of utilizing fuel: in society, as in the individual machine, failures in lubrication and transmission may be disastrous. It was important, for the further exploitation of the machine, that a social organization, appropriate to the technology itself, should have been invented. That the nineteenth century disclosed serious flaws in that organization—as it did in its financial twin, the joint stock company—does not lessen the importance of the original invention.

The clock and the printing press and the blast furnace were the giant inventions of the eotechnic phase, comparable to the steam engine in the period that followed, or the dynamo and the radio in the neotechnic phase. But they were surrounded by a multitude of inventions, too significant to be called minor, even when they fell short in performance of the inventor's expectations.

A good part of these inventions came to birth—or were further nourished—in the fecund mind of Leonardo da Vinci. Standing in the middle of this era, Leonardo summed up the technology of the artisans and military engineers who preceded him and released new

stores of scientific insight and inventive ingenuity: to catalog his
inventions and discoveries is almost to outline the structure of modern
technics. He was not alone in his own time: a military engineer
himself, he utilized to the full the common stock of knowledge that
was the property of his profession: nor was he altogether without
influence upon the period that followed, for it is probable that his
manuscripts were consulted and utilized by people who did not bother
particularly to record their obligations. But in his own person,
Leonardo embodied the forces of the period that was to follow.
He made the first scientific observations of the flight of birds, designed and built a flying machine, and designed the first parachute:
the conquest of space preoccupied him even though he was no more
successful than his obscure contemporary, G. B. Danti. Utilitarian
devices claimed his interest: he invented silk-winding machinery and
the alarm clock, he designed a power loom which was close to success: he invented the wheelbarrow and the lamp chimney and the
ship's log. Once he put before the Duke of Milan a project for the
mass production of standardized worker's dwellings. Even the motive
of amusement was not absent: he designed water shoes. As a mechanic he was incomparable: the antifriction roller bearing, the
universal joint, rope and belt drives, link chains, bevel and spiral
gears, the continuous motion lathe—all these were the work of his
powerful analytic mind. Indeed, his positive genius as technician
far outdoes his cold perfection as painter.

Even on the baser side of industrial exploitation Leonardo foreshadowed the forces that were to come. He was preoccupied not
merely with the desire for fame but for quick financial success:
"Early tomorrow, Jan. 2, 1496," he records in one of his notes,
"I shall make the leather belt and proceed to a trial. . . . One hundred times in each hour 400 needles will be finished, making 40,000
in an hour and 480,000 in 12 hours. Suppose we say 4000 thousands
which at 5 solidi per thousand gives 20,000 solidi: 1000 lira per
working day, and if one works 20 days in the month 60,000 ducats
the year." These wild dreams of freedom and power through a successful invention were to lure more than one daring mind, even
though the outcome were often to fail of realization as completely

as Leonardo's. Add to this Leonardo's contributions to warfare: the steam cannon, the organ gun, the submarine, and various detailed improvements upon the common devices of his time: inventions that represented an interest which, so far from dying out with the growth of industrialism, were rather substantiated and fortified by that growth. Even in the larger issue of Leonardo's life—the persistent warfare between the engineer and the artist—he typified most of the contradictions inherent in the new civilization, as it developed toward the Faustlike exploitation of the private ego and its satisfaction by means of financial and military and industrial power.

But Leonardo was not alone: both in his inventions and his anticipations he was surrounded by a gathering army of technicians and inventors. In 1535 the first diving bell was invented by Francesco del Marchi: in 1420 Joannes Fontana described a war-wagon or tank; and in 1518 the fire-engine is mentioned in the Augsburg Chronicles. In 1550 Palladio designed the first known suspension bridge in Western Europe while Leonardo, before him, had designed the drawbridge. In 1619 a tile making machine was invented; in 1680 the first power dredge was invented, and before the end of this century a French military man, De Gennes, had invented a power loom, while another Frenchman, the physician, Papin, had invented the steam engine and the steamboat. [For a fuller sense of the inventive richness of the eotechnic period, from the fifteenth to the eighteenth centuries, consult the List of Inventions.]

These are but samples from the great storehouse of eotechnic invention: seeds which came to life or lay dormant in dry soil or rocky crevices as wind and weather and chance dictated. Most of these inventions have been attributed to a later period, partly because they came to fruition then, partly because the first historians of the mechanical revolution, duly conscious of the vast strides that had been made in their own generation, were ignorant of the preparation and achievement that lay behind them, and were inclined at all events to belittle the preparatory period. Moreover, they were often not familiar with the manuscripts and books and artifacts that would have set them right. Thus it happens that England has sometimes been taken as the original home of inventions that had come into existence much earlier in Italy. So, too, the nineteenth century pinned on its own brows laurels that often enough belonged to the sixteenth and the seventeenth.

Since invention is almost never the sole work of a single inventor, however great a genius he may be, and since it is the product of the successive labors of innumerable men, working at various times and often toward various purposes, it is merely a figure of speech to attribute an invention to a single person: this is a convenient falsehood fostered by a spurious sense of patriotism and by the device of patent monopolies—a device that enables one man to claim special financial rewards for being the last link in the complicated social process that produced the invention. Any fully developed machine is a composite collective product: the present weaving machinery, according to Hobson, is a compound of about 800 inventions, while the present carding machinery is a compound of about 60 patents. This holds true for countries and generations as well: the joint stock of knowledge and technical skill transcends the boundaries of individual or national egos: and to forget that fact is not merely to enthrone superstition but to undermine the essential planetary basis of technology itself.

In calling attention to the scope and efficacy of eotechnic inventions one does not seek to belittle their debt to the past and to remoter regions—one merely wishes to show how much water had run under the bridge before people had become generally aware that a bridge had been built.

8: Weakness and Strength

The capital weakness of the eotechnic régime was not in the inefficiency of its power, still less in a lack of it; but in its irregularity. The dependence upon strong steady winds and upon the regular flow of water limited the spread and universalization of this economy, for there were districts in Europe that never fully benefited by it, and its dependence in both glass-making and metallurgy upon wood had, by the end of the eighteenth century, brought its powers to a low ebb. The forests of Russia and America might have delayed its collapse, as indeed they prolonged its reign within their own regions:

but they could not avert the steady dissipation of its fuel supply. Had the spoonwheel of the seventeenth century developed more rapidly into Fourneyron's efficient water-turbine, water might have remained the backbone of the power system until electricity had developed sufficiently to give it a wider area of use. But before this development could take place, the steam pumping engine had been invented. This engine was first used outside the mine, it is interesting to note, to raise water whose fall turned the conventional eotechnic waterwheel in hardware factories. As society became more closely co-ordinated on a basis of time, the interruption in its schedules through the irregularity of wind and water was a further defect: the wind-mill was finally defeated in Holland because it could not conform easily to labor regulations. And as distances increased and contracts in business emphasized the time-element, a more regular means of power became a financial necessity: delays and stoppages were costly.

But there were social weaknesses within the eotechnic régime that were equally grave. First of all, the new industries were outside the institutional controls of the old order. Glass-making, for example, by reason of the fact that it was always located in forested areas, tended to escape the restrictions of the town guilds: from the first it had a semi-capitalistic basis. Mining and iron-working, likewise, were almost from the beginning under a capitalistic system of production: even when mines were not worked by means of forced or servile labor, they were outside the control of the municipalities. Printing, again, was not subject to guild regulations; and even the textile industries escaped to the country: the factor who gave his name to the factory was a trader who farmed out the raw materials, and sometimes the necessary machines of production, and who bought up the product. The new industries, as Mantoux points out, tended to escape the manufacturing regulations of the guilds and even of the State itself—such as the English Statute of Apprentices of 1563: they grew up without social control. In other words, mechanical improvements flourished at the expense of the human improvements that had been strenuously introduced by the craft guilds; and the latter, in turn, were steadily losing force by reason of the growth of capitalistic monopolies which produced a steadily widening gap between masters and men. The machine had an anti-social bias: it tended by reason of its "progressive" character to the more naked forms of human exploitation.

Both the strength and the weakness of the eotechnic régime can in fact be witnessed in the technical development and the social dissolution and decay that took place in the textile industries, which were the backbone of the old economy.

Along with mining, the textile industries recorded the greatest number of improvements. While spinning with the distaff was carried on far into the seventeenth century, the spinning wheel had made its way into Europe from India by 1298. Within another century spinning mills and fulling mills had been introduced: by the sixteenth century, according to Usher, the fulling mills were also used as communal washing-machines: the fuller in his spare time did the village washing. Leonardo made the important invention of the flyer for spindles around 1490, and an authority on textiles, Mr. M. D. C. Crawford goes so far as to say that "without this inspired drawing we might have had no subsequent developments of textile machinery as we now know it." Johann Jurgen, a wood-carver of Brunswick, invented a partly automatic spinning wheel with a flier around 1530.

After Leonardo a succession of inventors worked on the power-loom. But the device that made it possible was Kay's flying shuttle, which greatly increased the productive capacity of the hand-loom weaver over eighty years before steam power was successfully applied to the automatic loom. This work was partly anticipated in the narrow-width ribbon loom, first invented in Danzig and then introduced into Holland; but the development of the power loom, through Bell and Monteith, was properly speaking a product of the paleotechnic phase, and Cartwright, the clergyman who usually gets full credit for its invention, played only an incidental rôle in the long chain of improvements that made it possible. While silk was spun by machinery in the fourteenth century, the first successful cotton spinning machine was not built until 1733 and patented in 1738, at a time when industry was still employing water power for prime movers. This series of inventions was in fact the final bequest of the eotechnic phase. Sombart marks the turning point of capitalism in the transfer of the center of gravity from the organic textile industries to the inorganic mining industries: that likewise marks the transition from the eotechnic to the paleotechnic economy.

One further set of inventions in the textile industry must be noted: the invention of knitting machinery in the sixteenth century. The origins of hand-knitting are obscure; if the art existed it played but a minor part before the fifteenth century. Knitting is not only perhaps the most distinctively European contribution to the textile industries but it was one of the first to be mechanized as the result of the invention of the knitting frame by another ingenious English clergyman. By taking advantage of the elasticity of yarns, knitting creates textiles which adapt themselves to the contours of the body and flex and contract with the movements of the muscles: while by adding to the amount of air-space within the yarn itself and between the strands, it increases warmth without adding to the weight. Knitted hose and undergarments—to say nothing of the wider use of the lighter washable cottons for body clothes—are all distinctly eotechnic contributions to comfort and cleanliness.

While the textile industries exhibited the steady advance of invention long before the introduction of the steam engine, they likewise witnessed the degradation of labor through the displacement of skill and through the breakdown of political control over the processes of production. The first characteristic is perhaps best seen in the industries where the division of the process could be carried farther than in the textile industries.

Manu-facture, that is, organized and partitioned handwork carried on in large establishments with or without power-machines, broke down the process of production into a series of specialized operations. Each one of these was carried on by a specialized worker whose facility was increased to the extent that his function was limited. This division was, in fact, a sort of empirical analysis of the working process, analyzing it out into a series of simplified human motions which could then be translated into mechanical operations. Once this analysis was performed, the rebuilding of the entire sequence of operations into a machine became more feasible. The mechanization of human labor was, in effect, the first step toward the humanization of the machine—humanization in the sense of giving the automaton some of the mechanical equivalents of life-likeness. The immediate effect of this division of process was a monstrous dehumanization: the worst drudgeries of craftsmanship can hardly be compared to it. Marx has summed up the process admirably.

"Whereas," Marx writes, "simple cooperation leaves the individual's methods of work substantially unaltered; manufacture revolutionizes these methods and cuts at the root of individual labor power. It transforms the worker into a cripple, a monster, by forcing him to develop some highly specialized dexterity at the cost of a world of productive impulses and faculties—much as in Argentina they slaughter a whole beast simply in order to get his hide or tallow. Not merely are the various partial operations allotted to different individuals; but the individual himself is split up, is transformed into the automatic motor of some partial operation. . . . To begin with the worker sells his labor power to capital because he himself lacks the material means requisite to the production of a commodity. But now his labor power actually renounces work unless it is sold to capital."

Here was both the process and the result which came about through the increased use of power and machinery in the eotechnic period. It marked the end of the guild system and the beginning of the wage worker. It marked the end of internal workshop discipline, administered by masters and journeymen through a system of apprenticeship, traditional teaching, and the corporate inspection of the product; while it indicated the beginning of an external discipline imposed by the worker and manufacturer in the interest of private profit—a system which lent itself to adulteration and to deteriorated standards of production almost as much as it lent itself to technical improvements. All this was a large step downward. In the textile industries the descent was rapid and violent during the eighteenth century.

In sum: as industry became more advanced from a mechanical point of view it at first became more backward from a human standpoint. Advanced agriculture, as practiced on the large estates toward the end of this period, sought to establish, as Arthur Young pointed out, the same standards in the field as had come to prevail in the workshop: specialization and division of process. If one wishes to view the eotechnic period at its best, one should perhaps behold it in the thirteenth century, before this process had set in: or at latest, at the end of the sixteenth century, when the ordinary worker, though still losing ground, losing freedom and self-control and substance, was unruly and resourceful—still capable of fighting or colonizing rather than ready to submit to the yoke of either becoming a machine or competing at sweated labor with the products of the machine. It remained for the nineteenth century to accomplish this final degradation.

But while one cannot ignore the defects of the eotechnic economy, including the fact that more powerful and accurate engines of destruction and exquisite apparatus for human torture were both put at the service of morbid ambitions and a corrupt ideology—while one cannot ignore these things one must not under-rate the real achievements. The new processes did save human labor and diminish—as the Swedish industrialist Polhem pointed out at the time— the amount and intensity of manual work. This result was achieved by the substitution of water-power for handwork, "with gains of 100 or even 1000 per cent in relative costs." It is easy to put a low estimate on the gains if one applies merely a quantitative measuring stick to them: if one compares the millions of horsepower now available to the thousands that then existed, if one compares the vast amount of goods poured forth by our factories with the modest output of the older workshops. But to judge the two economies correctly, one must also have a qualitative standard: one must ask not merely how much crude energy went into it, but how much of that went into the production of durable goods. The energy of the eotechnic régime did not vanish in smoke nor were its products thrown quickly on junk-heaps: by the seventeenth century it had transformed the woods and swamps of northern Europe into a continuous vista of wood and field, village and garden: an ordered human landscape replaced the bare meadows and the matted forests, while the social necessities of man had created hundreds of new cities, solidly built and commodiously arranged, cities whose spaciousness and order and beauty still challenge, even in their decay, the squalid anarchy of the new towns that succeeded them. In addition to the rivers, there were hundreds of miles of canals: in addition to the made lands of the north coastal area, there were harbors arranged for safety, and the beginnings of a lighthouse system. All these were solid achievements: works of art whose well-wrought forms stayed the process of entropy and postponed the final reckoning that all human things must make.

During this period the machine was adequately complemented by the utility: if the watermill made more power available the dyke and the drainage ditch created more usable soil. If the canal aided transport, the new cities aided social intercourse. In every department of activity there was equilibrium between the static and the dynamic, between the rural and the urban, between the vital and the mechanical. So it is not merely in the annual rate of converting energy or the annual rate of production that one must gauge the gains of the eotechnic period: many of its artifacts are still in use and still almost as good as new; and when one takes account of the longer span of time enjoyed by eotechnic products the balance tips back toward its own side of the arm. What it lacked in power, it made up for in time: its works had durability. Nor did the eotechnic period lack time any more than it lacked energy: far from moiling day and night to achieve as much as it did, it enjoyed in Catholic countries about a hundred complete holidays a year.

How rich the surplus of energies was by the seventeenth century one may partly judge by the high state of horticulture in Holland: when food is scarce one does not grow flowers to take its place. And wherever the new industry made its way during this period it directly enriched and improved the life of the community; for the services of art and culture, instead of being paralyzed by the increasing control over the environment, were given fuller sustenance. Can anything else account for the outburst of the arts during the Renascence, at a moment when the culture that supported them was so weak-spirited and the ostensible impulses so imitative and derivative?

The goal of the eotechnic civilization as a whole until it reached the decadence of the eighteenth century was not more power alone but a greater intensification of life: color, perfume, images, music, sexual ecstasy, as well as daring exploits in arms and thought and exploration. Fine images were everywhere: a field of tulips in bloom, the scent of new mown hay, the ripple of flesh under silk or the rondure of budding breasts: the rousing sting of the wind as the rain clouds scud over the seas, or the blue serenity of the sky and cloud, reflected with crystal clarity on the velvety surface of canal and pond and watercourse. One by one the senses were refined. Toward the end of this period the repetitious courses of the medieval dinner were analyzed out into the procession of foods that pass from the appetizer which rouses the necessary secretions to the sweet that signifies ultimate repletion. The touch, too, was refined: silks became commoner and the finest Dacca muslins from India took the place of coarse wools and linens: similarly the delicate smooth-surfaced Chinese porcelain supplemented the heavier Delft and Majolica and common earthenware.

Flowers in every garden improved the sensitiveness of the eye and the nose, making them quicker to take offense at the dungheap and the human ordure, and re-enforcing the general habits of household order and cleanliness that came in with eotechnic improvements. As early as Agricola's time he observes that "the place that Nature has provided with a river or stream can be made serviceable for many things; for water will never be wanting and can be carried through wooden pipes to baths in dwelling-houses." Refinement of smell was carried to such a pitch that it suggested Father Castel's *clavecin des odeurs.* One did not touch books or prints with dirty greasy hands: the well-thumbed books of the sixteenth and seventeenth centuries are still with us to prove it.

Re-enforcing the sense of cleanliness and this refinement of touch and taste, even in the kitchen, the first few rough iron pots gave way to copper pots and pans that were brought to a mirror-like polish by the industrious kitchen wench or housewife. But above all, during this period the eye was trained and refined: the delight of the eye even served other functions than pure vision by retarding them and giving the observer a chance to enter into them more fully. The wine-drinker gazed thoughtfully at the color of the wine before he supped it, and the lover's courtship became more intense, as well as more prolonged, as the visual pleasure of his beloved distracted him for a moment from the desire for possession. The wood-cut and the copper plate were popular arts during this period: even a great part of the vulgar work had affiliations to good form, and much of it had genuine distinction, while painting was one of the dominant expressions of the intellectual as well as the emotional life. Throughout life, alike for rich and poor, the spirit of play was understood and fostered. If the gospel of work took form during this period, it did not dominate it.

This great dilation of the senses, this more acute response to external stimuli, was one of the prime fruits of the eotechnic culture: it is still a vital part of the tradition of Western culture. Tempering the eotechnic tendency toward intellectual abstractionism, these sensual expressions formed a profound contrast to the contraction and starvation of the senses which had characterized the religious codes that preceded it, and was to characterize once more much of the doctrines and life of the nineteenth century. Culture and technics, though intimately related to each other through the activities of living men, often lie like non-conformable strata in geology, and, so to say, weather differently. During the greater part of the eotechnic period, however, they were in relative harmony. Except perhaps on the mine and the battlefield, they were both predominantly in the service of life. The rift between the mechanization and humanization, between power bent on its own aggrandizement and power directed toward wider human fulfillment had already appeared: but its consequences had still to become fully visible.

CHAPTER IV.　　THE PALEOTECHNIC PHASE

1: England's Belated Leadership

By the middle of the eighteenth century the fundamental industrial revolution, that which transformed our mode of thinking, our means of production, our manner of living, had been accomplished: the external forces of nature were harnessed and the mills and looms and spindles were working busily through Western Europe. The time had come to consolidate and systematize the great advances that had been made.

At this moment the eotechnic régime was shaken to its foundations. A new movement appeared in industrial society which had been gathering headway almost unnoticed from the fifteenth century on: after 1750 industry passed into a new phase, with a different source of power, different materials, different social objectives. This second revolution multiplied, vulgarized, and spread the methods and goods produced by the first: above all, it was directed toward the quantification of life, and its success could be gauged only in terms of the multiplication table.

For a whole century the second industrial revolution, which Geddes called the paleotechnic age, has received credit for many of the advances that were made during the centuries that preceded it. In contrast to the supposedly sudden and inexplicable outburst of inventions after 1760 the previous seven hundred years have often been treated as a stagnant period of small-scale petty handicraft production, feeble in power resources and barren of any significant accomplishments. How did this notion become popular? One reason, I think, is that the critical change that actually did take place during

the eighteenth century threw into shadow the older technical methods: but perhaps the main reason is that this change took place first
and most swiftly in England, and the observations of the new industrial methods, after Adam Smith—who was too early to appraise
the transformation—were made by economists who were ignorant
of the technical history of Western Europe, or who were inclined to
belittle its significance. The historians failed to appreciate the debt
of England's navy under Henry VIII to Italian shipbuilders, of her
mining industry to imported German miners, of her waterworks and
land-clearance schemes to Dutch engineers, and her silk spinning
mills to the Italian models which were copied by Thomas Lombe.

The fact is that England, throughout the Middle Ages, was one of
the backward countries of Europe: it was on the outskirts of the
great continental civilization and it shared in only a limited way in
the great industrial and civic development that took place in the
South from the tenth century onward. As a wool-raising center, in
the time of Henry VIII, England was a source of raw materials,
rather than a well-rounded agricultural and manufacturing country;
and with the destruction of the monasteries by the same monarch,
England's backwardness was only accentuated. It was not until the
sixteenth century that various traders and enterprisers began to develop mines and mills and glassworks on any considerable scale.
Few of the decisive inventions or improvements of the eotechnic
phase—one excepts knitting—had their home in England. England's
first great contribution to the new processes of thought and work came
through the marvellous galaxy of distinguished scientists it produced
in the seventeenth century: Gilbert, Napier, Boyle, Harvey, Newton,
and Hooke. Not until the eighteenth century did England participate
in any large degree in the eotechnic advances: the horticulture, the
landscape gardening, the canal building, even the factory organization of that period, correspond to developments that had taken place
from one to three centuries earlier in other parts of Europe.

Since the eotechnic régime had scarcely taken root in England,
there was less resistance there to new methods and new processes: the
break with the past came more easily, perhaps, because there was

less to break with. England's original backwardness helped to establish her leadership in the paleotechnic phase.

2: The New Barbarism

As we have seen, the earlier technical development had not involved a complete breach with the past. On the contrary, it had seized and appropriated and assimilated the technical innovations of other cultures, some very ancient, and the pattern of industry was wrought into the dominant pattern of life itself. Despite all the diligent mining for gold, silver, lead and tin in the sixteenth century, one could not call the civilization itself a mining civilization; and the handicraftsman's world did not change completely when he walked from the workshop to the church, or left the garden behind his house to wander out into the open fields beyond the city's walls.

Paleotechnic industry, on the other hand, arose out of the breakdown of European society and carried the process of disruption to a finish. There was a sharp shift in interest from life values to pecuniary values: the system of interests which only had been latent and which had been restricted in great measure to the merchant and leisure classes now pervaded every walk of life. It was no longer sufficient for industry to provide a livelihood: it must create an independent fortune: work was no longer a necessary part of living: it became an all-important end. Industry shifted to new regional centers in England: it tended to slip away from the established cities and to escape to decayed boroughs or to rural districts which were outside the field of regulation. Bleak valleys in Yorkshire that supplied water power, dirtier bleaker valleys in other parts of the land which disclosed seams of coal, became the environment of the new industrialism. A landless, traditionless proletariat, which had been steadily gathering since the sixteenth century, was drawn into these new areas and put to work in these new industries: if peasants were not handy, paupers were supplied by willing municipal authorities: if male adults could be dispensed with, women and children were used. These new mill villages and milltowns, barren of even the dead memorials of an older humaner culture, knew no other round and suggested no other outlet, than steady unremitting toil. The operations themselves were repetitive and monotonous; the environment was sordid; the life that was lived in these new centers was empty and barbarous to the last degree. Here the break with the past was complete. People lived and died within sight of the coal pit or the cotton mill in which they spent from fourteen to sixteen hours of their daily life, lived and died without either memory or hope, happy for the crusts that kept them alive or the sleep that brought them the brief uneasy solace of dreams.

Wages, never far above the level of subsistence, were driven down in the new industries by the competition of the machine. So low were they in the early part of the nineteenth century that in the textile trades they even for a while retarded the introduction of the power loom. As if the surplus of workers, ensured by the disfranchisement and pauperization of the agricultural workers, were not enough to re-enforce the Iron Law of Wages, there was an extraordinary rise in the birth-rate. The causes of this initial rise are still obscure; no present theory fully accounts for it. But one of the tangible motives was the fact that unemployed parents were forced to live upon the wages of the young they had begotten. From the chains of poverty and perpetual destitution there was no escape for the new mine worker or factory worker: the servility of the mine, deeply engrained in that occupation, spread to all the accessory employments. It needed both luck and cunning to escape those shackles.

Here was something almost without parallel in the history of civilization: not a lapse into barbarism through the enfeeblement of a higher civilization, but an upthrust into barbarism, aided by the very forces and interests which originally had been directed toward the conquest of the environment and the perfection of human culture. Where and under what conditions did this change take place? And how, when it represented in fact the lowest point in social development Europe had known since the Dark Ages did it come to be looked upon as a humane and beneficial advance? We must answer those questions.

The phase one here defines as paleotechnic reached its highest point, in terms of its own concepts and ends, in England in the middle of the nineteenth century: its cock-crow of triumph was the great industrial exhibition in the new Crystal Palace at Hyde Park in 1851: the first World Exposition, an apparent victory for free trade, free enterprise, free invention, and free access to all the world's markets by the country that boasted already that it was the workshop of the world. From around 1870 onwards the typical interests and preoccupations of the paleotechnic phase have been challenged by later developments in technics itself, and modified by various counterpoises in society. But like the eotechnic phase, it is still with us: indeed, in certain parts of the world, like Japan and China, it even passes for the new, the progressive, the modern, while in Russia an unfortunate residue of paleotechnic concepts and methods has helped misdirect, even partly cripple, the otherwise advanced economy projected by the disciples of Lenin. In the United States the paleotechnic régime did not get under way until the eighteen fifties, almost a century after England; and it reached its highest point at the beginning of the present century, whereas in Germany it dominated the years between 1870 and 1914, and, being carried to perhaps fuller and completer expression, has collapsed with greater rapidity there than in any other part of the world. France, except for its special coal and iron centers, escaped some of the worst defects of the period; while Holland, like Denmark and in part Switzerland, skipped almost directly from an eotechnic into a neotechnic economy, and except in ports like Rotterdam and in the mining districts, vigorously resisted the paleotechnic blight.

In short, one is dealing with a technical complex that cannot be strictly placed within a time belt; but if one takes 1700 as a beginning, 1870 as the high point of the upward curve, and 1900 as the start of an accelerating downward movement, one will have a sufficiently close approximation to fact. Without accepting any of the implications of Henry Adams's attempt to apply the phase rule of physics to the facts of history, one may grant an increasing rate of change to the processes of invention and technical improvement, at least up to the present; and if eight hundred years almost defines the eotechnic phase, one should expect a much shorter term for the paleotechnic one.

3: Carboniferous Capitalism

The great shift in population and industry that took place in the eighteenth century was due to the introduction of coal as a source of mechanical power, to the use of new means of making that power effective—the steam engine—and to new methods of smelting and working up iron. Out of this coal and iron complex, a new civilization developed.

Like so many other elements in the new technical world, the use of coal goes back a considerable distance in history. There is a reference to it in Theophrastus: in 320 B.C. it was used by smiths; while the Chinese not merely used coal for baking porcelain but even employed natural gas for illumination. Coal itself is a unique mineral: apart from the precious metals, it is one of the few unoxidized substances found in nature; at the same time it is one of the most easy to oxidize: weight for weight it is of course much more compact to store and transport than wood.

As early as 1234 the freemen of Newcastle were given a charter to dig for coal, and an ordinance attempting to regulate the coal nuisance in London dates from the fourteenth century. Five hundred years later coal was in general use as a fuel among glassmakers, brewers, distillers, sugar bakers, soap boilers, smiths, dyers, brickmakers, lime burners, founders, and calico printers. But in the meanwhile a more significant use had been found for coal: Dud Dudley at the beginning of the seventeenth century sought to substitute coal for charcoal in the production of iron: this aim was successfully accomplished by a Quaker, Abraham Darby, in 1709. By that invention the high-powered blast furnace became possible; but the method itself did not make its way to Coalbrookdale in Shropshire to Scotland and the North of England until the 1760's. The next development in the making of cast-iron awaited the introduction of a pump which should deliver to the furnace a more effective blast of air: this came with the invention of Watt's steam pump, and the demand for more iron, which followed, in turn increased the demand for coal.

Meanwhile, coal as a fuel for both domestic heating and power was started on a new career. By the end of the eighteenth century coal began to take the place of current sources of energy as an illuminant through Murdock's devices for producing illuminating gas. Wood, wind, water, beeswax, tallow, sperm-oil—all these were displaced steadily by coal and derivatives of coal, albeit an efficient type of burner, that produced by Welsbach, did not appear until electricity was ready to supplant gas for illumination. Coal, which could be mined long in advance of use, and which could be stored up, placed industry almost out of reach of seasonal influences and the caprices of the weather.

In the economy of the earth, the large-scale opening up of coal seams meant that industry was beginning to live for the first time on an accumulation of potential energy, derived from the ferns of the carboniferous period, instead of upon current income. In the abstract, mankind entered into the possession of a capital inheritance more splendid than all the wealth of the Indies; for even at the present rate of use it has been calculated that the present known supplies would last three thousand years. In the concrete, however, the prospects were more limited, and the exploitation of coal carried with it penalties not attached to the extraction of energy from growing plants or from wind and water. As long as the coal seams of England, Wales, the Ruhr, and the Alleghanies were deep and rich the limited terms of this new economy could be overlooked: but as soon as the first easy gains were realized the difficulties of keeping up the process became plain. For mining is a robber industry: the mine owner, as Messrs. Tryon and Eckel point out, is constantly consuming his capital, and as the surface measures are depleted the cost per unit of extracting minerals and ores becomes greater. The mine is the worst possible local base for a permanent civilization: for when the seams are exhausted, the individual mine must be closed down, leaving behind its debris and its deserted sheds and houses. The by-products are a befouled and disorderly environment; the end product is an exhausted one.

Now, the sudden accession of capital in the form of these vast coal fields put mankind in a fever of exploitation: coal and iron were the pivots upon which the other functions of society revolved. The activities of the nineteenth century were consumed by a series of rushes—the gold rushes, the iron rushes, the copper rushes, the petroleum rushes, the diamond rushes. The animus of mining affected the entire economic and social organism: this dominant mode of exploitation became the pattern for subordinate forms of industry. The reckless, get-rich-quick, devil-take-the-hindmost attitude of the mining rushes spread everywhere: the bonanza farms of the Middle West in the United States were exploited as if they were mines, and the forests were gutted out and mined in the same fashion as the minerals that lay in their hills. Mankind behaved like a drunken heir on a spree. And the damage to form and civilization through the prevalence of these new habits of disorderly exploitation and wasteful expenditure remained, whether or not the source of energy itself disappeared. The psychological results of carboniferous capitalism—the lowered morale, the expectation of getting something for nothing, the disregard for a balanced mode of production and consumption, the habituation to wreckage and debris as part of the normal human environment—all these results were plainly mischievous.

4: The Steam Engine

In all its broader aspects, paleotechnic industry rested on the mine: the products of the mine dominated its life and determined its characteristic inventions and improvements.

From the mine came the steam pump and presently the steam engine: ultimately the steam locomotive and so, by derivation, the steamboat. From the mine came the escalator, the elevator, which was first utilized elsewhere in the cotton factory, and the subway for urban transportation. The railroad likewise came directly from the mine: roads with wooden rails were laid down in Newcastle, England, in 1602: but they were common in the German mines a hundred years before, for they enabled the heavy ore carts to be moved easily over the rough and otherwise impassable surface of the mine. Around 1716 these wooden ways were capped with plates of malleable iron; and in 1767 cast iron bars were substituted. (Feldhaus notes that the invention of iron-clad wooden rails is illustrated at the time of the Hussite Wars around 1430: possibly the invention of a military engineer.) The combination of the railroad, the train of cars, and the locomotive, first used in the mines at the beginning of the nineteenth century, was applied to passenger transportation a generation later. Wherever the iron rails and wooden ties of this new system of locomotion went, the mine and the products of the mine went with it: indeed, the principal product carried by railroads is coal. The nineteenth century town became in effect—and indeed in appearance—an extension of the coal mine: The cost of transporting coal naturally increases with distance: hence the heavy industries tended to concentrate near the coal measures. To be cut off from the coal mine was to be cut off from the source of paloetechnic civilization.

In 1791, less than a generation after Watt had perfected the steam engine, Dr. Erasmus Darwin, whose poetic fancies were to become the leading ideas of the next century, apostrophized the new powers in the following verses:

> Soon shall thy arm, unconquered steam, afar
> Drag the slow barge, or drive the rapid car;
> Or on wide-waving wings expanded bear
> The flying chariot through the fields of air.
> Fair crews triumphant, leaning from above,
> Shall wave their flutt'ring kerchiefs as they move
> Or warrior bands alarm the gaping crowd,
> And armies shrink beneath the shadowy cloud.

His perceptions were quick and his anticipations were just. The technical history of the next hundred years was directly or indirectly the history of steam.

The need for more efficient mining which could reach the deeper seams prompted the effort to devise a more powerful pump than human labor or horse could work, and more regular and more accessible than wind or water mills: this was necessary to clear the galleries of water. The translation of Hero's Pneumatics, which contains devices for using steam, was published in Europe in 1575, and a series of inventors in the sixteenth century, Porta, Cardan, De Caus made various suggestions for using the power of steam to perform work. A century later the second Marquis of Worcester busied himself with the invention of a steam pumping engine (1630), thus transforming the instrument from a scientific toy into a practical mechanism. In 1633 the Marquis was granted a patent for his "water-commanding" engine, and he purposed to develop a water works for supplying water to the inhabitants of London. Nothing came of this; but the work was carried further by Thomas Savery whose device, called The Miner's Friend, was first publicized in 1698.

Dr. Papin, in France, had been working on the same lines: he described his engine as a "new means to create considerable motive power at low prices": the purpose was clear enough. Following up Papin's work, Newcomen, in 1712, erected an improved type of pumping engine. While the Newcomen engine was clumsy and inefficient, since it lost enormous quantities of heat in effecting condensation, it exceeded in power any single earlier prime mover, and through the application of steam power at the very source of energy, the coal mine itself, it was possible to sink the mines deeper and still keep them free of water. The main lines of the invention were laid down before Watt came upon the scene. It was his mission, not to invent the steam engine, but to raise considerably its efficiency by creating a separate condensing chamber and by utilizing the expansive pressure of the steam itself. Watt worked on the steam engine from 1765 on, applied for a patent in 1769, and between 1775 and 1800 erected 289 engines in England. His earlier steam engines were all pumps. Not until 1781 did Watt devote himself to inventing a rotary prime mover; and the answer to this problem was the great double-action fifty horsepower engine that his firm installed in the Albion Flour Mill in 1786, following the ten horsepower engine he first made for use in a brewery in London. In less than twenty years, so great was the demand for power, he installed 84 engines in cotton mills, 9 in wool and worsted mills, 18 in canal-works and 17 in breweries.

Watt's improvement of the steam engine in turn required improvements in the metallurgical arts. The machine work of his day in England was extremely inaccurate, and in boring cylinders for his engine he was obliged to "tolerate errors in his cylinders amounting to the thickness of a little finger in a cylinder 28 inches in diameter." So the demand for better engines, leading to Wilkinson's boring machine about 1776, and to Maudslay's numerous inventions and sim-plifications a generation later—including his perfection of the French slide rest for the lathe—gave a great stimulus to the machine crafts. Incidentally, the Albion Mills, designed by Rennie, were not merely the first to use steam for grinding wheat, but are supposed to have been the first important establishment in which every piece of the plant and equipment, axles, wheels, pinions, and shafts, was made of metal.

In more than one department, then, the 1780's mark the definite crystallization of the paleotechnic complex: Murdock's steam carriage, Cort's reverberatory furnace, Wilkinson's iron boat, Cartwright's power loom, and Jouffroy's and Fitch's steamboats, the latter with a screw propeller, date back to this decade.

The whole technique of wood had now to be perfected in the more difficult, refractory material—iron. The change from eotechnic to paleotechnic of course passed through transitional stages; but it could not remain at a halfway point. Though in America and Russia wood might, for example, be used right up to the third quarter of the nineteenth century for locomotives and steamboats, the need for coal developed with the larger and larger demands for fuel that the universalization of the machine carried with it. The very fact that Watt's steam engine consumed about eight and a half pounds of coal per horsepower, in comparison with Smeaton's atmospheric engine, which had used almost sixteen pounds, only increased the demand for more of Watt's kind, and widened the area of exploitation. The water-turbine was not perfected till 1832: in the intervening two generations steam had won supremacy, and it remained the *symbol* of increased efficiency. Even in Holland the efficient steam engine was presently introduced to assist in the Zuyder Zee reclamation: once the new scale, the new magnitudes, the new regularities were established, wind and water power could not without further aid compete with steam.

But note an important difference: the steam engine tended toward monopoly and concentration. Wind and water power were free; but coal was expensive and the steam engine itself was a costly investment; so, too, were the machines that it turned. Twenty-four hour operations, which characterized the mine and the blast furnace, now came into other industries which had heretofore respected the limitations of day and night. Moved by a desire to earn every possible sum on their investments, the textile manufacturers lengthened the working day: and whereas in England in the fifteenth century it had been fourteen or fifteen hours long in mid-summer with from two and a half to three hours allowed for recreation and meals, in the new milltowns it was frequently sixteen hours long all the year round, with a single hour off for dinner. Operated by the steam engine, lighted by gas, the new mills could work for twenty-four hours. Why not the worker? The steam engine was pacemaker.

Since the steam engine requires constant care on the part of the stoker and engineer, steam power was more efficient in large units than in small ones: instead of a score of small units, working when required, one large engine was kept in constant motion. Thus steam power fostered the tendency toward large industrial plants already present in the subdivision of the manufacturing process. Great size, forced by the nature of the steam engine, became in turn a symbol of efficiency. The industrial leaders not only accepted concentration and magnitude as a fact of operation, conditioned by the steam engine: they came to believe in it by itself, as a mark of progress. With the big steam engine, the big factory, the big bonanza farm, the big blast furnace, efficiency was supposed to exist in direct ratio to size. Bigger was another way of saying better.

But the steam engine tended toward concentration and bigness in still another way. Though the railroad increased travel distances and the amount of locomotion and transportation, it worked within relatively narrow regional limits: the poor performance of the railroad on grades over two per cent caused the new lines to follow the watercourses and valley bottoms. This tended to drain the population out of the back country, that had been served during the eotechnic phase by high roads and canals: with the integration of the railroad system and the growth of international markets, population tended to heap up in the great terminal cities, the junctions, the port towns. The main line express services tended to further this concentration and the feeder lines and cross country services ran down, died out, or were deliberately extirpated: to travel across country it was often necessary to go twice the distance through a central town and back again, hairpinwise.

Though the steam carriage was invented and put into use on the old coaching roads in England before the railroad, it never successfully challenged it: for a British act of Parliament drove it off the roads as soon as the railroad appeared on the scene. Steam power thus increased the areas of cities; it also increased the tendency of the new urban communities to coalesce along the main line of transportation and travel. That purely physical massing of population to which Patrick Geddes gave the name conurbation, was a direct product of the coal-and-iron régime. It must be distinguished carefully from the social formation of the city, to which it bears a casual resemblance by reason of its concentration of buildings and people. The prosperity of these new areas was measured in terms of the size of their new factories, the size of the population, the current rate of growth. In every way, then, the steam engine accentuated and deepened that quantification of life which had been taking place slowly and in every department during the three centuries that had preceded its introduction. By 1852 the railroad had reached the East Indies: by 1872 Japan and by 1876 China. Wherever it went it carried the methods and ideas of this mining civilization along with it.

5: Blood and Iron

Iron and coal dominated the paleotechnic period. Their color spread everywhere, from grey to black: the black boots, the black stove-pipe hat, the black coach or carriage, the black iron frame of the hearth, the black cooking pots and pans and stoves. Was it mourning? Was it protective coloration? Was it mere depression of the senses? No matter what the original color of the paleotechnic milieu might be, it was soon reduced, by reason of the soot and cinders that accompanied its activities, to its characteristic tones, grey, dirty brown, black. The center of the new industrialism in England was appropriately called the Black Country: by 1850 there was a similar blackness around the Pittsburgh district in America, and presently there was another in the Ruhr and around Lille.

Iron became the universal material. One went to sleep in an iron bed and washed one's face in the morning in an iron washbowl: one practiced gymnastics with the aid of iron dumb-bells or other iron weight-lifting apparatus; one played billiards on an iron billiard table, made by Messrs. Sharp and Roberts; one sat behind an iron locomotive and drove to the city on iron rails, passing over an iron bridge and arriving at an iron-covered railroad station: in America, after 1847, the front of the office-building might even be made of cast iron. In the most typical of Victorian utopias, that of J. S. Buckingham, the ideal city is built almost entirely of iron.

Although the Italians had designed iron bridges in the sixteenth century, the first to be built in England was in 1779, across the River Severn: the first iron dome was put on the Halles des Blés in Paris in 1817; the first iron ship was built in 1787, and the first iron steamship in 1821. So deep was the faith in iron during the paleotechnic period that it was not merely a favorite form of medicine, chosen as much for its magical association with strength as for any tangible benefits, but it was likewise offered for sale, if not actually used, for cuffs and collars to be worn by men, while, with the development of spring steel, iron even replaced whalebone in the apparatus used by the women of the period to deform their breasts, pelvises and hips. If the widest and most advantageous use of iron was in warfare, there was no part of existence, nevertheless, that was not touched directly or indirectly by the new material.

The cheaper, more efficient production of iron was indeed a direct result of the tremendous military demand for it. The first notable improvement in the production of iron, after the Darby process for making cast iron and the Huntsman process for making crucible steel was that made by Henry Cort, an English naval agent: he took out a patent for his puddling process in 1784 and made a timely contribution not merely to the success of England's iron industry in the export trade but to the victory of British arms during the Napoleonic wars. In 1856 Henry Bessemer, an Englishman, took out the patent for decarbonizing cast iron in his egg-shaped converter to make steel: a process slightly antedated by the independent invention of a Kentucky ironmaster, William Kelly. Thanks to Bessemer and the later Siemens-Martin process for making steel, the artillery arm flourished in warfare as never before: and after this period the ironclad or the steelclad warship, using long-range guns, became one of the most effective consumers of the national revenue in existence—as well as one of the most deadly weapons of war. Cheap iron and steel made it feasible to equip larger armies and navies than ever before: bigger cannon, bigger warships, more complicated equipment; while the new railroad system made it possible to put more men in the field and to put them in constant communication with the base of supplies at ever greater distances: war became a department of large-scale mass production.

In the very midst of celebrating the triumphs of peace and internationalism in 1851, the paleotechnic régime was preparing for a series of more lethal wars in which, as a result of modern methods of production and transport entire nations would finally become involved: the American Civil War, the Franco-Prussian War, most deadly and vicious of all, the World War. Nourished by war, the armament industries, whose plants were over-swollen through railroad building and past wars, sought new markets: in America, they found an outlet in the steel-framed building; but in the long run they were forced back on the more reliable industry of war, and they loyally served their stockholders by inciting competitive fears and rivalries among the nations: the notorious part recently played by the American steel manufacturers in wrecking the International Arms Conference of 1927 was only typical of a thousand less publicized moves during the previous century.

Bloodshed kept pace with iron production: in essence, the entire paloetechnic period was ruled, from beginning to end, by the policy of blood and iron. Its brutal contempt for life was equalled only by the almost priestly ritual it developed in preparation for inflicting death. Its "peace" was indeed the peace that passeth understanding: what was it but latent warfare?

What, then, is the nature of this material that exercised such a powerful effect upon the affairs of men? The use of meteoric iron possibly goes back very far in history: there is record of iron derived from the ordinary ores as far back as 1000 B.C., but the rapid

oxidation of iron may have wiped out traces of a much earlier utilization. Iron was associated in Egypt with Set, God of the waste
and desert, an object of fear; and through iron's close ties with
the military arts this association remains a not inappropriate one.

Iron's principal virtue lies in its combination of great strength
and malleability. While varying amounts of carbon change its characteristics, from toughness to brittleness, as steel or wrought iron it
has greater strength than any of the other common metals; and since,
in suitable cross section, an I-beam of iron is as strong as a solid
block, it matches its strength with relative lightness and transportability as compared for example with stone. But not merely is iron
strong under compression, like many varieties of stone: unlike stone,
it is strong in tension and when used in chains and cables, as the
Chinese were the first to use it, its characteristic properties come out
perhaps most clearly. One must pay for these excellent qualities by
working iron under a more intense heat than copper, zinc, or tin:
whereas steel melts at 1800 degrees Centigrade, and cast iron at
1500, copper has a melting point of 1100 and certain types of bronze
only half that heat: so that the casting of bronze long preceded the
casting of iron. On a large scale, iron-making demands power production: hence, while wrought iron dates back at least 2500 years,
cast-iron was not invented until the fourteenth century when the
water-driven bellows finally made the high temperature needed in
the blast furnace possible. To handle iron in large masses, conveying
it, rolling it, hammering it, all the accessory machinery must be
brought to an advanced stage of development. Though the ancients
produced hard implements of copper by hammering it cold, the cold
rolling of steel awaited advanced types of power machinery.
Nasmyth's steam hammer, invented in 1838, was one of the final
steps toward iron working in the grand style which made possible
the titanic machines and utilities of the later half of the nineteenth
century.

But iron has defects almost commensurate with its virtues. In its
usual impure state it is subject to fairly rapid oxidation, and until
the rustless steel alloys were discovered in the neotechnic period it
was necessary to cover iron with at least a film of non-oxidizing mate-

rial. Left to itself, iron rusts away: without constant lubrication bearings become jammed and without constant painting the iron ships and bridges and sheds would in the space of a generation become dangerously weakened: unless constant care is assured, the stone viaducts of the Romans, for example, are superior for long-time use. Again: iron is subject to changes in temperature: allowances must be made for expansion and contraction in summer and winter and during different parts of the same day: and without a protective covering of a fire-resistant material, the iron loses its strength so rapidly under heat that the soundest structure would become a mass of warped and twisted metal. But if iron oxidizes too easily, it has at least this compensating attribute: next to aluminum it is the commonest metal on the earth's crust. Unfortunately, the commonness and cheapness of iron, together with the fact that it was used according to rule-of-thumb prescription long before its properties were scientifically known, fostered a certain crudeness in its utilization: allowing for ignorance by erring on the side of safety, the designers used over-size members in their iron structures which did not sufficiently embrace the esthetic advantages—to say nothing of economic gain—possible through lightness and through the closer adaptation of structure to function. Hence the paradox: between 1775 and 1875 there was technological backwardness in the most advanced part of technology. If iron was cheap and if power was plentiful, why should the engineer waste his talents attempting to use less of either? By any paleotechnic standard, there was no answer to this question. Much of the iron that the period boasted was dead weight.

6: The Destruction of Environment

The first mark of paleotechnic industry was the pollution of the air. Disregarding Benjamin Franklin's happy suggestion that coal smoke, being unburnt carbon, should be utilized a second time in the furnace, the new manufacturers erected steam engines and factory chimneys without any effort to conserve energy by burning up thoroughly the products of the first combustion; nor did they at first attempt to utilize the by-products of the coke-ovens or burn up the gases produced in the blast-furnace. For all its boasts of improvement, the steam engine was only ten per cent efficient: ninety per cent of the heat created escaped in radiation, and a good part of the fuel went up the flue. Just as the noisy clank of Watt's original engine was maintained, against his own desire to do away with it, as a pleasing mark of power and efficiency, so the smoking factory chimney, which polluted the air and wasted energy, whose pall of smoke increased the number and thickness of natural fogs and shut off still more sunlight—this emblem of a crude, imperfect technics became the boasted symbol of prosperity. And here the concentration of paleotechnic industry added to the evils of the process itself. The pollution and dirt of a small iron works situated in the open country could be absorbed or carried away without difficulty. When twenty large iron works were grouped together, concentrating their effluvia and their waste-products, a wholesale deterioration of the environment inevitably followed.

How serious a loss was occasioned by these paleotechnic habits one can see even today, and one can put it in terms that even paleotects can understand: the annual cost of keeping Pittsburgh clean because of smoke has been estimated at $1,500,000 for extra laundry work, $750,000 for extra general cleaning, and $360,000 for extra curtain cleaning: an estimate which does not include losses due to the corrosion of buildings, to extra cost of light during periods of smog, and the losses occasioned by the lowering of health and vitality through interference with the sun's rays. The hydrochloric acid evolved by the Le Blanc process for manufacturing sodium carbonate was wasted until an act of the British Parliament in 1863, incited by the corrosive action of the gas on the surrounding vegetation and metal work, compelled its conservation. Need one add that the chlorine in the "waste-product" was turned to highly profitable commercial uses as a bleaching powder?

In this paleotechnic world the realities were money, prices, capital, shares: the environment itself, like most of human existence, was treated as an abstraction. Air and sunlight, because of their deplorable lack of value in exchange, had no reality at all. Andrew Ure, the great British apologist for Victorian capitalism, was aghast at the excellent physician who testified before Sadler's Factory Investigating Commission on the basis of experiments made by Dr. Edwards in Paris with tadpoles, that sunlight was essential to the growth of children: a belief which he backed up—a century before the effect of sunlight in preventing rickets was established—by pointing to the absence of deformities of growth, such as were common in milltowns, among the Mexicans and Peruvians, regularly exposed to sunlight. In response to this Ure proudly exhibited the illustration of a factory room without windows as an example of the excellent gas-lighting which served as a substitute for the sun!

The values of the paleotechnic economy were topsy-turvy. Its abstractions were reverenced as "hard facts" and ultimate realities; whereas the realities of existence were treated by the Gradgrinds and Bounderbys as abstractions, as sentimental fancies, even as aberrations. So this period was marked throughout the Western World by the widespread perversion and destruction of environment: the tactics of mining and the debris of the mine spread everywhere. The current annual wastage through smoke in the United States is huge— one estimate is as high as approximately $200,000,000. In an all too literal sense, the paleotechnic economy had money to burn.

In the new chemical industries that sprang up during this period no serious effort was made to control either air pollution or stream pollution, nor was any effort made to separate such industries from the dwelling-quarters of the town. From the soda works, the ammonia works, the cement-making works, the gas plant, there emerged dust, fumes, effluvia, sometimes noxious for human organisms. In 1930 the upper Meuse district in Belgium was in a state of panic because a heavy fog resulted in widespread choking and in the death of 65 people: on careful examination it turned out that there had been only a particularly heavy concentration of the *usual* poison gases, chiefly sulphurous anhydride. Even where the chemical factories were not conspicuously present, the railroad distributed smut and dirt: the reek of coal was the very incense of the new industrialism. A clear sky in an industrial district was the sign of a strike or a lockout or an industrial depression.

If atmospheric sewage was the first mark of paleotechnic industry, stream pollution was the second. The dumping of the industrial and chemical waste-products into the streams was a characteristic mark of the new order. Wherever the factories went, the streams became foul and poisonous: the fish died or were forced, like the Hudson shad, to migrate, and the water became unfit for either drinking or bathing. In many cases the refuse so wantonly disposed of was in fact capable of being used: but the whole method of industry was so short-sighted and so unscientific that the full utilization of by-products did not concern anyone for the first century or so. What the streams could not transport away remained in piles and hillocks on the outskirts of the industrial plant, unless it could be used to fill in the water-courses or the swamps on the new sites of the industrial city. These forms of industrial pollution of course go back very far in the history of paleotechnic industry: Agricola makes mention of them, and they remain to this day one of the most durable attributes of the mining economy.

But with the new concentration of industry in the industrial city there was still a third source of stream pollution. This was from human excrement, recklessly dumped into the rivers and tidal waters without any preliminary treatment, to say nothing of attempts to conserve the valuable nitrogenous elements for fertilizer. The smaller rivers, like the Thames and later the Chicago River became little less than open sewers. Lacking the first elements of cleanliness, lacking even a water supply, lacking sanitary regulations of any kind, lacking the open spaces and gardens of the early medieval city, which made cruder forms of sewage disposal possible, the new industrial towns became breeding places for disease: typhoid bacteria filtered through the soil from privy and open sewer into the wells from which the poorer classes got their water, or they were pumped out of the river which served equally as a reservoir for drinking water and a sewage outlet: sometimes, before the chlorine treatment was introduced, the municipal waterworks were the chief source of infection. Diseases of dirt and diseases of darkness flourished: smallpox, typhus, typhoid, rickets, tuberculosis. In the very hospitals, the prevalent dirt counteracted the mechanical advances of surgery: a great part of those who survived the surgeon's scalpel succumbed to "hospital fever." Sir Frederick Treves remembered how the surgeons of Guy's Hospital boasted of the incrustations of blood and dirt on their operating coats, as a mark of long practice! If that was surgical cleanliness, what could one expect of the impoverished workers in the new slums?

But there were other types of environmental degradation besides these forms of pollution. Foremost among these was that resulting from the regional specialization of industry. Natural regional specializations exist by reason of strong differences in climate and geological formation and topography: under natural conditions, no one attempts to grow coffee in Iceland. But the new specialization was based, not upon conforming to regional opportunities, but upon concentrating upon a single aspect of industry and pushing this to the exclusion of every other form of art and work. Thus England, the home of the new specialization, turned all its resources and energy and man-power into mechanical industry and permitted agriculture to languish: similarly, within the new industrial complex, one locality specialized in steel and another in cotton, with no attempt at diversification of manufacture. The result was a poor and constricted social life and a precarious industry. By reason of specialization a variety of regional opportunities were neglected, and the amount of wasteful cross-haulage in commodities that could be produced with equal efficiency in any locality was increased; while the shutting down of the single industry meant the collapse of the entire local community. Above all, the psychological and social stimulus derived from cultivating numerous different occupations and different modes of thought and living disappeared. Result: an insecure industry, a lop-sided social life, an impoverishment of intellectual resources, and often a physically depleted environment. This intensive regional specialization at first brought huge pecuniary profits to the masters of industry; but the price it exacted was too high. Even in terms of mechanical efficiency the process was a doubtful one, because it was a barrier against that borrowing from foreign processes which is one of the principal means of effecting new inventions and creating industries. While when one considers the environment as an element in human ecology, the sacrifice of its varied potentialities to mechanical industries alone was highly inimical to human welfare: the usurpation of park sites and bathing sites by the new steel works and coke-ovens, the reckless placement of railroad yards with no respect to any fact except cheapness and convenience for the railroad itself, the destruction of forests, and the building up of solid masses of brick and paving stone without regard for the special qualities of site and soil—all these were forms of environmental destruction and waste. The cost of this indifference to the environment as a human resource—who can measure it? But who can doubt that it offsets a large part of the otherwise real gains in producing cheap textiles and transporting surplus foods?

7: The Degradation of the Worker

Kant's doctrine, that every human being should be treated as an end, not as a means, was formulated precisely at the moment when mechanical industry had begun to treat the worker solely as a means —a means to cheaper mechanical production. Human beings were dealt with in the same spirit of brutality as the landscape: labor was a resource to be exploited, to be mined, to be exhausted, and finally to be discarded. Responsibility for the worker's life and health ended with the cash-payment for the day's labor.

The poor propagated like flies, reached industrial maturity—ten or twelve years of age—promptly, served their term in the new textile mills or the mines, and died inexpensively. During the early paleotechnic period their expectation of life was twenty years less than that of the middle classes. For a number of centuries the degradation of labor had been going on steadily in Europe; at the end of the eighteenth century, thanks to the shrewdness and near-sighted rapacity of the English industrialists, it reached its nadir in England. In other countries, where the paleotechnic system entered later, the same brutality emerged: the English merely set the pace. What were the causes at work?

By the middle of the eighteenth century the handicraft worker had been reduced, in the new industries, into a competitor with the machine. But there was one weak spot in the system: the nature of human beings themselves: for at first they rebelled at the feverish pace, the

rigid discipline, the dismal monotony of their tasks. The main difficulty, as Ure pointed out, did not lie so much in the invention of an
effective self-acting mechanism as in the "distribution of the different
members of the apparatus into one cooperative body, in impelling
each organ with its appropriate delicacy and speed, and above all, in
training human beings to renounce their desultory habits of work
and to identify themselves with the unvarying regularity of the complex automaton." "By the infirmity of human nature," wrote Ure
again, "it happens that the more skillful the workman, the more self-
willed and intractable he is apt to become, and of course the less fit
and component of the mechanical system in which . . . he may do
great damage to the whole."

The first requirement for the factory system, then, was the castra-
tion of skill. The second was the discipline of starvation. The third
was the closing up of alternative occupations by means of land-
monopoly and dis-education.

In actual operation, these three requirements were met in reverse
order. Poverty and land monopoly kept the workers in the locality
that needed them and removed the possibility of their improving their
position by migration: while exclusion from craft apprenticeship,
together with specialization in subdivided and partitioned mechanical
functions, unfitted the machine-worker for the career of pioneer or
farmer, even though he might have the opportunity to move into the
free lands in the newer parts of the world. Reduced to the function
of a cog, the new worker could not operate without being joined to a
machine. Since the workers lacked the capitalists' incentives of
gain and social opportunity, the only things that kept them bound
to the machine were starvation, ignorance, and fear. These three
conditions were the foundations of industrial discipline, and they
were retained by the directing classes even though the poverty of the
worker undermined and periodically ruined the system of mass production which the new factory discipline promoted. Therein lay one
of the inherent "contradictions" of the capitalist scheme of production.

It remained for Richard Arkwright, at the beginning of the paleotechnic development, to put the finishing touches upon the factory

system itself: perhaps the most remarkable piece of regimentation, all things considered, that the last thousand years have seen.

Arkwright, indeed, was a sort of archetypal figure of the new order: while he is often credited, like so many other successful capitalists, with being a great inventor, the fact is that he was never guilty of a single original invention: he appropriated the work of less astute men. His factories were located in different parts of England, and in order to supervise them he had to travel with Napoleonic diligence, in a post-chaise, driven at top speed: he worked far into the night, on wheels as well as at his desk. Arkwright's great contribution to his personal success and to the factory system at large was the elaboration of a code of factory discipline: three hundred years after Prince Maurice had transformed the military arts, Arkwright perfected the industrial army. He put an end to the easy, happy-go-lucky habits that had held over from the past: he forced the one-time independent handicraftsman to "renounce his old prerogative of stopping when he pleases, because," as Ure remarks, "he would thereby throw the whole establishment into disorder."

Following upon the earlier improvements of Wyatt and Kay, the enterpriser in the textile industries had a new weapon of discipline in his hands. The machines were becoming so automatic that the worker himself, instead of performing the work, became a machine-tender, who merely corrected failures in automatic operation, like a breaking of the threads. This could be done by a woman as easily as by a man, and by an eight-year-old child as well as by an adult, provided discipline were harsh enough. As if the competition of children were not enough to enforce low wages and general submission, there was still another police-agent: the threat of a new invention which would eliminate the worker altogether.

From the beginning, technological improvement was the manufacturer's answer to labor insubordination, or, as the invaluable Ure reminded his readers, new inventions "confirmed the great doctrine already propounded that when capital enlists science into its service the refractory hand of labor will be taught docility." Nasmyth put this fact in its most benign light when he held, according to Smiles, that strikes were more productive of good than of evil, since they served to stimulate invention. "In the case of many of our most potent self-acting tools and machines, manufacturers could not be induced to adopt them until compelled to do so by strikes. This was the case of the self-acting mule, the wool-combing machine, the planing-machine, the slotting-machine, Nasmyth's steam-arm, and many others."

At the opening of the period, in 1770, a writer had projected a new scheme for providing for paupers. He called it a House of Terror: it was to be a place where paupers would be confined at work for fourteen hours a day and kept in hand by a starvation diet. Within a generation, this House of Terror had become the typical paleotechnic factory: in fact the ideal, as Marx well says, paled before the reality.

Industrial diseases naturally flourished in this environment: the use of lead glaze in the potteries, phosphorus in the match-making industry, the failure to use protective masks in the numerous grinding operations, particularly in the cutlery industry, increased to enormous proportions the fatal forms of industrial poisoning or injury: mass consumption of china, matches, and cutlery resulted in a steady destruction of life. As the pace of production increased in certain trades, the dangers to health and safety in the industrial process itself increased: in glass-making, for example, the lungs were overtaxed, in other industries the increased fatigue resulted in careless motions and the maceration of a hand or the amputation of a leg might follow.

With the sudden increase of population that marked the opening years of the paleotechnic period, labor appeared as a new natural resource: a lucky find for the labor-prospector and labor-miner. Small wonder that the ruling classes flushed with moral indignation when they found that Francis Place and his followers had endeavored to propagate a knowledge of contraceptives among the Manchester operatives in the eighteen-twenties: these philanthropic radicals were threatening an otherwise inexhaustible supply of raw material. And in so far as the workers were diseased, crippled, stupefied, and reduced to apathy and dejection by the paleotechnic environment they were only, up to a certain point, so much the better adapted to the new routine of factory and mill. For the highest standards of factory

efficiency were achieved with the aid of only partly used human organisms—in short, of defectives.

With the large scale organization of the factory it became necessary that the operatives should at least be able to read notices, and
from 1832 onwards measures for providing education for the child
laborers were introduced in England. But in order to unify the whole
system, the characteristic limitations of the House of Terror were
introduced as far as possible into the school: silence, absence of motion, complete passivity, response only upon the application of an
outer stimulus, rote learning, verbal parroting, piece-work acquisition of knowledge—these gave the school the happy attributes of jail
and factory combined. Only a rare spirit could escape this discipline,
or battle successfully against this sordid environment. As the habituation became more complete, the possibility of escaping to other occupations and other environments became more limited.

One final element in the degradation of the worker must be noted:
the maniacal intensity of work. Marx attributed the lengthening of the
working day in the paleotechnic period to the capitalist's desire to
extract extra surplus value from the laborer: as long as values in use
predominated, he pointed out, there was no incentive to industrial
slavery and overwork: but as soon as labor became a commodity, the
capitalist sought to obtain as large a share of it as possible for himself at the smallest expense. But while the desire for gain was perhaps
the impulse uppermost in lengthening the worker's day—as it happened, a mistaken method even from the most limited point of view—
one must still explain the sudden intensity of the desire itself. This
was not a result of capitalist production's unfolding itself according
to an inner dialectic of development: the desire for gain was a causal
factor in that development. What lay behind its sudden impetus and
fierce intensity was the new contempt for any other mode of life or
form of expression except that associated with the machine. The
esoteric natural philosophy of the seventeenth century had finally
become the popular doctrine of the nineteenth. The gospel of work
was the positive side of the incapacity for art, play, amusement, or
pure craftsmanship which attended the shriveling up of the cultural
and religious values of the past. In the pursuit of gain, the ironmas-

ters and textile masters drove themselves almost as hard as they drove their workers: they scrimped and stinted and starved themselves at the beginning, out of avarice and the will-to-power, as the workers themselves did out of sheer necessity. The lust for power made the Bounderbys despise a humane life: but they despised it for themselves almost as heartily as they despised it for their wage-slaves. If the laborers were crippled by the doctrine, so were the masters.

For a new type of personality had emerged, a walking abstraction: the Economic Man. Living men imitated this penny-in-the-slot automaton, this creature of bare rationalism. These new economic men sacrificed their digestion, the interests of parenthood, their sexual life, their health, most of the normal pleasures and delights of civilized existence to the untrammeled pursuit of power and money. Nothing retarded them; nothing diverted them . . . except finally the realization that they had more money than they could use, and more power than they could intelligently exercise. Then came belated repentance: Robert Owen founds a utopian co-operative colony, Nobel, the explosives manufacturer, a peace foundation, Carnegie free libraries, Rockefeller medical institutes. Those whose repentance took a more private form became the victims of their mistresses, their tailors, their art dealers. Outside the industrial system, the Economic Man was in a state of neurotic maladjustment. These successful neurotics looked upon the arts as unmanly forms of escape from work and business enterprise: but what was their one-sided, maniacal concentration upon work but a much more disastrous escape from life itself? In only the most limited sense were the great industrialists better off than the workers they degraded: jailer and prisoner were both, so to say, inmates of the same House of Terror.

Yet though the actual results of the new industrialism were to increase the burdens of the ordinary worker, the ideology that fostered it was directed toward his release. The central elements in that ideology were two principles that had operated like dynamite upon the solid rock of feudalism and special privilege: the principle of utility and the principle of democracy. Instead of justifying their existence by reason of tradition and custom, the institutions of society were forced to justify themselves by their actual use. It was in the

name of social improvement that many obsolete arrangements that
had lingered on from the past were swept away, and it was likewise
by reason of their putative utility to mankind at large that the most
humane and enlightened minds of the early nineteenth century welcomed machines and sanctioned their introduction. Meanwhile, the
eighteenth century had turned the Christian notion of the equality of
all men in Heaven into an equality of all men on earth: they were
not to achieve it by conversion and death and immortality, but were
supposed to be "born free and equal." While the bourgeoisie interpreted these terms to their own advantage, the notion of democracy
nevertheless served as a psychological rationalization for machine industry: for the mass production of cheap goods merely carried the
principle of democracy on to the material plane, and the machine
could be justified because it favored the process of vulgarization.
This notion took hold very slowly in Europe; but in America, where
class barriers were not so solid, it worked out into a levelling upward
of the standard of expenditure. Had this levelling meant a genuine
equalization of the standard of life, it would have been a beneficent
one: but in reality it worked out spottily, following the lines most
favorable for profits, and thus often levelling downward, undermining
taste and judgment, lowering quality, multiplying inferior goods.

8: The Starvation of Life

The degradation of the worker was the central point in that more
widespread starvation of life which took place during the paleotechnic
régime, and which still continues in those many areas and occupations
where paleotechnic habits predominate.

In the depauperate homes of the workers in Birmingham and Leeds
and Glasgow, in New York and Philadelphia and Pittsburgh, in Hamburg and Elberfeld-Barmen and Lille and Lyons, and in similar centers from Bombay to Moscow, rickety and undernourished children
grew up: dirt and squalor were the constant facts of their environment. Shut off from the country by miles of paved streets, the most
common sights of field and farm might be strange to them: the sight
of violets, buttercups, day-lilies, the smell of mint, honeysuckle, the
locust trees, the raw earth opened by the plow, the warm hay piled

up in the sun, or the fishy tang of beach and salt-marsh. Overcast by the smoke-pall, the sky itself might be shut out and the sunlight diminished; even the stars at night became dim.

The essential pattern set by paleotechnic industry in England, with its great technical lead and its sedate, well-disciplined operatives, was repeated in every new region, as the machine girdled the globe.

Under the stress of competition, adulterants in food became a commonplace of Victorian industry: flour was supplemented with plaster, pepper with wood, rancid bacon was treated with boric acid, milk was kept from souring with embalming fluid, and thousands of medical nostrums flourished under the protection of patents, bilge-water or poison whose sole efficacy resided in the auto-hypnotism produced by the glowing lies on their labels. Stale and rancid food degraded the sense of taste and upset the digestion: gin, rum, whisky, strong tobacco made the palate less sensitive and befuddled the senses: but drink still remained the "quickest way of getting out of Manchester." Religion ceased in large groups to be the opiate of the poor: indeed the mines and the textile mills often lacked even the barest elements of the older Christian culture: and it would be more nearly true to say that opiates became the religion of the poor.

Add to the lack of light a lack of color: except for the advertisements on the hoardings, the prevailing tones were dingy ones: in a murky atmosphere even the shadows lose their rich ultramarine or violet colors. The rhythm of movement disappeared: within the factory the quick staccato of the machine displaced the organic rhythms, measured to song, that characterized the old workshop, as Bücher has pointed out: while the dejected and the outcast shuffled along the streets in Cities of Dreadful Night, and the sharp athletic movements of the sword dances and the morris dances disappeared in the surviving dances of the working classes, who began to imitate clumsily the graceful boredom of the idle and the leisured.

Sex, above all, was starved and degraded in this environment. In the mines and factories an indiscriminate sexual intercourse of the most brutish kind was the only relief from the tedium and drudgery of the day: in some of the English mines the women pulling the carts even worked completely naked—dirty, wild, and degraded as only the worst slaves of antiquity had been. Among the agricultural population in England sexual experience before marriage was a period of experimental grace before settling down: among the new industrial workers, it was often preliminary to abortion, as contemporary evidence proves. The organization of the early factories, which threw girls and boys into the same sleeping quarters, also gave power to the overseers of the children which they frequently abused: sadisms and perversions of every kind were common. Home life was crowded out of existence; the very ability to cook disappeared among the women workers.

Even among the more prosperous middle classes, sex lost both its intensity and its priapic sting. A cold rape followed the prudent continences and avoidances of the pre-marital state of women. The secrets of sexual stimulation and sexual pleasure were confined to the specialists in the brothels, and garbled knowledge about the possibilities of intercourse were conveyed by well-meaning amateurs or by quacks whose books on sexology acted as an additional bait, frequently, for their patent medicines. The sight of the naked body, so necessary for its proud exercise and dilation, was discreetly prohibited even in the form of undraped statues: moralists looked upon it as a lewd distraction that would take the mind off work and undermine the systematic inhibitions of machine industry. Sex had no industrial value. The ideal paleotechnic figure did not even have legs, to say nothing of breasts and sexual organs: even the bustle disguised and deformed the rich curve of the buttocks in the act of making them monstrous.

This starvation of the senses, this restriction and depletion of the physical body, created a race of invalids: people who knew only partial health, partial physical strength, partial sexual potency: it was the rural types, far from the paleotechnic environment, the country squire and the parson and the agricultural laborer, who had in the life insurance tables the possibility of a long life and a healthy one. Ironically enough, the dominant figures in the new struggle for existence lacked biological survival value: biologically the balance of power was in the countryside, and it was only by faking the statistics

—that is, by failing to correct them for age-groups—that the weaknesses of the new industrial towns could be concealed.

With the starvation of the senses went a general starvation of the mind: mere literacy, the ability to read signs, shop notices, newspapers, took the place of that general sensory and motor training that went with the handicraft and the agricultural industries. In vain did the educators of the period, like Schreber in Germany with his projects for Schrebergärten as necessary elements in an integral education, and like Spencer in England with his praise of leisure, idleness, and pleasant sport, attempt to combat this desiccation of the mind and this drying up of life at the roots. The manual training that was introduced was as abstract as drill; the art fostered by South Kensington was more dead and dull than the untutored products of the machine.

The eye, the ear, the touch, starved and battered by the external environment, took refuge in the filtered medium of print; and the sad constraint of the blind applied to all the avenues of experience. The museum took the place of concrete reality; the guidebook took the place of the museum; the criticism took the place of the picture; the written description took the place of the building, the scene in nature, the adventure, the living act. This exaggerates and caricatures the paleotechnic state of mind; but it does not essentially falsify it. Could it have been otherwise? The new environment did not lend itself to first hand exploration and reception. To take it at second hand, to put at least a psychological distance between the observer and the horrors and deformities observed, was really to make the best of it. The starvation and diminution of life was universal: a certain dullness and irresponsiveness, in short, a state of partial anesthesia, became a condition of survival. At the very height of England's industrial squalor, when the houses for the working classes were frequently built beside open sewers and when rows of them were being built back to back—at that very moment complacent scholars writing in middle-class libraries could dwell upon the "filth" and "dirt" and "ignorance" of the Middle Ages, as compared with the enlightenment and cleanliness of their own.

How was that belief possible? One must pause for a second to examine its origin. For one cannot understand the technics, unless one appreciates its debt to the mythology it had conjured up.

9: The Doctrine of Progress

The mechanism that produced the conceit and the self-complacence of the paleotechnic period was in fact beautifully simple. In the eighteenth century the notion of Progress had been elevated into a cardinal doctrine of the educated classes. Man, according to the philosophers and rationalists, was climbing steadily out of the mire of superstition, ignorance, savagery, into a world that was to become ever more polished, humane and rational—the world of the Paris salons before the hailstorm of revolution broke the windowpanes and drove the talkers to the cellar. Tools and instruments and laws and institutions had all been improved: instead of being moved by instincts and governed by force, men were capable of being moved and governed by reason. The student at the university had more mathematical knowledge than Euclid; and so, too, did the middle class man, surrounded by his new comforts, have more wealth than Charlemagne. In the nature of progress, the world would go on forever and forever in the same direction, becoming more humane, more comfortable, more peaceful, more smooth to travel in, and above all, much more rich.

This picture of a steady, persistent, straight-line, and almost uniform improvement throughout history had all the parochialism of the eighteenth century: for despite Rousseau's passionate conviction that the advance in the arts and sciences had depraved morals, the advocates of Progress regarded their own period—which was in fact a low one measured by almost any standard except scientific thought and raw energy—as the natural peak of humanity's ascent to date. With the rapid improvement of machines, the vague eighteenth century doctrine received new confirmation in the nineteenth century. The laws of progress became self-evident: were not new machines being invented every year? Were they not transformed by successive modifications? Did not chimneys draw better, were not houses warmer, had not railroads been invented?

Here was a convenient measuring stick for historical comparison.

Assuming that progress was a reality, if the cities of the nineteenth century were dirty, the cities of the thirteenth century must have been six centuries dirtier: for had not the world become constantly cleaner? If the hospitals of the early nineteenth century were overcrowded pest-houses, then those of the fifteenth century must have been even more deadly. If the workers of the new factory towns were ignorant and superstitious, then the workers who produced Chartres and Bamberg must have been more stupid and unenlightened. If the greater part of the population were still destitute despite the prosperity of the textile trades and the hardware trades, then the workers of the handicraft period must have been more impoverished. The fact that the cities of the thirteenth century were far brighter and cleaner and better ordered than the new Victorian towns: the fact that medieval hospitals were more spacious and more sanitary than their Victorian successors: the fact that in many parts of Europe the medieval worker had demonstrably a far higher standard of living than the paleotechnic drudge, tied triumphantly to a semi-automatic machine—these facts did not even occur to the exponents of Progress as possibilities for investigation. They were ruled out automatically by the theory itself.

Plainly, by taking some low point of human development in the past, one might over a limited period of time point to a real advance. But if one began with a high point—for example, the fact that German miners in the sixteenth century frequently worked in three shifts of only eight hours each—the facts of progress, when one surveyed the mines of the nineteenth century, were non-existent. Or if one began with the constant feudal strife of fourteenth century Europe, the peace that prevailed over great areas of Western Europe between 1815 and 1914 was a great gain. But if one compared the amount of destruction caused by a hundred years of the most murderous warfare in the Middle Ages with what took place in four short years during the World War, precisely because of such great instruments of technological progress as modern artillery, steel tanks, poison gas, bombs and flame throwers, picric acid and T.N.T., the result was a step backward.

Value, in the doctrine of progress, was reduced to a time-calculation: value was in fact *movement in time*. To be old-fashioned or to be "out of date" was to lack value. Progress was the equivalent in history of mechanical motion through space: it was after beholding a thundering railroad train that Tennyson exclaimed, with exquisite aptness, "Let the great world spin forever down the ringing grooves of change." The machine was displacing every other source of value partly because the machine was by its nature the most progressive element in the new economy.

What remained valid in the notion of progress were two things that had no essential connection with human improvement. First: the fact of life, with its birth, development, renewal, decay, which one might generalize, in such a fashion as to include the whole universe, as the fact of change, motion, transformation of energy. Second: the social fact of accumulation: that is the tendency to augment and conserve those parts of the social heritage which lend themselves to transmission through time. No society can escape the fact of change or evade the duty of selective accumulation. Unfortunately change and accumulation work in both directions: energies may be dissipated, institutions may decay, and societies may pile up evils and burdens as well as goods and benefits. To assume that a later point in development necessarily brings a higher kind of society is merely to confuse the neutral quality of complexity or maturity with improvement. To assume that a later point in time *necessarily* carries a greater accumulation of values is to forget the recurrent facts of barbarism and degradation.

Unlike the organic patterns of movement through space and time, the cycle of growth and decay, the balanced motion of the dancer, the statement and return of the musical composition, progress was motion toward infinity, motion without completion or end, motion for motion's sake. One could not have too much progress; it could not come too rapidly; it could not spread too widely; and it could not destroy the "unprogressive" elements in society too swiftly and ruthlessly: for progress was a good in itself independent of direction or end. In the name of progress, the limited but balanced economy of the Hindu village, with its local potter, its local spinners and weavers, its local smith, was overthrown for the sake of providing a market for the potteries of the Five Towns and the textiles of Manchester and the superfluous hardware of Birmingham. The result was impoverished villages in India, hideous and destitute towns in England, and a great wastage in tonnage and man-power in plying the oceans between: but at all events a victory for progress.

Life was judged by the extent to which it ministered to progress, progress was not judged by the extent to which it ministered to life. The last possibility would have been fatal to admit: it would have transported the problem from the cosmic plane to a human one. What paleotect dared ask himself whether labor-saving, money-grubbing, power-acquiring, space-annihilating, thing-producing devices were in fact producing an equivalent expansion and enrichment of life? That question would have been the ultimate heresy. The men who asked it, the Ruskins, the Nietzsches, the Melvilles, were in fact treated as heretics and cast out of this society: in more than one case, they were condemned to an exacerbating solitude that reached the limit of madness.

10: The Struggle for Existence

But progress had an economic side: at bottom it was little less than an elaborate rationalizing of the dominant economic conditions. For Progress was possible only through increased production: production grew in volume only through larger sales: these in turn were an incentive to mechanical improvements and fresh inventions which ministered to new desires and made people conscious of new necessities. So the struggle for the market became the dominant motive in a progressive existence.

The laborer sold himself to the highest bidder in the labor market. His work was not an exhibition of personal pride and skill but a commodity, whose value varied with the quantity of other laborers who were available for performing the same task. For a while the professions, like law and medicine, still maintained a qualitative standard: but their traditions were insidiously undermined by the more general practices of the market. Similarly, the manufacturer sold his product in the commercial market. Buying cheap and selling dear, he had no other standard than that of large profits: at the height of this economy John Bright defended the adulteration of goods in the British House of Commons as a necessary incident of competitive sale.

To widen the margin between the costs of production and the return from sales in a competitive market, the manufacturer depressed wages, lengthened hours, speeded up motions, shortened the worker's period of rest, deprived him of recreation and education, robbed him in youth of the opportunities for growth, in maturity of the benefits of family life, and in old age of his security and peace. So unscrupulous was the competition that in the early part of the period, the manufacturers even defrauded their own class: the mines that used Watt's steam engine refused to pay him the royalties they owed, and Shuttle Clubs were formed by the manufacturers to assist members sued by Kay for royalties on his invention.

This struggle for the market was finally given a philosophic name: it was called the struggle for existence. Wage worker competed against wage worker for bare subsistence; the unskilled competed against the skilled; women and children competed against the male heads of families. Along with this horizontal struggle between the different elements in the working class, there was a vertical struggle that rent society in two: the class struggle, the struggle between the possessors and the dispossessed. These universal struggles served as basis for the new mythology which complemented and extended the more optimistic theory of progress.

In his essay on population the Reverend T. R. Malthus shrewdly generalized the actual state of England in the midst of the disorders that attended the new industry. He stated that population tended to expand more rapidly than the food supply, and that it avoided starvation only through a limitation by means of the positive check of continence, or the negative checks of misery, disease, and war. In the course of the struggle for food, the upper classes, with their thrift and foresight and superior mentality emerged from the ruck of mankind. With this image in mind, and with Malthus's Essay on Population as the definite stimulus to their thoughts, two British biologists, Charles Darwin and Alfred Wallace, projected the intense struggle for the market upon the world of life in general.

Another philosopher of industrialism, just as characteristically a railroad engineer by profession as Spinoza had been a lens grinder, coined a phrase that touched off the whole process: to the struggle for existence and the process of natural selection Spencer appended the results: "the survival of the fittest." The phrase itself was a tautology; for survival was taken as the proof of fitness: but that did not decrease its usefulness.

This new ideology arose out of the new social order, not out of Darwin's able biological work. His scientific study of modifications, variations, and the processes of sexual selection were neither furthered nor explained by a theory which accounted not for the occurrence of new organic adaptations, but merely for a possible mechanism whereby certain forms had been weeded out after the survivors had been favorably modified. Moreover, there were the demonstrable facts of commensalism and symbiosis, to say nothing of ecological partnership, of which Darwin himself was fully conscious, to modify the Victorian nightmare of a nature red in tooth and claw.

The point is, however, that in paleotechnic society the weaker were indeed driven to the wall and mutual aid had almost disappeared. The Malthus-Darwin doctrine explained the dominance of the new bourgeoisie, people without taste, imagination, intellect, moral scruples, general culture or even elementary bowels of compassion, who rose to the surface precisely because they fitted an environment that had no place and no use for any of these humane attributes. Only anti-social qualities had survival value. Only people who valued machines more than men were capable under these conditions of governing men to their own profit and advantage.

11: Class and Nation

The struggle between the possessing classes and the working classes during this period assumed a new form, because the system of production and exchange and the common intellectual milieu had all profoundly altered. This struggle was closely observed and for the first time accurately appraised by Friedrich Engels and Karl Marx. Just as Darwin had extended the competition of the market to the entire world of life, so did Engels and Marx extend the contemporary class struggle to the entire history of society.

But there was a significant difference between the new class struggles and the slave uprisings, the peasant rebellions, the local conflicts between masters and journeymen that had occurred previously in Europe. The new struggle was continuous, the old had been sporadic. Except for the medieval utopian movements—such as the Lollards— the earlier conflicts had been, in the main, struggles over abuses in a system which both master and worker accepted: the appeal of the worker was to an antecedent right or privilege that had been grossly violated. The new struggle was over the system itself: it was an attempt on the workers' part to modify the system of free wage competition and free contract that left the worker, a helpless atom, free to starve or cut his own throat if he did not accept the conditions the industrialists offered.

From the standpoint of the paleotechnic worker, the goal of the struggle was control of the labor market: he sought for power as a bargainer, obtaining a slightly larger share of the costs of production, or, if you will, the profits of sale. But he did not, in general, seek responsible participation as a worker in the business of production: he was not ready to be an autonomous partner in the new collective mechanism, in which the least cog was as important to the process as a whole as the engineers and scientists who had devised it and who controlled it. Here one marks the great gap between handicraft and the early machine economy. Under the first system the worker was on his way to being a journeyman; the journeyman, broadened by travel to other centers, and inducted into the mysteries of his craft, was capable, not merely of bargaining with his employer, but of *taking his place*. The class conflict was lessened by the fact that the masters could not take away the workers' tools of production, which were personal, nor could they decrease his actual pleasure of craftsmanship. Not until specialization and expropriation had given the employer a special advantage did the conflict begin to take on its paleotechnic form. Under the capitalist system the worker could achieve security and mastery only by leaving his class. The consumer's cooperative movement was a partial exception to this on the side of consumption: far more important ultimately than the spectacular wage-battles that were fought during this period; but it did not touch the organization of the factory itself.

Unfortunately, on the terms of the class struggle, there was no means of preparing the worker for the final results of his conquest. The struggle was in itself an education for warfare, not for industrial management and production. The battle was constant and bitter, and it was conducted without mercy on the part of the exploiting classes, who used the utmost brutality that the police and the soldiery were capable of, on occasion, to break the resistance of the workers. In the course of this war one or another part of the proletariat—chiefly the more skilled occupations—made definite gains in wages and hours, and they shook off the more degrading forms of wage-slavery and sweating: but the fundamental condition remained unaltered. Meanwhile, the machine process itself, with its matter-of-fact procedure, its automatism, its impersonality, its reliance upon the specialized services and intricate technological studies of the engineer, was getting further and further beyond the worker's unaided power of intellectual apprehension or political control.

Marx's original prediction that the class struggle would be fought out on strict class lines between an impoverished international proletariat and an equally coherent international bourgeoisie was falsified by two unexpected conditions. One was the growth of the middle classes and the small industries: instead of being automatically wiped out they showed unexpected resistance and staying power. In a crisis, the big industries with their vast over-capitalization and their enormous overhead, were less capable of adjusting themselves to the situation than the smaller ones. In order to make the market more secure, there were even fitful attempts to raise the standard of consumption among the workers themselves: so the sharp lines necessary for successful warfare only emerged in periods of depression. The second fact was the new alignment of forces between country and country, which tended to undermine the internationalism of capital and disrupt the unity of the proletariat. When Marx wrote in the eighteen fifties Nationalism seemed to him, as it seemed to Cobden, to be a dying movement: events showed that, on the contrary, it had taken a new lease on life.

With the massing of the population into national states which continued during the nineteenth century, the national struggle cut at right angles to the class struggle. After the French revolution war, which was once the sport of dynasties, became the major industrial occupation of whole peoples: "democratic" conscription made this possible.

The struggle for political power, always limited in the past by financial weakness, technical restrictions, the indifference and hostility of the underlying population, now became a struggle between states for the command of exploitable areas: the mines of Lorraine, the diamond fields of South Africa, the South American markets, possible sources of supply or possible outlets for products that could not be absorbed by the depressed proletariat of the industrial countries, or, finally, possible fields for investment for the surplus of capital heaped up in the "progressive" countries.

"The present," exclaimed Ure in 1835, "is distinguished from every preceding age by an universal ardor of enterprise in arts and manufactures. Nationals, convinced at length that war is always a losing game, have converted their swords and muskets into factory implements, and now contend with each other in the bloodless but still formidable strife of trade. They no longer send troops to fight on distant fields, but fabrics to drive before them those of their old adversaries in arms, to take possession of a foreign market. To impair the resources of a rival at home, by underselling his wares abroad, is the new belligerent system, in pursuance of which every nerve and sinew of the people are put upon the strain." Unfortunately the sublimation was not complete: economic rivalries added fuel to national hates and gave a pseudo-rational face to the most violently irrational motives.

Even the leading utopias of the paleotechnic phase were nationalist and militarist: Cabet's Icaria, which was contemporary with the liberal revolutions of 1848, was a masterpiece of warlike regimentation in every detail of life, whilst Bellamy, in 1888, took the organization of the army, on a basis of compulsory service, as the pattern for all industrial activities. The intensity of these nationalist struggles, aided by the more tribal instincts, somewhat weakened the effect of the class struggles. But they were alike in this respect: neither the state as conceived by the followers of Austin, nor the proletarian class as conceived by the followers of Marx, were organic entities or true social groups: they were both arbitrary collections of individuals, held together not by common functions, but by a common collective symbol of loyalty and hate. This collective symbol had a magical office: it was willed into existence by magical formulae and incantations, and kept alive by a collective ritual. So long as the ritual was piously maintained the subjective nature of its premises could be ignored. But the "nation" had this advantage over the "class": it could conjure up more primitive responses, for it played, not on material advantage, but on naïve hates and manias and death-wishes. After 1850 nationalism became the drill master of the restless proletariat, and the latter worked out its sense of inferiority and defeat by identification with the all-powerful State.

12: The Empire of Muddle

The quantity of goods produced by the machine was supposed to be automatically regulated by the law of supply and demand: commodities, like water, were supposed to seek their own level: in the long run, only so much goods would be produced as could be sold at a profit. The lessening of profits automatically, according to this theory, closed the valve of production; while the increase of profits automatically opened it and even would lead to the construction of new feeders. Producing the necessaries of life, was, however, merely a by-product of making profits. Since there was more money to be made in textiles for foreign markets than in sound workers' houses for domestic use, more profit in beer and gin than in unadulterated bread, the elementary necessities of shelter—and sometimes even food—were scandalously neglected. Ure, the lyric poet of the textile industries, readily confessed that "to the production of food and domestic accommodation not many automatic inventions have been applied, or seem to be extensively applicable." As prophecy this proved absurd; but as a description of the current limitations, it was correct.

The shortage of housing for the workers, the congestion of domestic quarters, the erection of vile insanitary barracks to serve as substitutes for decent human shelter—these were universal characteristics of the paleotechnic régime. Fortunately, the terrible incidence of disease in the poorer quarters of the cities awakened the attention of health officers, and in the name of sanitation and public health various measures were taken, dating in England to Shaftesbury's "model" housing acts in 1851, to alleviate the worst conditions by restrictive legislation, compulsory slum repair, and even an insignificant modicum of slum clearance and improved housing. Some of the best examples, from the eighteenth century on, appeared in the colliery villages of England, possibly as a result of their semi-feudal traditions, to be followed in the 1860's by Krupp's workers' housing at Essen. Slowly, a small number of the worst evils were wiped away, despite the fact that the new laws were in opposition to the holy principles of free competitive enterprise in the production of illth.

The jockeying for profits without any regard for the stable ordering of production had two unfortunate results. For one thing, it undermined agriculture. As long as food supplies and materials could be obtained cheaply from some far part of the earth, even at the expense of the speedy exhaustion of the soils that were being recklessly cropped for cotton and wheat, no effort was made to keep agriculture and industry in equipoise. The countryside, reduced in general to the margin of subsistence, was further depressed by the drift of population into the apparently thriving factory towns, with infant mortality rates that often rose as high as 300 or more per thousand live births. The application of machines to sowing, reaping, threshing, instituted on a large scale with the multitude of new reapers invented at the beginning of the century—McCormick's was merely one of many—only hastened the pace of this development.

The second effect was even more disastrous. It divided the world into areas of machine production and areas of foods and raw materials: this made the existence of the over-industrialized countries more precarious, the further they were cut off from their rural base of supplies: hence the beginning of strenuous naval competition. Not merely did the existence of the coal-agglomerations themselves depend upon their ability to command water from distant streams and lakes, and food from distant fields and farms: but continued production depended upon the ability to bribe or browbeat other parts of the earth into accepting their industrial products. The Civil War in America, by cutting off the supply of cotton, reduced to a state of extreme penury the brave and honest textile workers of Lancashire. And the fear of repeating such events, in other industries beside cotton, was responsible in good part for the panicky imperialism and armament competition that developed throughout the world after 1870. As paleotechnic industry was founded originally upon systematic child slavery, so it was dependent for its continued growth upon a forced outlet for its goods.

Unfortunately for the countries that relied upon this process to go on indefinitely, the original consuming areas—the new or the "backward" countries—speedily took possession of the common heritage in science and technics and began to produce machined goods for themselves. That tendency became universal by the eighties. It was temporarily limited by the fact that England, which long retained its technical superiority in weaving and spinning, could use 7 operatives per thousand spindles in 1837 and only 3 operatives per thousand in 1887, while Germany, its nearest competitor at the second date still used from 7½ to 9, while Bombay required 25. But in the long run neither England nor the "advanced countries" could hold the lead: for the new machine system was a universal one. Therewith one of the main props of paleotechnic industry was displaced.

The hit-or-miss tactics of the market place pervaded the entire social structure. The leaders of industry were for the most part empirics: boasting that they were "practical" men they prided themselves on their technical ignorance and naïvety. Solvay, who made a fortune out of the Solvay soda process, knew nothing about chemistry; neither did Krupp, the discoverer of cast-steel; Hancock, one of the early experimenters with India rubber was equally ignorant. Bessemer, the inventor of many things besides the Bessemer process of making steel, at first merely stumbled on his great invention through the accident of using iron with a low phosphorus content: it was only the failure of his method with the continental ores that had a high phosphorus content that led him to consider the chemistry of the process.

Within the industrial plant scientific knowledge was at a discount. The practical man, contemptuous of theory, scornful of exact training, ignorant of science, was uppermost. Trade secrets, sometimes important, sometimes merely childish empiricism, retarded the co-operative extension of knowledge which has been the basis of all our major technical advances; whilst the system of patent monopolies was used by astute business men to drive improvements out of the market, if they threatened to upset existing financial values, or to delay their introduction—as the automatic telephone was delayed—until the original rights to the patent had expired. Right down to the World War an unwillingness to avail itself of scientific knowledge or to promote scientific research characterized paleotechnic industry throughout the world. Perhaps the only large exception to this, the German dye industry, was due to its close connection with the poisons and explosives necessary for military purposes.

While free competition prevailed between individual manufacturers, planned production for industry as a whole was impossible: each manufacturer remained his own judge, on the basis of limited knowledge and information, of the amount of goods he could profitably produce and dispose of. The labor market itself was based on absence of plan: it was, in fact, by means of a constant surplus of unemployed workers, who were never systematically integrated into industry, that wages could be kept low. This excess of the unemployed in "normal and prosperous" times was essential to competitive production. The location of industries was unplanned: accident, pecuniary advantage, habit, gravitation toward the surplus labor market, were as important as the tangible advantages from a technical standpoint. The machine—the outcome of man's impulse to conquer his environment and to canalize his random impulses into orderly activities—produced during the paleotechnic phase the systematic negation of all its characteristics: nothing less than the empire of muddle. What was, indeed, the boasted "mobility of labor" but the breakdown of stable social relations and the disorganization of family life?

The state of paleotechnic society may be described, ideally, as one of wardom. Its typical organs, from mine to factory, from blast-furnace to slum, from slum to battlefield, were at the service of death. Competition: struggle for existence: domination and submission: extinction. With war at once the main stimulus, the underlying basis, and the direct destination of this society, the normal motives and reactions of human beings were narrowed down to the desire for domination and to the fear of annihilation—the fear of poverty, the fear of unemployment, the fear of losing class status, the fear of starvation, the fear of mutilation and death. When war finally came, it was welcomed with open arms, for it relieved the intolerable suspense: the shock of reality, however grim, was more bearable than the constant menace of spectres, worked up and paraded forth by the journalist and the politician. The mine and the battlefield underlay all the paleotechnic activities; and the practices they stimulated led to the widespread exploitation of fear.

The rich feared the poor and the poor feared the rent collector: the middle classes feared the plagues that came from the vile insanitary quarters of the industrial city and the poor feared, with justice, the dirty hospitals to which they were taken. Toward the latter part of the period religion adopted the uniform of war: singing Onward Christian Soldiers, the converted marched with defiant humility in military dress and order: imperialist salvation. The school was regimented like an army, and the army camp became the universal school: teacher and pupil feared each other, even as did capitalist and worker. Walls, barred windows, barbed wire fences surrounded the factory as well as the jail. Women feared to bear children and men feared to beget them: the fear of syphilis and gonorrhea tainted sexual intercourse: behind the diseases themselves lurked Ghosts: the spectre of locomotor ataxia, paresis, insanity, blind children, crippled legs, and the only known remedy for syphilis, till salvarsan, was itself a poison. The drab prisonlike houses, the palisades of dull streets, the treeless backyards filled with rubbish, the unbroken rooftops, with never a gap for park or playground, underlined this environment of death. A mine explosion, a railway wreck, a tenement house fire, a military assault upon a group of strikers, or finally the more potent outbreak of war—these were but appropriate punctuation marks. Exploited for power and profit, the destination of most of the goods made by the machine was either the rubbish heap or the battlefield. If the landlords and other monop- olists enjoyed an unearned increment from the massing of population and the collective efficiency of the machine, the net result for society at large might be characterized as the unearned excrement.

13: Power and Time

During the paleotechnic period the changes that were manifested in every department of technics rested for the most part on one central fact: the increase of energy. Size, speed, quantity, the multi- plication of machines, were all reflections of the new means of utiliz- ing fuel and the enlargement of the available stock of fuel itself. Power was at last dissociated from its natural human and geographic limitations: from the caprices of the weather, from the irregularities of the rainfall and wind, from the energy intake in the form of food which definitely restricts the output of men and animals.

Power, however, cannot be dissociated from another factor in work, namely time. The chief use of power during the paleotechnic period was to decrease the time during which any given quantity of work can be performed. That much of the time so saved was frittered away in disordered production, in stoppages derived from the weaknesses of the social institutions that accompanied the factory, and in unem- ployment is a fact which diminished the putative efficiency of the new régime. Vast were the labors performed by the steam engine and its accessories; but vast, likewise, were the losses that accom- panied them. Measured by effective work, that is, by human effort transformed into direct subsistence or into durable works of art and technics, the relative gains of the new industry were pitifully small. Other civilizations with a smaller output of power and a larger ex- penditure of time had equalled and possibly surpassed the paleo- technic period in real efficiency.

With the enormous increase in power a new tempo had entered production: the regimentation of time, which had been sporadic and fitful now began to influence the entire Western World. The symptom of this change was the mass production of cheap watches: first begun in Switzerland, and then undertaken on a large scale in Waterbury, Connecticut, in the eighteen-fifties.

Time-saving now became an important part of labor-saving. And as time was accumulated and put by, it was reinvested, like money capital, in new forms of exploitation. From now on filling time and killing time became important considerations: the early paleotechnic employers even stole time from their workers by blowing the factory whistle a quarter of an hour earlier in the morning, or by moving the hands of the clock around more swiftly during the lunch period: where the occupation permitted, the worker often reciprocated when the employer's back was turned. Time, in short, was a commodity in the sense that money had become a commodity. Time as pure duration, time dedicated to contemplation and reverie, time divorced from mechanical operations, was treated as a heinous waste. The paleotechnic world did not heed Wordsworth's Expostulation and Reply: it had no mind to sit upon an old gray stone and dream its time away.

Just as, on one hand, the filling up of time-compartments became a duty, so the necessity of "cutting things short" made itself manifest, too. Poe attributed the vogue of the short-story in the forties to the need for brief snatches of relaxation in the routine of a busy day. The immense growth of periodical literature during this period, following the cheap, large-scale production of the steam-driven printing press (1814) was likewise a mark of the increasing mechanical division of time. While the three-volume novel served the sober domestic habits of the Victorian middle classes, the periodical— quarterly, monthly, daily, and finally almost hourly—served the bulk of the popular needs. Human pregnancies still lasted nine months; but the tempo of almost everything else in life was speeded, the span was contracted, and the limits were arbitrarily clipped, not in terms of the function and activity, but in terms of a mechanical system of time accountancy. Mechanical periodicity took the place of organic and functional periodicity in every department of life where the usurpation was possible.

The spread of rapid transportation occasioned a change in the method of time-keeping itself. Sun time, which varies a minute every eight miles as one travels from east to west, could no longer be observed. Instead of a local time based upon the sun, it was necessary to have a conventional time belt, and to change abruptly by a whole hour when one entered the next time belt. Standard time was imposed by the transcontinental railroads themselves in 1875 in the United States, ten years before the regulations for standard time were officially promulgated at a World Congress. This carried to a conclusion that standardization of time that had begun with the foundation of the Greenwich observatory two hundred years before, and had been carried further, first on the sea, by comparing ship's chronometers with Greenwich time. The entire planet was now divided off into a series of time-belts. This orchestrated actions over wider areas than had ever been able to move simultaneously before.

Mechanical time now became second nature: the acceleration of the tempo became a new imperative for industry and "progress." To reduce the time on a given job, whether the work was a source of pleasure, or pain, or to quicken movement through space, whether the traveler journeyed for enjoyment or profit, was looked upon as a sufficient end in itself. Some of the specific fears as to the results of this acceleration were absurd, as in the notion that flight through space at twenty miles an hour on the railroad would cause heart disease and undermine the human frame; but in its more general application, this alteration of the tempo from the organic period, which cannot be greatly quickened without maladjustment of function, to the mechanical period, which can be stretched out or intensified, was indeed made too lightly and thoughtlessly.

Apart from the primitive physical delight in motion for its own sake, this acceleration of the tempo could not be justified except in terms of pecuniary rewards. For power and time, the two components of mechanical work, are in human terms only a function of purpose. They have no more significance, apart from human purpose, than has the sunlight that falls in the solitude of the Sahara desert. During the paleotechnic period, the increase of power and the acceleration of movement became ends in themselves: ends that justified themselves apart from their human consequences.

Technologically, the department in which paleotechnic industry rose to its greatest eminence was not the cotton mill but the railroad system. The success of this new invention is all the more remarkable because so little of the earlier technique of the stage-coach could be carried over into the new means of transportation. The railroad was the first industry to benefit by the use of electricity; for the telegraph made possible a long distance signalling system and remote control; and it was in the railroad that the routing through of production and the timing and inter-relationship of the various parts of production took place more than a generation before similar tables and schedules and forecasts made their way into industry as a whole. The invention of the necessary devices to ensure regularity and safety, from the air-brake and the vestibule car to the automatic switch and automatic signal system, and the perfection of the system for routing goods and traffic at varying rates of speed and under varying weather conditions from point to point, was one of the superb technical and administrative achievements of the nineteenth century. That there were various curbs on the efficiency of the system as a whole goes without saying: financial piracy, lack of rational planning of industries and cities, failure to achieve unified operation of continental trunk lines. But within the social limitations of the period, the railroad was both the most characteristic and the most efficient form of technics.

14: The Esthetic Compensation

But paleotechnic industry was not without an ideal aspect. The very bleakness of the new environment provoked esthetic compensations. The eye, deprived of sunlight and color, discovered a new world in twilight, fog, smoke, tonal distinctions. The haze of the factory town exercised its own visual magic: the ugly bodies of human beings, the sordid factories and rubbish heaps, disappeared in the fog, and instead of the harsh realities one encountered under the sun, there was a veil of tender lavenders, grays, pearly yellows, wistful blues.

It was an English painter, J. W. M. Turner, working in the very
heyday of the paleotechnic régime, who left the fashionable classic
landscape with its neat parklike scenery and its artful ruins to create
pictures, during the later part of his career, that had only two
subjects: Fog and Light. Turner was perhaps the first painter to
absorb and directly express the characteristic effects of the new
industrialism: his painting of the steam-locomotive, emerging through
the rain, was perhaps the first lyric inspired by the steam engine.

The smoking factory chimney had helped to create this dense
atmosphere; and by means of the atmosphere one escaped, in vision,
some of the worst effects of the factory chimney. In painting even
the acrid smells disappeared, and only the illusion of loveliness
remained. At a distance, through the mist, the Doulton pottery works
in Lambeth, with their piously misprized decoration, are almost as
stimulating as any of the pictures in the Tate Gallery. Whistler,
from his studio on the Chelsea Embankment, overlooking the factory
district of Battersea, expressed himself through this fog and mist
without the help of light: the finest gradations of tone disclosed and
defined the barges, the outline of a bridge, the distant shore: in the
fog, a row of street lamps shone like tiny moons on a summer night.

But Turner, not merely reacting to the fog, but reacting against it,
turned from the garbage-strewn streets of Covent Market, from the
blackened factories and the darkened London slums, to the purity
of light itself. In a series of pictures he painted a hymn to the wonder
of light, such a hymn as a blind man might sing on finding his eyesight, a paean to light emerging from night and fog and smoke and
conquering the world. It was the very lack of sun, the lack of color,
the starvation within the industrial towns for the sight of rural scenes,
that sharpened the art of landscape painting during this period, and
gave birth to its chief collective triumph, the work of the Barbizon
school and the later impressionists, Monet, Sisley, Pissarro, and most
characteristic if not most original of all, Vincent Van Gogh.

Van Gogh knew the paleotechnic city in its most complete gloom,
the foul, bedraggled, gas-lighted London of the seventies: he also
knew the very source of its dark energies, places like the mines at
La Borinage where he had lived with the miners. In his early pictures

he absorbed and courageously faced the most sinister parts of his environment: he painted the gnarled bodies of the miners, the almost animal stupor of their faces, bent over the bare dinner of potatoes, the eternal blacks, grays, dark blues, and soiled yellows of their poverty-stricken homes. Van Gogh identified himself with this sombre, forbidding routine: then, going to France, which had never entirely succumbed to the steam engine and large-scale production, which still retained its agricultural villages and its petty handicrafts, he found himself quickened to revolt against the deformities and deprivations of the new industrialism. In the clear air of Provence, Van Gogh beheld the visual world with a sense of intoxication deepened by the bleak denial he had known so long: the senses, no longer blanketed and muffled by smoke and dirt, responded in shrill ecstasy. The fog lifted: the blind saw: color returned.

Though the chromatic analysis of the impressionists was derived directly from Chevreul's scientific researches on color, their vision was unbelievable to their contemporaries: they were denounced as impostors because the colors they painted were not dulled by studio walls, subdued by fog, mellowed by age, smoke, varnish: because the green of their grass was yellow in the intensity of sunlight, the snow pink, and the shadows on the white walls lavender. Because the natural world was not sober, the paleotects thought the artists were drunk.

While color and light absorbed the new painters, music became both more narrow and more intense in reaction against the new environment. The workshop song, the street cries of the tinker, the dustman, the pedlar, the flower vendor, the chanties of the sailor hauling the ropes, the traditional songs of the field, the wine-press, the taproom, were slowly dying out during this period: at the same time, the power to create new ones was disappearing. Labor was orchestrated by the number of revolutions per minute, rather than by the rhythm of song or chant or tattoo. The ballad, with its old religious, military, or tragic contents, was thinned out into the sentimental popular song, watered even in its eroticism: its pathos became bathos: only as literature for the cultivated classes, in the poems of Coleridge and Wordsworth and Morris, did the ballad survive. It is scarcely possible to mention in the same breath "Mary Hamilton's to the Kirk Gane" and, let us say, The Baggage Car Ahead. Song and poesy ceased to be folk possessions: they became "literary," professionalized, segregated. No one thought any longer of asking the servants to come into the living room to take part in madrigal or ballad. What happened to poetry had happened likewise to pure music. But music, in the creation of the new orchestra, and in the scope and power and movement of the new symphonies, became in a singularly representative way the ideal counterpart of industrial society.

The baroque orchestra had been built up on the sonority and volume of stringed instruments. Meanwhile mechanical invention had added enormously to the range of sound and the qualities of tone that could be produced: it even made the ear alive to new sounds and new rhythms. The thin little clavichord became the massive machine known as the piano, with its great sounding board, and its extended keyboard: similarly, a series of instruments was introduced by Adolph Sax, the inventor of the saxophone, around 1840, between the wood-winds and the old brasses. All the instruments were now scientifically calibrated: the production of sound became, within limits, standardized and predictable. And with the increase in the number of instruments, the division of labor within the orchestra corresponded to that of the factory: the division of the process itself became noticeable in the newer symphonies. The leader was the superintendent and production manager, in charge of the manufacture and assembly of the product, namely the piece of music, while the composer corresponded to the inventor, engineer, and designer, who had to calculate on paper, with the aid of such minor instruments as the piano, the nature of the ultimate product—working out its last detail before a single step was made in the factory. For difficult compositions, new instruments were sometimes invented or old ones resurrected; but in the orchestra the collective efficiency, the collective harmony, the functional division of labor, the loyal cooperative interplay between the leaders and the led, produced a collective unison greater than that which was achieved, in all probability, within any single factory. For one thing, the rhythm was more subtle;

and the timing of the successive operations was perfected in the symphony orchestra long before anything like the same efficient routine came about in the factory.

Here, then, in the constitution of the orchestra, was the ideal pattern of the new society. It was achieved in art before it was approached in technics. As for the products made possible by the orchestra, the symphonies of Beethoven and Brahms or the re-orchestrated music of Bach, they have the distinction of being the most perfect works of art produced during the paleotechnic period: no poem, no painting, expresses such depth and energy of spirit, gathering resources from the very elements of life that were stifling and deforming the existing society, as completely as the new symphonies. The visual world of the Renascence had been almost obliterated: in France alone, which had not altogether succumbed either to decay or to progress, did this world remain alive in the succession of painters between Delacroix and Renoir. But what was lost in the other arts, what had disappeared almost completely in architecture, was recovered in music. Tempo, rhythm, tone, harmony, melody, polyphony, counterpoint, even dissonance and atonality, were all utilized freely to create a new ideal world, where the tragic destiny, the dim longings, the heroic destinies of men could be entertained once more. Cramped by its new pragmatic routines, driven from the marketplace and the factory, the human spirit rose to a new supremacy in the concert hall. Its greatest structures were built of sound and vanished in the act of being produced. If only a small part of the population listened to these works of art or had any insight into their meaning, they nevertheless had at least a glimpse of another heaven than Coketown's. The music gave more solid nourishment and warmth than Coketown's spoiled and adulterated foods, its shoddy clothes, its jerrybuilt houses.

Apart from painting and music, one looks almost in vain among the cottons of Manchester, the ceramics of Burslem and Limoges, or the hardware of Solingen and Sheffield, for objects fine enough to be placed on even the most obscure shelves of a museum. Although the best English sculptor of the period, Alfred Stevens, was commissioned to make designs for Sheffield cutlery, his work was an exception. Disgusted with the ugliness of its own products, the paleotechnic period turned to past ages for models of authenticated art. This movement began with the realization that the art produced by the machines for the great exposition of 1851 was beneath contempt. Under the patronage of Prince Albert, the school and museum at South Kensington were founded, in order to improve taste and design: the result was merely to eviscerate what vitality its ugliness possessed. Similar efforts in the German speaking countries, under the leadership of Gottfried Semper, and in France and Italy and the United States, produced no better results. For the moment handicraft, as re-introduced by De Morgan, La Farge, and William Morris, provided the only live alternative to dead machine designs. The arts were degraded to the level of Victorian ladies' fancy work: a triviality, a waste of time.

Naturally, human life as a whole did not stop short during this period. Many people still lived, if with difficulty, for other ends than profit, power, and comfort: certainly these ends were not within reach of the millions of men and women who composed the working classes. Perhaps most of the poets and novelists and painters were distressed by the new order and defied it in a hundred ways: above all, by existing as poets and novelists and painters, useless creatures, whose confrontation of life in its many-sided unity was looked upon by the Gradgrinds as a wanton escape from the "realities" of their abstract accountancy. Thackeray deliberately cast his works in a pre-industrial environment, in order to evade the new issues. Carlyle, preaching the gospel of work, denounced the actualities of Victorian work. Dickens satirized the stock-promoter, the Manchester individualist, the utilitarian, the blustering self-made man: Balzac and Zola, painting the new financial order with a documentary realism, left no question as to its degradation and nastiness. Other artists turned with Morris and the Pre-Raphaelites back to the Middle Ages, where Overbeck and Hoffmann in Germany, and Chateaubriand and Hugo in France, had preceded them: still others turned with Browning to Renascence Italy, with Doughty to primitive Arabia, with Melville and Gauguin to the South Seas, with Thoreau to the primeval woods, with Tolstoy to the peasants. What did they seek? A few simple things not to be found between the railroad terminal and the factory: plain animal self-respect, color in the outer environment and emotional depth in the inner landscape, a life lived for its own values, instead of a life on the make. Peasants and savages had retained some of these qualities: and to recover them became one of the main duties of those who sought to supplement the iron fare of industrialism.

15: Mechanical Triumphs

The human gains in the paleotechnic phase were small: perhaps for the mass of the population non-existent: the progressive and utilitarian John Stuart Mill, was at one here with the most bitter critic of the new régime, John Ruskin. But a multitude of detailed advances were made in technics itself. Not merely did the inventors and machine-makers of the paleotechnic phase improve tools and refine the whole apparatus of mechanical production, but its scientists and philosophers, its poets and artists, helped lay the foundation for a more humane culture than that which had prevailed even during the eotechnic period. Though science was only sporadically applied to industrial production, most notably perhaps, through Euler and Camus, in the improvement of gears, the pursuit of science went on steadily: the great advances made during the seventeenth century were matched once more in the middle of the nineteenth in the conceptual reorganization of every department of scientific thought— advances to which we attach the names of von Meyer, Mendelev, Faraday, Clerk-Maxwell, Claude Bernard, Johannes Müller, Darwin, Mendel, Willard Gibbs, Mach, Quetelet, Marx, and Comte, to mention only some of the outstanding figures. Through this scientific work, technics itself entered a new phase, whose characteristics we shall examine in the next chapter. *The essential continuity of science and technics remains a reality through all their shifts and phases.*

The technical gains made during this phase were tremendous: it was an era of mechanical realization when, at last, the ability of the tool-makers and machine-makers had caught up with the demands of the inventor. During this period the principal machine tools were perfected, including the drill, the planer, and the lathe: power-propelled vehicles were created and their speeds were steadily increased: the rotary press came into existence: the capacity to produce, maniplate and transport vast masses of metal was enlarged: and many of the chief mechanical instruments of surgery—including the stethoscope and the ophthalmoscope—were invented or perfected, albeit one of the most notable advances in instrumentation, the use of the obstetrical forceps, was a French invention during the eotechnic phase. The extent of the gains can be made most clear if one confines attention roughly to the first hundred years. Iron production increased from 17,000 tons in 1740 to 2,100,000 tons in 1850. With the invention in 1804 of a machine for dressing the cotton warps with starch to prevent breaking, the power loom for cotton weaving at last became practical: Horrocks' invention of a successful loom in 1803 and its improvement in 1813 transformed the cotton industry. Because of the cheapness of hand workers—as late as 1834 it was estimated that there were 45,000 to 50,000 in Scotland alone and about 200,000 in England—power loom weaving came in slowly: while in 1823 there were only 10,000 steam looms in Great Britain in 1865 there were 400,000. These two industries serve as a fairly accurate index of paleotechnic productivity.

Apart from the mass-production of clothes and the mass-distribution of foods, the great achievements of the paleotechnic phase were not in the end-products but in the intermediate machines and utilities. Above all, there was one department that was peculiarly its possession: the use of iron on a great scale. Here the engineers and workers were on familiar ground, and here, in the iron steamship, in the iron bridge, in the skeleton tower, and in the machine-tools and machines, they recorded their most decisive triumphs.

Both the iron bridge and the iron ship have a brief history. While numerous designs for iron bridges were made in Italy by Leonardo and his contemporaries, the first iron bridge in England was not built till the end of the eighteenth century. The problems to be worked out in the use of structural iron were all unfamiliar ones, and while the engineer had recourse to mathematical assistance in making and checking his calculations, the actual technique was in advance of the mathematical expression. Here was a field for ingenuity, daring experiment, bold departures.

In the course of less than a century the ironmakers and the structural engineers reached an astonishing perfection. The size of the steamship increased speedily from the tiny Clermont, 133 feet long, and 60 tons gross, to the Great Eastern, finished in 1858, the monster of the Atlantic, with decks 691 feet long, 22,500 tons gross, capable of generating 1600 H.P. in its screw engines and 1000 H.P. in its paddle-wheel engines. The regularity of performance also increased: by 1874 the City of Chester crossed the ocean regularly in eight days and between 1 and 12 hours over, on eight successive voyages. The rate of speed increased in crossing the Atlantic from the twenty-six days made by the Savannah in 1819 to the seven days and twenty hours made in 1866. This rate of increase tended to slacken during the next seventy years: a fact equally true of railroad transportation. What held for speed held likewise for size, as the big steamships lost by their bulk ease of handling in harbors and as they reached the depths of the channel in safe harbors. The Great Eastern was five times as big as the Clermont: the biggest steamship today is less than twice as big as the Great Eastern. The speed of transatlantic travel in 1866 was over three times as fast as in 1819 (47 years) but the present rate is less than twice as fast as 1866 (67 years). This holds true in numerous departments of technics: acceleration and quantification and multiplication went on faster in the early paleotechnic phase than they have gone on since in the same province.

An early mastery was likewise achieved in the building of iron structures. Perhaps the greatest monument of the period was the Crystal Palace in England: a timeless building which binds together the eotechnic phase, with its invention of the glass hothouse, the paleotechnic, with its use of the glass-covered railroad shed, and the neotechnic, with its fresh appreciation of sun and glass and structural lightness. But the bridges were the more typical monuments: not forgetting Telford's iron chain suspension bridge over Menai straits (1819-1825); the Brooklyn Bridge, begun in 1869 and the Firth of Forth bridge, a great cantilever construction, begun in 1867, were perhaps the most complete esthetic consummations of the new industrial technique. Here the quantity of the material, even the element of size itself, had a part in the esthetic result, emphasizing the difficulty of the task and the victory of the solution. In these magnificent works the sloppy empiric habits of thought, the catch-penny economies of the textile manufacturers, were displaced: such methods, though they played a scandalous part in contributing to the disasters of the early railroad and the early American river-steamboat, were at last sloughed off: an objective standard of performance was set and achieved. Lord Kelvin was consulted by the Glasgow shipbuilders in the working out of their difficult technical problems: these machines and structures revealed an honest, justifiable pride in confronting hard conditions and conquering obdurate materials. What Ruskin said in praise of the old wooden ships of the line applies even more to their greater iron counterparts in the merchant trade: it will bear repeating. "Take it all in all, a ship of the line is the most honorable thing that man, as a gregarious animal, has produced. By himself, unhelped, he can do better things than ships of the line; he can make poems and pictures, and other such concentrations of what is best in him. But as a being living in flocks and hammering out, with alternate strokes and mutual agreement, what is necessary for him in these flocks to get or produce, the ship of the line is his first work. Into that he has put as much of his human patience, common sense, forethought, experimental philosophy, self-control, habits of order and obedience, thoroughly wrought hard work, defiance of brute elements, careless courage, careful patriotism, and calm expectation of the judgment of God, as can well be put' into a space 300 feet long by 80 feet broad. And I am thankful to have lived in an age when I could see this thing so done."

This period of daring experimentation in iron structures reached its climax in the early skyscrapers of Chicago, and in Eiffel's great bridges and viaducts: the famous Eiffel Tower of 1888 overtopped these in height but not in mastery.

Ship-building and bridge-building, moreover, were extremely complex tasks: they required a degree of inter-relation and co-ordination that few industries, except possibly railroads, approached. These structures called forth all the latent military virtues of the régime and used them to good purpose: men risked their lives with superb nonchalance every day, smelting the iron, hammering and riveting the steel, working on narrow platforms and slender beams; and there was little distinction in the course of production between the engineer, the foremen, and the common workers: each had his share in the common task; each faced the danger. When the Brooklyn Bridge was being built, it was the Master Mechanic, not a common workman, who first tested the carriage that was used to string the cable. William Morris characterized the new steamships, with true insight, as the Cathedrals of the Industrial Age. He was right. They brought forth a fuller orchestration of the arts and sciences than any other work that the paleotects were engaged upon, and the final product was a miracle of compactness, speed, power, inter-relation, and esthetic unity. The steamer and the bridge were the new symphonies in steel. Hard grim men produced them: wage slaves or taskmasters. But like the Egyptian stone carver many thousand years before they knew the joy of creative effort. The arts of the drawing room wilted in comparison. The masculine reek of the forge was a sweeter perfume than any the ladies affected.

In back of all these efforts was a new race of artists: the English toolmakers of the late eighteenth and the early nineteenth century. These toolmakers sprang by necessity out of two dissimilar habitats: the machine works of Bolton and Watt and the wood-working shop of Joseph Bramah. In looking around for a workman to carry out a newly patented lock, Bramah seized on Henry Maudslay, a bright young mechanic who had begun work in the Woolwich Arsenal. Maudslay became not merely one of the most skilled mechanics of all time: his passion for exact work led him to bring order into the making of the essential parts of machines, above all, machine-screws. Up to this time screws had been usually cut by hand: they were difficult to make and expensive and were used as little as possible: no system was observed as to pitch or form of the threads. Every bolt and nut, as Smiles remarks, was a sort of specialty in itself. Maudslay's screw-cutting lathe was one of the decisive pieces of standardization that made the modern machine possible. He carried the spirit of the artist into every department of machine making: standardizing, refining, reducing to exact dimensions. Thanks to Maudslay interior angles, instead of being in the sharp form of an L were curved. Maudslay was used by M. I. Brunel to make his tackle-block machine; and out of his workshop, trained by his exact methods, came an apostolic succession of mechanics: Nasmyth, who invented the steam hammer, Whitworth, who perfected the rifle and the cannon, Roberts, Muirs, and Lewis. Another great mechanic of the time, Clement, also trained by Bramah, worked on Babbage's calculating machine, between 1823 and 1842—the most refined and intricate mechanism, according to Roe, that had so far been produced.

These men spared no effort in their machine-work: they worked toward perfection, without attempting to meet the cheaper competition of inferior craftsmen. There were, of course, men of similar stamp in America, France, and Germany: but for the finest work the English toolmakers commanded an international market. Their productions, ultimately, made the steamship and the iron bridge possible. The remark of an old workman of Maudslay's can well bear repetition: "It was a pleasure to see him handle a tool of any kind, but he was *quite splendid* with an eighteen inch file." That was the tribute of a competent critic to an excellent artist. And it is in machines that one must seek the most original examples of directly paleotechnic art.

16: The Paleotechnic Passage

The paleotechnic phase, then, did two things. It explored the blind alleys, the ultimate abysses, of a quantitative conception of life, stimulated by the will-to-power and regulated only by the conflict of one power-unit—an individual, a class, a state—with another power-unit. And in the mass-production of goods it showed that mechanical improvements alone were not sufficient to produce socially valuable results—or even the highest degree of industrial efficiency.

The ultimate outcome over this over-stressed power ideology and this constant struggle was the World War—that period of senseless strife which came to a head in 1914 and is still being fought out by the frustrated populations that have come under the machine system. This process can have no other end than an impotent victory: the extinction of both sides together, or the suicide of the successful nation at the very moment that it has finished slaughtering its victim. Though for convenience I have talked of the paleotechnic phase in its past tense, it is still with us, and the methods and habits of thought it has produced still rule a great part of mankind. If they are not supplanted, the very basis of technics itself may be undermined, and our relapse into barbarism will go on at a speed directly proportional to the complication and refinement of our present technological inheritance.

But the truly significant part of the paleotechnic phase lay not in what it produced but in what it led to: it was a period of transition, a busy, congested, rubbish-strewn avenue between the eotechnic and the neotechnic economies. Institutions do not affect human life only directly: they also affect it by reason of the contrary reactions they produce. While humanly speaking the paleotechnic phase was a disastrous interlude, it helped by its very disorder to intensify the search for order, and by its special forms of brutality to clarify the goals of humane living. Action and reaction were equal—and in opposite directions.

CHAPTER V.   THE NEOTECHNIC PHASE

1: The Beginnings of Neotechnics

The neotechnic phase represents a third definite development in the machine during the last thousand years. It is a true mutation: it differs from the paleotechnic phase almost as white differs from black. But on the other hand, it bears the same relation to the eotechnic phase as the adult form does to the baby.

During the neotechnic phase, the conceptions, the anticipations, the imperious visions of Roger Bacon, Leonardo, Lord Verulam, Porta, Glanvill, and the other philosophers and technicians of that day at last found a local habitation. The first hasty sketches of the fifteenth century were now turned into working drawings: the first guesses were now re-enforced with a technique of verification: the first crude machines were at last carried to perfection in the exquisite mechanical technology of the new age, which gave to motors and turbines properties that had but a century earlier belonged almost exclusively to the clock. The superb animal audacity of Cellini, about to cast his difficult Perseus, or the scarcely less daring work of Michelangelo, constructing the dome of St. Peter's, was replaced by a patient co-operative experimentalism: a whole society was now prepared to do what had heretofore been the burden of solitary individuals.

Now, while the neotechnic phase is a definite physical and social complex, one cannot define it as a period, partly because it has not yet developed its own form and organization, partly because we are still in the midst of it and cannot see its details in their ultimate relationships, and partly because it has not displaced the older

régime with anything like the speed and decisiveness that characterized the transformation of the eotechnic order in the late eighteenth
century. Emerging from the paleotechnic order, the neotechnic institutions have nevertheless in many cases compromised with it, given
way before it, lost their identity by reason of the weight of vested
interests that continued to support the obsolete instruments and the
anti-social aims of the middle industrial era. *Paleotechnic ideals
still largely dominate the industry and the politics of the Western
World:* the class struggles and the national struggles are still pushed
with relentless vigor. While eotechnic practices linger on as civilizing
influences, in gardens and parks and painting and music and the
theater, the paleotechnic remains a barbarizing influence. To deny
this would be to cling to a fool's paradise. In the seventies Melville
framed a question in fumbling verse whose significance has deepened
with the intervening years:

> *. . . Arts are tools;*
> *But tools, they say, are to the strong:*
> *Is Satan weak? Weak is the wrong?*
> *No blessed augury overrules:*
> *Your arts advanced in faith's decay:*
> *You are but drilling the new Hun*
> *Whose growl even now can some dismay.*

To the extent that neotechnic industry has failed to transform the
coal-and-iron complex, to the extent that it has failed to secure an
adequate foundation for its humaner technology in the community
as a whole, to the extent that it has lent its heightened powers to
the miner, the financier, the militarist, the possibilities of disruption
and chaos have increased.

But the beginnings of the neotechnic phase can nevertheless be
approximately fixed. The first definite change, which increased the
efficiency of prime movers enormously, multiplying it from three
to nine times, was the perfection of the water-turbine by Fourneyron
in 1832. This came at the end of a long series of studies, begun
empirically in the development of the spoon-wheel in the sixteenth
century and carried on scientifically by a series of investigators,

notably Euler in the middle of the eighteenth century. Burdin, Fourneyron's master, had made a series of improvements in the turbine type of water-wheel—a development for which one may perhaps thank France's relative backwardness in paleotechnic industry—and Fourneyron built a single turbine of 50 H.P. as early as 1832. With this, one must associate a series of important scientific discoveries made by Faraday during the same decade. One of these was his isolation of benzine: a liquid that made possible the commercial utilization of rubber. The other was his work on electromagnetic currents, beginning with his discovery in 1831 that a conductor cutting the lines of force of a magnet created a difference in potential: shortly after he made this purely scientific discovery, he received an anonymous letter suggesting that the principle might be applied to the creation of great machines. Coming on top of the important work done by Volta, Galvani, Oersted, Ohm, and Ampére, Faraday's work on electricity, coupled with Joseph Henry's exactly contemporary research on the electro-magnet, erected a new basis for the conversion and distribution of energy and for most of the decisive neotechnic inventions.

By 1850 a good part of the fundamental scientific discoveries and inventions of the new phase had been made: the electric cell, the storage cell, the dynamo, the motor, the electric lamp, the spectroscope, the doctrine of the conservation of energy. Between 1875 and 1900 the detailed application of these inventions to industrial processes was carried out in the electric power station and the telephone and the radio telegraph. Finally, a series of complementary inventions, the phonograph, the moving picture, the gasoline engine, the steam turbine, the airplane, were all sketched in, if not perfected, by 1900: these in turn effected a radical transformation of the power plant and the factory, and they had further effects in suggesting new principles for the design of cities and for the utilization of the environment as a whole. By 1910 a definite counter-march against paleotechnic methods began in industry itself.

The outlines of the process were blurred by the explosion of the World War and by the sordid disorders and reversions and compensations that followed it. Though the instruments of a neotechnic civilization are now at hand, and though many definite signs of an integration are not lacking, one cannot say confidently that a single region, much less our Western Civilization as a whole, has entirely embraced the neotechnic complex: for the necessary social institutions and the explicit social purposes requisite even for complete technological fulfillment are lacking. The gains in technics are never registered automatically in society: they require equally adroit inventions and adaptations in politics; and the careless habit of attributing to mechanical improvements a direct rôle as instruments of culture and civilization puts a demand upon the machine to which it cannot respond. Lacking a cooperative social intelligence and good-will, our most refined technics promises no more for society's improvement than an electric bulb would promise to a monkey in the midst of a jungle.

True: the industrial world produced during the nineteenth century is either technologically obsolete or socially dead. But unfortunately, its maggoty corpse has produced organisms which in turn may debilitate or possibly kill the new order that should take its place: perhaps leave it a hopeless cripple. One of the first steps, however, toward combating such disastrous results is to realize that even technically the Machine Age does not form a continuous and harmonious unit, that there is a deep gap between the paleotechnic and neotechnic phases, and that the habits of mind and the tactics we have carried over from the old order are obstacles in the way of our developing the new.

2: The Importance of Science

The detailed history of the steam engine, the railroad, the textile mill, the iron ship, could be written without more than passing reference to the scientific work of the period. For these devices were made possible largely by the method of empirical practice, by trial and selection: many lives were lost by the explosion of steamboilers before the safety-valve was generally adopted. And though all these inventions would have been the better for science, they came into existence, for the most part, without its direct aid. It was the practical men in the mines, the factories, the machine shops and the clockmakers' shops and the locksmiths' shops or the curious amateurs with a turn for manipulating materials and imagining new processes, who made them possible. Perhaps the only scientific work that steadily and systematically affected the paleotechnic design was the analysis of the elements of mechanical motion itself.

With the neotechnic phase, two facts of critical importance become plain. First, the scientific method, whose chief advances had been in mathematics and the physical sciences, took possession of other domains of experience: the living organism and human society also became the objects of systematic investigation, and though the work done in these departments was handicapped by the temptation to take over the categories of thought, the modes of investigation, and the special apparatus of quantitative abstraction developed for the isolated physical world, the extension of science here was to have a particularly important effect upon technics. Physiology became for the nineteenth century what mechanics had been for the seventeenth: instead of mechanism forming a pattern for life, living organisms began to form a pattern for mechanism. Whereas the mine dominated the paleotechnic period, it was the vineyard and the farm and the physiological laboratory that directed many of the most fruitful investigations and contributed to some of the most radical inventions and discoveries of the neotechnic phase.

Similarly, the study of human life and society profited by the same impulses toward order and clarity. Here the paleotechnic phase had succeeded only in giving rise to the abstract series of rationalizations and apologies which bore the name of political economy: a body of doctrine that had almost no relation to the actual organization of production and consumption or to the real needs and interests and habits of human society. Even Karl Marx, in criticizing these doctrines, succumbed to their misleading verbalisms: so that whereas Das Kapital is full of great historic intuitions, its description of price and value remains as prescientific as Ricardo's. The abstractions of economics, instead of being isolates and derivatives of reality, were in fact mythological constructions whose only justification would be in the impulses they excited and the actions they prompted Following Vico, Condorcet, Herder and G. F. Hegel, who were philosophers of history, Comte, Quetelet, and Le Play laid down the new science of sociology; while on the heels of the abstract psychologists from Locke and Hume onward, the new observers of human nature, Bain, Herbart, Darwin, Spencer, and Fechner integrated psychology with biology and studied the mental processes as a function of all animal behavior.

In short, the concepts of science, hitherto associated largely with the cosmic, the inorganic, the "mechanical" were now applied to every phase of human experience and every manifestation of life. The analysis of matter and motion, which had greatly simplified the original tasks of science, now ceased to exhaust the circle of scientific interests: men sought for an underlying order and logic of events which would embrace more complex manifestations. The Ionian philosophers had long ago had a clue to the importance of order itself in the constitution of the universe. But in the visible chaos of Victorian society Newlands' original formulation of the periodic table as the Law of Octaves was rejected, not because it was insufficient, but because Nature was deemed unlikely to arrange the elements in such a regular horizontal and vertical pattern.

During the neotechnic phase, the sense of order became much more pervasive and fundamental. The blind whirl of atoms no longer seemed adequate even as a metaphorical description of the universe. During this phase, the hard and fast nature of matter itself underwent a change: it became penetrable to newly discovered electric impulses, and even the alchemist's original guess about the transmutation of the elements was turned, through the discovery of radium, into a reality. The image changed from "solid matter" to "flowing energy."

Second only to the more comprehensive attack of the scientific method upon aspects of existence hitherto only feebly touched by it, was the direct application of scientific knowledge to technics and the conduct of life. In the neotechnic phase, the main initiative comes, not from the ingenious inventor, but from the scientist who establishes the general law: the invention is a derivative product. It was Henry who in essentials invented the telegraph, not Morse; it was Faraday who invented the dynamo, not Siemens; it was Oersted who invented the electric motor, not Jacobi; it was Clerk-Maxwell and Hertz who invented the radio telegraph, not Marconi and De Forest. The translation of the scientific knowledge into practical instruments was a mere incident in the process of invention. While distinguished individual inventors like Edison, Baekeland and Sperry remained, the new inventive genius worked on the materials provided by science.

Out of this habit grew a new phenomenon: deliberate and systematic invention. Here was a new material: problem—find a new use for it. Or here was a necessary utility: problem—find the theoretic formula which would permit it to be produced. The ocean cable was finally laid only when Lord Kelvin had contributed the necessary scientific analysis of the problem it presented: the thrust of the propeller shaft on the steamer was finally taken up without clumsy and expensive mechanical devices, only when Michell worked out the behavior of viscous fluids: long distance telephony was made possible only by systematic research by Pupin and others in the Bell Laboratories on the several elements in the problem. Isolated inspiration and empirical fumbling came to count less and less in invention. In a whole series of characteristic neotechnic inventions the thought was father to the wish. And typically, this thought is a collective product.

While the independent theoretic mind was still, naturally, immensely stimulated by the suggestions and needs of practical life, as Carnot had been stirred to his researches on heat by the steam engine, as the chemist, Louis Pasteur, was stirred to bacteriological research by the predicament of the vintners, brewers, and silkworm growers, the fact was that a liberated scientific curiosity might at any moment prove as valuable as the most factual pragmatic research. Indeed, this freedom, this remoteness, this contemplative isolation, so foreign to the push of practical success and the lure of immediate applications, began to fill up a general reservoir of ideas, which flowed over, as if by gravity, into practical affairs. The possibilities for human life could be gauged by the height of the reservoir itself, rather than by the pressure the derivative stream might show at any moment. And though science had been impelled, from the beginning, by practical needs and by the desire for magical controls, quite as much perhaps as by the will-to-order, it came during the nineteenth century to act as a counterweight to the passionate desire to reduce all existence to terms of immediate profit and success. The scientists of the first order, a Faraday, a Clerk-Maxwell, a Gibbs, were untouched by pragmatic sanctions: for them science existed, as the arts exist, not simply as a means of exploiting nature, but as a mode of life: good for the states of mind they produce as well as for the external conditions they change.

Other civilizations reached a certain stage of technical perfection and stopt there: they could only repeat the old patterns. Technics in its traditional forms provided no means of continuing its own growth. Science, by joining on to technics, raised so to say the ceiling of technical achievement and widened its potential cruising area. In the interpretation and application of science a new group of men appeared, or rather, an old profession took on new importance. Intermediate between the industrialist, the common workman, and the scientific investigator came the engineer.

We have seen how engineering as an art goes back to antiquity, and how the engineer began to develop as a separate entity as a result of military enterprise from the fourteenth century onward, designing fortifications, canals, and weapons of assault. The first great school devised for the training of engineers was the Ecole Polytechnique, founded in Paris in 1794 in the midst of the revolution: the school at St. Etienne, the Berlin Polytechnic and Rensselaer (1824) came shortly after it: but it was only in the middle of the nineteenth century that South Kensington, Stevens, Zürich, and other schools followed. The new engineers must master all the problems involved in the development of the new machines and utilities, and in the application of the new forms of energy: the range of the profession must in all its specialized branches be as wide as Leonardo's had been in its primitive relatively undifferentiated state.

Already in 1825 Auguste Comte could say:

"It is easy to recognize in the scientific body as it now exists a certain number of *engineers* distinct from men of science properly so-called. This important class arose of necessity when Theory and Practice, which set out from such distant points, had approached sufficiently to give each other the hand. It is this that makes its distinctive character still so undefined. As to characteristic doctrines fitted to constitute the special existence of the class of engineers, their true nature cannot be easily indicated because their rudiments only exist. . . . The establishment of the class of engineers in its proper characteristics is the more important because this class will, without doubt, constitute the direct and necessary instrument of coalition between men of science and industrialists by which alone the new social order can commence." (Comte: Fourth Essay, 1825.)

The situation to which Comte looked forward did not become possible until the neotechnic phase itself had begun to emerge. As the methods of exact analysis and controlled observation began to penetrate every department of activity, the concept of the engineer broadened to the more general notion of technician. More and more, each of the arts sought for itself a basis in exact knowledge. The infusion of exact, scientific methods into every department of work and action, from architecture to education, to some extent increased the scope and power of the mechanical world-picture that had been built up in the seventeenth century: for technicians tended to take the world of the physical scientist as the most real section of experience, because it happened, on the whole, to be the most measurable; and they were sometimes satisfied with superficial investigations as long as they exhibited the general form of the exact sciences. The specialized, one-sided, factual education of the engineer, the absence of humanistic interests in both the school of engineering itself and the environment into which the engineer was thrust, only accentuated these limitations. Those interests to which Thomas Mann teasingly introduced his half-baked nautical engineer in The Magic Mountain, the interests of philosophy, religion, politics, and love, were absent from the utilitarian world: but in the long run, the broader basis of the neotechnic economy itself was to have an effect, and the restoration of the humanities in the California Institute of Technology and the Stevens Institute was a significant step toward repairing the breach that was opened in the seventeenth century. Unlike the paleotechnic economy, which had grown so exclusively out of the mine, the neotechnic economy was applicable at every point in the valley section—as important in its bacteriology for the farmer as in its psychology for the teacher.

3: New Sources of Energy

The neotechnic phase was marked, to begin with, by the conquest of a new form of energy: electricity. The lodestone and the properties of amber when rubbed were both known to the Greeks; but the first modern treatise on electricity dates back to Dr. John Gilbert's *De Magnete,* published in 1600. Dr. Gilbert related frictional elec-tricity to magnetism, and after him the redoubtable burgomaster of Magdeburg, Otto von Guericke, he of the Magdeburg hemi-spheres, recognized the phenomenon of repulsion, as well as attrac-tion, while Leibniz apparently was the first to observe the electric spark. In the eighteenth century, with the invention of the Leyden jar, and with Franklin's discovery that lightning and electricity were one, the experimental work in this field began to take shape. By 1840 the preliminary scientific exploration was done, thanks to Oersted, Ohm, and above all, to Faraday; and in 1838 Joseph Henry had even observed the inductive effects at a distance from a Leyden jar: the first hint of radio communication.

Technics did not lag behind science. By 1838 Professor Jacobi, at St. Petersburg, had succeeded in propelling a boat on the Neva at four miles an hour by means of an "electro-magnetic engine," David-son on the Edinburgh and Glasgow Railway achieved the same speed; while in 1849 Professor Page attained a speed of 19 miles per hour on a car on the Baltimore and Washington Railroad. The electric arc light was patented in 1846 and applied to the lighthouse at Dungeness, England, in 1862. Meanwhile, a dozen forms of the elec-tric telegraph had been invented: by 1839 Morse and Steinheil had made possible instantaneous communication over long distances, using grounded wires at either end. The practical development of the dynamo by Werner Siemens (1866) and the alternator by Nikola Tesla (1887) were the two necessary steps in the substi-tution of electricity for steam: the central power station and dis-tribution system, invented by Edison (1882) presently developed.

In the application of power, electricity effected revolutionary

changes: these touched the location and the concentration of industries and the detailed organization of the factory—as well as a multitude of inter-related services and institutions. The metallurgical
industries were transformed and certain industries like rubber production were stimulated. Let us look more closely at some of these
changes.

During the paleotechnic phase, industry depended completely
upon the coal mine as a source of power. Heavy industries were
compelled to settle close to the mine itself, or to cheap means of
transportation by means of the canal and the railroad. Electricity,
on the other hand, can be developed by energy from a large number
of sources: not merely coal, but the rapidly running river, the falls,
the swift tidal estuary are available for energy; so are the direct
rays of the sunlight (7000 H.P. per sun-acre) for the sun-batteries
that have been built in Egypt; so too is the windmill, when accumula-
tors are provided. Inaccessible mountain areas, like those in the Alps,
the Tyrol, Norway, the Rockies, interior Africa, became for the
first time potential sources of power and potential sites for modern
industry: the harnessing of water-power, thanks to the supreme efficiency of the water-turbine, which rates around 90 per cent, opened
up new sources of energy and new areas for colonization—areas
more irregular in topography and often more salubrious in climate
than the valley-bottoms and lowlands of the earlier eras. Because of
the enormous vested interest in coal measures, the cheaper sources
of energy have not received sufficient systematic attention upon the
part of inventors: but the present utilization of solar energy in
agriculture— about 0.13 per cent of the total amount of solar energy
received—presents a challenge to the scientific engineer; while the
possibility of using differences of temperature between the upper
and lower levels of sea water in the tropics offers still another prospect for escaping servitude to coal.

The availability of water-power for producing energy, finally,
changes the potential distribution of modern industry throughout the
planet, and reduces the peculiar industrial dominance that Europe
and the United States held under the coal-and-iron régime. For Asia
and South America are almost as well endowed with water-power—

over fifty million horsepower each—as the older industrial regions, and Africa has three times as much as either Europe or North America. Even within Europe and the United States a shifting of the industrial center of gravity is taking place: thus the leadership in hydro-electric power development has gone to Italy, France, Norway, Switzerland and Sweden in the order named, and a similar shift is taking place toward the two great spinal mountain-systems of the United States. The coal measures are no longer the exclusive measures of industrial power.

Unlike coal in long distance transportation, or like steam in local distribution, electricity is much easier to transmit without heavy losses of energy and higher costs. Wires carrying high tension alternating currents can cut across mountains which no road vehicle can pass over; and once an electric power utility is established the rate of deterioriation is slow. Moreover, electricity is readily convertible into various forms: the motor, to do mechanical work, the electric lamp, to light, the electric radiator, to heat, the x-ray tube and the ultra-violet light, to penetrate and explore, and the selenium cell, to effect automatic control.

While small dynamos are less efficient than large dynamos, the difference in performance between the two is much less than that between the big steam-engine and the small steam-engine. When the water-turbine can be used, the advantage of being able to use electricity with high efficiency in all sizes and power-ratings becomes plain: if there is not a sufficiently heavy head of water to operate a large alternator, excellent work can nevertheless be done for a small industrial unit, like a farm, by harnessing a small brook or stream and using only a few horsepower; and by means of a small auxiliary gasoline engine continuous operation can be assured despite seasonal fluctuations in the flow of the water. The water turbine has the great advantage of being automatic: once installed, the costs of production are almost nil, since no fireman or attendant is necessary. With larger central power stations there are other advantages. Not all power need be absorbed by the local area: by a system of interlinked stations, surplus power may be transmitted over long distances, and in case of a breakdown in a single plant the supply

itself will remain adequate by turning on the current from the asso·
ciated plants.

4: The Displacement of the Proletariat

The typical productive units of the paleotechnic period were
afflicted with giantism: they increased in size and agglomerated together without attempting to scale size to efficiency. In part this
grew out of the defective system of communication which antedated
the telephone: this confined efficient administration to a single manufacturing plant and made it difficult to disperse the several units,
whether or not they were needed on a single site. It was likewise
abetted by the difficulties of economic power production with small
steam engines: so the engineers tended to crowd as many productive units as possible on the same shaft, or within the range of
steam pressure through pipes limited enough to avoid excessive
condensation losses. The driving of the individual machines in the
plant from a single shaft made it necessary to spot the machines
along the shafting, without close adjustment to the topographical
needs of the work itself: there were friction losses in the belting, and
the jungle of belts offered special dangers to the workers: in addition to these defects, the shafting and belting limited the use of
local transport by means of travelling cranes.

The introduction of the electric motor worked a transformation
within the plant itself. For the electric motor created flexibility in
the design of the factory: not merely could individual units be placed
where they were wanted, and not merely could they be designed for
the particular work needed: but the direct drive, which increased
the efficiency of the motor, also made it possible to alter the layout
of the plant itself as needed. The installation of motors removed the
belts which cut off light and lowered efficiency, and opened the way
for the rearrangement of machines in functional units without regard
for the shafts and aisles of the old-fashioned factory: each unit
could work at its own rate of speed, and start and stop to suit its
own needs, without power losses through the operation of the plant
as a whole. According to the calculations of a German engineer, this
has raised the performance fifty per cent in efficiency. Where large

units were handled, the automatic servicing of the machines through travelling cranes now became simple. All these developments have come about during the last forty years; and it goes without saying that it is only in the more advanced plants that all these refinements and economies in operation have been embraced.

With the use of electricity, as Henry Ford has pointed out, small units of production can nevertheless be utilized by large units of administration, for efficient administration depends upon record-keeping, charting, routing, and communication, and not necessarily upon a local overseership. In a word, the size of the productive unit is no longer determined by the local requirements of either the steam engine or the managerial staff: it is a function of the operation itself. But the efficiency of small units worked by electric motors utilizing current either from local turbines or from a central power plant has given small-scale industry a new lease on life: on a purely technical basis it can, for the first time since the introduction of the steam engine, compete on even terms with the larger unit. Even domestic production has become possible again through the use of electricity: for if the domestic grain grinder is less efficient, from a purely mechanical standpoint, than the huge flour mills of Minneapolis, it permits a nicer timing of production to need, so that it is no longer necessary to consume bolted white flours because whole wheat flours deteriorate more quickly and spoil if they are ground too long before they are sold and used. To be efficient, the small plant need not remain in continuous operation nor need it produce gigantic quantities of foodstuffs and goods for a distant market: it can respond to local demand and supply; it can operate on an irregular basis, since the overhead for permanent staff and equipment is proportionately smaller; it can take advantage of smaller wastes of time and energy in transportation, and by face to face contact it can cut out the inevitable red-tape of even efficient large organizations.

As an element in large-scale standardized industry, making products for a continental market, the small plant can now survive. "There is no point," as Henry Ford says, "in centralizing manufacturing unless it results in economies. If we, for instance, centered our entire production in Detroit we should have to employ about

6,000,000 people. . . . A product that is used all over the country ought to be made all over the country, in order to distribute buying power more evenly. For many years we have followed the policy of making in our branches whatever parts they were able to make for the area they served. A good manufacturer who makes himself a specialist will closely control his production and is to be preferred over a branch." And again Ford says: "In our first experimenting . . . we thought that we had to have the machine lines with their assembly and also the final assembly all under one roof, but as we grew in understanding we learned that the making of each part was a separate business in itself, and to be made wherever it could be made the most efficiently, and that the final assembly line could be anywhere. This gave us the first evidence of the flexibility of modern production, as well as indication of the savings that might be made in cutting down unnecessary shipping."

Even without the use of electric power the small workshop, because of some of the above facts, has survived all over the world, in defiance of the confident expectations of the early Victorian economists, marvelling over the mechanical efficiency of the monster textile mills: with electricity, the advantages of size from any point of view, except in possible special operations like the production of iron, becomes questionable. In the production of steel from scrap iron the electric furnace may be used economically for operations on a much smaller scale than the blast-furnace permits. Moreover, the weakest part mechanically of automatic production lies in the expense and hand-labor involved in preparation for shipment. To the extent that a local market and a direct service does away with these operations it removes a costly and completely uneducative form of work. Bigger no longer automatically means better: flexibility of the power unit, closer adaptation of means to ends, nicer timing of operation, are the new marks of efficient industry. So far as concentration may remain, it is largely a phenomenon of the market, rather than of technics: promoted by astute financiers who see in the large organization an easier mechanism for their manipulations of credit, for their inflation of capital values, for their monopolistic controls.

The electric power plant is not merely the driving force in the new technology: it is likewise perhaps one of the most characteristic end-products; for it is in itself an exhibition of that complete automatism to which, as Mr. Fred Henderson and Mr. Walter Polakov have ably demonstrated, our modern system of power production tends. From the movement of coal off the railroad truck or the coal barge, by means of a travelling crane, operated by a single man, to the stoking of the coal in the furnace by a mechanical feeder, power machinery takes the place of human energy: the worker, instead of being a source of work, becomes an observer and regulator of the performance of the machines—a supervisor of production rather than an active agent. Indeed the direct control of the local worker is the same in principle as the remote control of the management itself, supervising, through reports and charts, the flow of power and goods through the entire plant.

The qualities the new worker needs are alertness, responsiveness, an intelligent grasp of the operative parts: in short, he must be an all-round mechanic rather than a specialized hand. Short of complete automatism, this process is still a dangerous one for the worker: for partial automatism had been reached in the textile plants in England by the eighteen-fifties without any great release of the human spirit. But with complete automatism freedom of movement and initiative return for that small part of the original working force now needed to operate the plant. Incidentally, it is interesting to note that one of the most important labor-saving and drudgery-saving devices, the mechanical firing of boilers, was invented at the height of the paleotechnic period: in 1845. But it did not begin to spread rapidly in power plants until 1920, by which time coal had begun to feel competition from automatic oil burners. (Another great economy invented in the same year [1845], the use of blast-furnace gases for fuel, did not come in till much later.)

In all the neotechnic industries that produce completely standardized goods, automatism in operation is the goal toward which they tend. But, as Barnett points out, "the displacing power of machines varies widely. One man on the stone-planer is capable of producing as much as eight men can produce by hand. One man on the semi-automatic bottle machine can make as many as four hand-blowers. A linotype operator can set up as much matter as four hand-compositors. The Owens bottle machine in its latest form is capable of an output per operative equal to that of eighteen hand-blowers." To which one may add that in the automatic telephone exchange the number of operators has been reduced about eighty per cent, and in an American textile plant a single worker can look after 1200 spindles. While the deadliest form of high-paced, piece-meal, unvaried labor still remain in many so-called advance industries, like the straight-line assemblage of Ford cars, a form of work as dehumanized and as backward as any practiced in the worst manufacturing processes of the eighteenth century—while this is true, in the really neotechnic industries and processes the worker has been almost eliminated.

Power production and automatic machines have steadily been diminishing the worker's importance in factory production. Two million workers were cast out between 1919 and 1929 in the United States, while production itself actually increased. Less than a tenth of the population of the United States is sufficient to produce the bulk of its manufactured goods and its mechanical services. Benjamin Franklin figured that in his day the spread of work and the elimination of the kept classes would enable all the necessary production to be accomplished with an annual toll of five hours per worker per day. Even with our vast increase in consumptive standards, both in intermediate machines and utilities and in final goods, a fragment of that time would probably suffice for a neotechnic industry, if it were organized efficiently on a basis of steady, full-time production.

Parallel to the advances of electricity and metallurgy from 1870 onward were the advances that took place in chemistry. Indeed, the emergence of the chemical industries after 1870 is one of the definite signs of the neotechnic order, since the advance beyond the age-old empirical methods used, for example, in distilling and in the manufacture of soap naturally was limited by the pace of science itself. Chemistry not merely assumed a relatively larger share in every phase of industrial production from metallurgy to the fabrication of artificial silk: but the chemical industries themselves, by their very nature, exhibited the characteristic neotechnic features a whole generation before mechanical industry showed them. Here Mataré's figures, though they are almost a generation old, are still significant: in the advanced mechanical industries only 2.8 per cent of the entire personnel were technicians: in the old-fashioned chemical industries, such as vinegar works and breweries, there were 2.9 per cent; but in the more recent chemical industries, dyes, starch products, gas works, and so forth, 7.1 per cent of the personnel were technicians. Similarly, the processes themselves tend to be automatic, and the percentage of workers employed is smaller than even in advanced machine industries, while workers who supervise them must have similar capacities to those at the remote control boards of a power station or a steamship. Here, as in neotechnic industry generally, advances in production increase the number of trained technicians in the laboratory and decrease the number of human robots in the plant. In short, one witnesses in the chemical processes —apart from the ultimate packaging and boxing—the general change that characterizes all genuinely neotechnic industry: the displacement of the proletariat.

That these gains in automatism and power have not yet been assimilated by society is plain; and I shall revert to the problem here presented in the final chapter.

5: Neotechnic Materials

Just as one associates the wind and water power of the eotechnic economy with the use of wood and glass, and the coal of the paleotechnic period with iron, so does electricity bring into wide industrial use its own specific materials: in particular, the new alloys, the rare earths, and the lighter metals. At the same time, it creates a new series of synthetic compounds that supplement paper, glass and wood: celluloid, vulcanite, bakelite and the synthetic resins, with special properties of unbreakability, electrical resistance, imperviousness to acids, or elasticity.

Among the metals, electricity places a premium upon those that have a high degree of conductivity: copper and aluminum. Area for area, copper is almost twice as good a conductor as aluminum but weight for weight aluminum is superior to any other metal, even silver, while iron and nickel are practically useless except where resistance is needed, as for example in electric heating. Perhaps the most distinctively neotechnic metal is aluminum, for it was discovered in 1825 by the Dane, Oersted, one of the fruitful early experimenters with electricity, and it remained a mere curiosity of the laboratory through the high paleotechnic period. It was not until 1886, the decade that saw the invention of the motion picture and the discovery of the Hertzian wave, that patents for making aluminum commercially were taken out. One need not wonder at aluminum's slow development: for the commercial process of extraction is dependent upon the use of large quantities of electric energy: the principal cost of reducing the aluminum ore by the electrolytic process is the use of from ten to twelve kilowatt hours of energy for every pound of metal recovered. Hence the industry must naturally attach itself to a cheap source of electric power.

Aluminum is the third most abundant element on the earth's crust, following oxygen and silicon; but at present it is manufactured chiefly from its hydrated oxide, bauxite. If the extraction of aluminum from clay is not yet commercially feasible, no one can doubt that an effective means will eventually be found: hence the supply of aluminum is practically inexhaustible, all the more because its slow oxidation permits society to build up steadily a reserve of scrap metal. This entire development has taken place over a period of little more than forty years, those same forty years that saw the introduction of central power plants and multiple motor installations in factories; and while copper production in the last twenty years has increased a good fifty per cent, aluminum production has increased during the same period 316 per cent. Everything from typewriter frames to airplanes, from cooking vessels to furniture, can now be made of aluminum and its stronger alloys. With aluminum, a new standard of lightness is set: a dead weight is lifted from all forms of locomotion, and the new aluminum cars for railroads can attain a higher speed with a smaller output of power. If one of the great achievements of the paleotechnic period was the translation of clumsy wooden machines into stronger and more accurate iron ones, one of the chief tasks of the neotechnic period is to translate heavy iron forms into lighter aluminum ones. And just as the technique of water-power and electricity had an effect in reorganizng even the coal-consumption and steam-production of power plants, so the lightness of aluminum is a challenge to more careful and more accurate design in such machines and utilities as still use iron and steel. The gross over-sizing of standard dimensions, with an excessive factor of safety based upon a judicious allowance for ignorance, is intolerable in the finer design of airplanes; and the calculations of the airplane engineer must in the end react back upon the design of bridges, cranes, steel-buildings: in fact, such a reaction is already in evidence. Instead of bigness and heaviness being a happy distinction, these qualities are now recognized as handicaps: lightness and compactness are the emergent qualities of the neotechnic era.

The use of the rare metals and the metallic earths is another characteristic advance of this phase: tantalum, tungsten, thorium, and cerium in lamps, iridium and platinum in mechanical contact points —the tips of fountain pens or the attachments in removable dentures—and of nickel, vanadium, tungsten, manganese and chromium in steel. Selenium, whose electrical resistance varies inversely with the intensity of light, was another metal which sprang into wide use with electricity: automatic counting devices and electric door-openers are both possible by reason of this physical property.

As a result of systematic experiment in metallurgy a revolution took place here comparable to that which was involved in the change from the steam-engine to the dynamo. For the rare metals now have a special place in industry, and their careful use tends to promote habits of thrift even in the exploitation of the commoner minerals. Thus the production of rustless steel will decrease the erosion of steel and add to the metal worth redeeming from the scrapheap. Already the supply of steel is so large and its conservation has at last become so important that over half the burden of the open hearth furnaces in the United States is scrap metal—and the open hearth process now takes care of 80 per cent of the domestic steel production. The rare elements, most of which were undiscovered until the nineteenth century, cease to be curiosities or to have, like gold, chiefly a decorative or honorific value: their importance is

out of all proportion to their bulk. The significance of minute quan·
tities—which we shall note again in physiology and medicine—is
characteristic of the entire metallurgy and technics of the new phase.
One might say, for dramatic emphasis, that paleotechnics regarded
only the figures to the left of the decimal, whereas neotechnics is preoccupied with those to the right.

There is still another important consequence of this new complex.
While certain products of the neotechnic phase, like glass, copper,
and aluminum, exist like iron in great quantities, there are other
important materials—asbestos, mica, cobalt, radium, uranium,
thorium, helium, cerium, molybdenum, tungsten—which are exceed-
ingly rare, or which are strictly limited in their distribution. Mica,
for example, has unique properties that make it indispensable in
the electrical industry: its regular cleavage, great flexibility, elasticity, transparency, non-conductivity of heat and electricity and general resistance to decomposition make it the best possible material
for radio condensers, magnetos, spark plugs, and other necessary
instruments: but while it has a fairly wide distribution there are important parts of the earth that are completely without it. Manganese,
one of the most important alloys for hard steel, is concentrated chiefly
in India, Russia, Brazil and the Gold Coast of Africa. With tungsten,
seventy per cent of the supply comes from South America and 9.3
per cent from the United States; as for chromite, almost half the
present supply comes from South Rhodesia, 12.6 per cent from New
Caledonia, and 10.2 per cent from India. The rubber supply, similarly, is still limited to certain tropical or sub-tropical areas, notably
Brazil and the Malayan archipelago.

Note the importance of these facts in the scheme of world commodity flow. Both eotechnic and paleotechnic industry could be carried on within the framework of European society: England, Germany, France, the leading countries, had a sufficient supply of wind,
wood, water, limestone, coal, iron ore; so did the United States.
Under the neotechnic régime their independence and their self-
sufficiency are gone. *They must either organize and safeguard and
conserve a worldwide basis of supply, or run the risk of going destitute and relapse into a lower and cruder technology.* The basis of

the material elements in the new industry is neither national nor continental but planetary: this is equally true, of course, of its technological and scientific heritage. A laboratory in Tokio or Calcutta may produce a theory or an invention which will entirely alter the possibilities of life for a fishing community in Norway. Under these conditions, no country and no continent can surround itself with a wall without wrecking the essential, international basis of its technology: so if the neotechnic economy is to survive, it has no other alternative than to organize industry and its polity on a worldwide scale. Isolation and national hostilities are forms of deliberate technological suicide. The geographical distribution of the rare earths and metals by itself almost establishes that fact.

One of the greatest of neotechnic advances is associated with the chemical utilization of coal. Coal tar, once the unfortunate refuse of the paleotechnic type of beehive coke oven, became an important source of wealth: from each ton of coal "the by-product oven produces approximately 1500 pounds of coke, 111,360 cubic feet of gas, 12 gallons of tar, 25 pounds of ammonium sulphate, and 4 gallons of light oils." Through the breakdown of coal tar itself the chemist has produced a host of new medicines, dyes, resins, and even perfumes. As with advances in mechanization, it has tended to provide greater freedom from local conditions, from the accidents of supply and the caprices of nature: though a plague in silkworms might reduce the output of natural silk, artificial silk, which was first successfully created in the eighties, could partly take its place.

But while chemistry set itself the task of imitating or reconstructing the organic—ironically its first great triumph was Wohler's production of urea in 1825—certain organic compounds for the first time became important in industry: so that one cannot without severe qualification accept Sombart's characterization of modern industry as the supplanting of organic materials with inorganic ones. The greatest of these natural products was rubber, out of which the Indians of the Amazon had, by the sixteenth century, created shoes, clothes, and hot water bottles, to say nothing of balls and syringes. The development of rubber is exactly contemporary to that of electricity, even as cotton in Western Europe exactly parallels the steam engine, for it was Faraday's isolation of benzine, and the later use of naphtha, that made its manufacture possible elsewhere than at its place of origin. The manifold uses of rubber, for insulation, for phonograph records, for tires, for soles and heels of shoes, for rainproof clothing, for hygienic accessories, for the surgeon's gloves, for balls used in play give it a unique place in modern life. Its elasticity and impermeability and its insulating qualities make it a valuable substitute, on occasion, for fibre, metal, and glass, despite its low melting point. Rubber constitutes one of the great capital stocks of industry, and reclaimed rubber, according to Zimmerman, formed from 35 to 51 per cent of the total rubber production in the United States between 1925 and 1930. The use of corn and cane stalks for composite building materials and for paper illustrates another principle: the attempt to live on current energy income, instead of on capital in the form of trees and mineral deposits.

Almost all these new applications date since 1850; most of them came after 1875; while the great achievements in colloidal chemistry have come only within our own generation. We owe these materials and resources quite as much to fine instruments and laboratory apparatus as we do to power-machinery. Plainly, Marx was in error when he said that machines told more about the system of production that characterized an epoch than its utensils and utilities did: for it would be impossible to describe the neotechnic phase without taking into account various triumphs in chemistry and bacteriology in which machines played but a minor part. Perhaps the most important single instrument that the later neotechnic period has created is the three-element oscillator—or amplifier—developed by De Forest out of the Fleming valve: a piece of apparatus in which the only moving parts are electric charges. The movement of limbs is more obvious than the process of osmosis: but they are equally important in human life; and so too the relatively static operations of chemistry are as important to our technology as the more obvious engines of speed and movement. Today our industry owes a heavy debt to chemistry: tomorrow it may incur an even heavier debt to physiology and biology: already, in fact, it begins to be apparent.

6: Power and Mobility

Only second in importance to the discovery and utilization of electricity was the improvement that took place in the steam engine and the internal combustion engine. At the end of the eighteenth century Dr. Erasmus Darwin, who anticipated so many of the scientific and technical discoveries of the nineteenth century, predicted that the internal combustion engine would be more useful than the steam engine in solving the problem of flight. Petroleum, which was known and used by the ancients, and which was exploited in America as a quack Indian medicine, was tapped by drilling wells, for the first time in the modern period, in 1859: after that it was rapidly exploited. The value of the lighter distillates as fuels was equalled only by that of the heavier oils as lubricants.

From the eighteenth century onward the gas engine was the subject of numerous experiments: even the use of powdered explosives, on the analogy of cannon-fire, was tried; and the gas engine was finally perfected by Otto in 1876. With the improvement of the internal combustion engine a vast new source of power was opened up, fully equal to the old coal beds in importance, even if doomed to be consumed at a possibly more rapid rate. But the main point about fuel oil (used by the later Diesel engine) and gasoline was their relative lightness and transportability. Not merely could they be conveyed from well to market by permanent pipe-lines but, since they were liquids, and since the vaporizations and combustion of the fuel left little residue in comparison with coal, they could be stowed away easily, in odds and ends of space where coal could not be placed or reached: being fed by gravity or pressure the engine had no need for a stoker.

The effect of introducing liquid fuel and of mechanical stokers for coal, in electric power plants, and on steamships, was to emancipate a race of galley slaves, the stokers, those miserable driven men whose cruel work Eugene O'Neill properly took as the symbol of proletarian oppression in his drama, The Hairy Ape. Meanwhile, the efficiency of the steam engine was raised: the invention of Parson's steam turbine in 1884 increased the efficiency of the steam engine from ten or twelve for the old reciprocating engine to a good thirty per cent for the turbine, and the later use of mercury vapor instead of steam in turbines raised this to 41.5 per cent. How rapid was the advance in efficiency may be gauged from the average consumption of coal in power stations: it dropped from 3.2 pounds per kilowatt hour in 1913 to 1.34 pounds in 1928. These improvements made possible the electrification of railroads even where cheap water power could not be secured.

The steam engine and the internal combustion engine raced neck and neck: in 1892, by utilizing a more scientific mode of combustion, through the compression of air alone, Diesel invented an improved type of oil engine which has been built in units as large as 15,000 brake-horsepower, as in the generating plant at Hamburg. The development of the smaller internal combustion engine during the eighties and nineties was equally important for the perfection of the automobile and the airplane.

Neotechnic transportation awaited this new form of power, in which all the weight should be represented by the fuel itself, instead of carrying, like the steam engine, the additional burden of water. With the new automobile, power and movement were no longer chained to the railroad line: a single vehicle could travel as fast as a train of cars: again the smaller unit was as efficient as the larger one. (I put aside the technical question as to whether, with oil as fuel, the steam engine might not have competed effectively with the internal combustion engine, and whether it may not, in an improved and simplified form, re-enter the field.)

The social effects of the automobile and the airplane did not begin to show themselves on any broad scale until around 1910: the flight of Blériot across the English channel in 1909 and the introduction of the cheap, mass-produced motor car by Henry Ford were significant turning points.

But what happened here, unfortunately, is what happened in almost every department of industrial life. The new machines followed, not their own pattern, but the pattern laid down by previous economic and technical structures. While the new motor car was called a horseless carriage it had no other point of resemblance than the fact that it ran on wheels: it was a high-powered locomotive, equivalent to from five to a hundred horses in power, capable of safe speeds up to sixty miles an hour, as soon as the cord tire was invented, and having a daily cruising radius of two to three hundred miles. This private locomotive was set to running on the old-fashioned dirt roads or macadam highways that had been designed for the horse and wagon; and though after 1910 these highways were widened and concrete took the place of lighter materials for the surface, the pattern of the transportation lines remained what it had been in the past. All the mistakes that had been made in the railroad building period were made again with this new type of locomotive. Main highways cut through the center of towns, despite the congestion, the friction, the noise, and the dangers that attended this old paleotechnic practice. Treating the motor car solely as a mechanical object, its introducers made no attempt to introduce appropriate utilities which would realize its potential benefits.

Had anyone asked in cold blood—as Professor Morris Cohen has suggested—whether this new form of transportation would be worth the yearly sacrifice of 30,000 lives in the United States alone, to say nothing of the injured and the maimed, the answer would doubtless have been No. But the motor car was pumped onto the market at an accelerating rate, by business men and industrialists who looked for improvements only in the mechanical realm, and who had no flair for inventions on any other plane. Mr. Benton MacKaye has demonstrated that fast transportation, safe transportation and pedestrian movement, and sound community building are parts of a single process: the motor car demanded for long distance transportation the Townless Highway, with stations for entrance and exit at regular intervals and with overpasses and underpasses for major cross traffic arteries: similarly, for local transportation, it demanded the Highwayless Town, in which no neighborhood community would be split apart by major arteries or invaded by the noise of through traffic.

Even from the standpoint of speed by itself, the solution does not rest solely with the automotive engineer. A car capable of fifty miles an hour on a well-planned road system is a faster car than one that can do a hundred miles an hour, caught in the muddle and congestion of an old-fashioned highway net, and so reduced to twenty miles per hour. The rating of a car at the factory, in terms of speed and horsepower, has very little to do with its actual efficiency: in short, the motor car is as inefficient without its appropriate utilities as the electric power plant would be if the conducting units were iron wire rather than copper. Developed by a society so preoccupied with purely mechanical problems and purely mechanical solutions—themselves determined largely by speed in achieving financial rewards to the investing classes—the motor car has never attained anything like its potential efficiency except here and there in the remoter rural regions. Cheapness and quantity production, combined with the extravagant re-building of old-fashioned highway systems—with here and there honorable exceptions, as in New Jersey, Michigan and Westchester County, New York—have only increased the inefficiency of motor cars in use. The losses from congestion, both in the crowded and hopelessly entangled metropolises, and along the roads by means of which people attempt to escape the cities on holidays, are incalculably large in countries which, like the United States and England, have taken over the motor car most heedlessly and complacently.

This weakness in the development of neotechnic transportation has come out during the last generation in still another relationship: the geographic distribution of the population. Both the motor car and the airplane have a special advantage over the ordinary steam locomotives: the second can fly over areas that are impassable to any other mode of transportation, and the first can take easily grades which are prohibitive to the ordinary steam locomotive. By means of the motor car the upland areas, where electric power can be cheaply produced, and where the railroad enters at a considerable disadvantage can be thrown open to commerce, industry, and population. These uplands are likewise often the most salubrious seat of living, with their fine scenery, their bracing ionized air, their range of recreation, from mountain-climbing and fishing to swimming and ice-skating. Here is, I must emphasize, the special habitat of the neotechnic civilization, as the low coastal areas were for the eotechnic phase, and the valley bottoms and coal beds were for the paleotechnic period. Population nevertheless, instead of being released into these new centers of living, has continued in many countries to flow into the metropolitan centers of industry and finance: the motor car served to facilitate this congestion instead of dispelling it. In addition, because of the very spread of the overgrown centers the flying fields could be placed only at the extreme outskirts of the bigger cities, on such remaining land as had not been built upon or chopped into suburban subdivisions: so that the saving in time through the swift-ness and short-cuts of airplane travel is often counter-balanced, on short flights, by the length of time it takes to reach the center of the big city from the flying fields on the outskirts.

7: The Paradox of Communication

Communication between human beings begins with the immediate physiological expressions of personal contact, from the howlings and cooings and head-turnings of the infant to the more abstract gestures and signs and sounds out of which language, in its fulness, develops. With hieroglyphics, painting, drawing, the written alphabet, there grew up during the historic period a series of abstract forms of expression which deepened and made more reflective and pregnant the intercourse of men. The lapse of time between expression and reception had something of the effect that the arrest of action pro-duced in making thought itself possible.

With the invention of the telegraph a series of inventions began to bridge the gap in time between communication and response de-spite the handicaps of space: first the telegraph, then the telephone, then the wireless telegraph, then the wireless telephone, and finally television. As a result, communication is now on the point of return-ing, with the aid of mechanical devices, to that instantaneous reaction of person to person with which it began; but the possibilities of this immediate meeting, instead of being limited by space and time, will be limited only by the amount of energy available and the mechanical perfection and accessibility of the apparatus. When the radio tele-phone is supplemented by television communication will differ from direct intercourse only to the extent that immediate physical contact will be impossible: the hand of sympathy will not actually grasp the recipient's hand, nor the raised fist fall upon the provoking head.

What will be the outcome? Obviously, a widened range of intercourse: more numerous contacts: more numerous demands on attention and time. But unfortunately, the possibility of this type of immediate intercourse on a worldwide basis does not necessarily mean a less trivial or a less parochial personality. For over against the convenience of instantaneous communication is the fact that the great economical abstractions of writing, reading, and drawing, the media of reflective thought and deliberate action, will be weakened. Men often tend to be more socialized at a distance, than they are in their immediate, limited, and local selves: their intercourse sometimes proceeds best, like barter among savage peoples, when neither group is visible to the other. That the breadth and too-frequent repetition of personal intercourse may be socially inefficient is already plain through the abuse of the telephone: a dozen five minute conversations can frequently be reduced in essentials to a dozen notes whose reading, writing, and answering takes less time and effort and nervous energy than the more personal calls. With the telephone the flow of interest and attention, instead of being self-directed, is at the mercy of any strange person who seeks to divert it to his own purposes.

One is faced here with a magnified form of a danger common to all inventions: a tendency to use them whether or not the occasion demands. Thus our forefathers used iron sheets for the fronts of buildings, despite the fact that iron is a notorious conductor of heat: thus people gave up learning the violin, the guitar, and the piano when the phonograph was introduced, despite the fact that the passive listening to records is not in the slightest degree the equivalent of active performance; thus the introduction of anesthetics increased fatalities from superfluous operations. The lifting of restrictions upon close human intercourse has been, in its first stages, as dangerous as the flow of populations into new lands: it has increased the areas of friction. Similarly, it has mobilized and hastened mass-reactions, like those which occur on the eve of a war, and it has increased the dangers of international conflict. To ignore these facts would be to paint a very falsely over-optimistic picture of the present economy.

Nevertheless, instantaneous personal communication over long distances is one of the outstanding marks of the neotechnic phase: it is the mechanical symbol of those world-wide cooperations of thought and feeling which must emerge, finally, if our whole civiliza-tion is not to sink into ruin. The new avenues of communication have the characteristic features and advantages of the new technics; for they imply, among other things, the use of mechanical apparatus to duplicate and further organic operations: in the long run, they promise not to displace the human being but to re-focus him and enlarge his capacities. But there is a proviso attached to this promise: namely, that the culture of the personality shall parallel in refine-ment the mechanical development of the machine. Perhaps the great-est social effect of radio-communication, so far, has been a political one: the restoration of direct contact between the leader and the group. Plato defined the limits of the size of a city as the number of people who could hear the voice of a single orator: today those limits do not define a city but a civilization. Wherever neotechnic instruments exist and a common language is used there are now the elements of almost as close a political unity as that which once was possible in the tiniest cities of Attica. The possibilities for good and evil here are immense: the secondary personal contact with voice and image may increase the amount of mass regimentation, all the more because the opportunity for individual members reacting directly upon the leader himself, as in a local meeting, becomes farther and farther removed. At the present moment, as with so many other neo-technic benefits, the dangers of the radio and the talking picture seem greater than the benefits. *As with all instruments of multiplication the critical question is as to the function and quality of the object one is multiplying.* There is no satisfactory answer to this on the basis of technics alone: certainly nothing to indicate, as the earlier exponents of instantaneous communication seem pretty uniformly to have thought, that the results will automatically be favorable to the com-munity.

8: The New Permanent Record

Man's culture depends for its transmission in time upon the per-
ment record: the building, the monument, the inscribed word. During
the early neotechnic phase, vast changes were made here, as important as those brought about five hundred years earlier through
the invention of wood-engraving, copper-etching, and printing. The
black-and-white image, the color-image, the sound, and the moving
image were translated into permanent records, which could be
manifolded, by mechanical and chemical means. In the invention
of the camera, the phonograph, and the moving picture the interplay
of science and mechanical dexterity, which has already been stressed,
was again manifested.

While all these new forms of permanent record were first employed chiefly for amusement, and while the interest behind them
was esthetic rather than narrowly utilitarian, they had important uses
in science, and they even reacted upon our conceptual world as
well. The photograph, to begin with, served as an independent objective check upon observation. The value of a scientific experiment
lies partly in the fact that it is repeatable and thus verifiable by independent observers: but in the case of astronomical observations, for
example, the slowness and fallibility of the eye can be supplemented
by the camera, and the photograph gives the effect of repetition to
what was, perhaps, a unique event, never to be observed again. In
the same fashion, the camera gives an almost instantaneous cross-
section of history—arresting images in their flight through time. In
the case of architecture this mechanical copying on paper led to
unfortunately similar artifices in actual buildings, and instead of
enriching the mind left a trail of arrested images in the form of
buildings all over the landscape. For history is non-repeatable, and
the only thing that can be rescued from history is the note that one
takes and preserves at some moment of its evolution. To divorce an
object from its integral time-sequence is to rob it of its complete
meaning, although it makes it possible to grasp spatial relations
which may otherwise defy observation. Indeed, the very value of
the camera as a reproducing device is to present a memorandum,

as it were, of that which cannot in any other fashion be reproduced.

In a world of flux and change, the camera gave a means of combating the ordinary processes of deterioration and decay, not by "restoration" or "reproduction" but by holding in convenient form the lean image of men, places, buildings, landscapes: thus serving as an extension of the collective memory. The moving picture, carrying a succession of images through time, widened the scope of the camera and essentially altered its function; for it could telescope the slow movement of growth, or prolong the fast movement of jumping, and it could keep in steady focus events which could not otherwise be held in consciousness with the same intensity and fixity. Heretofore records had been confined to snatches of time, or, when they sought to move with time itself, they were reduced to abstractions. Now they could become continuous images of the events they represented. So the flow of time ceased to be representable by the successive mechanical ticks of the clock: its equivalent—and Bergson was quick to seize this image—was the motion picture reel.

One may perhaps over-rate the changes in human behavior that followed the invention of these new devices; but one or two suggest themselves. Whereas in the eotechnic phase one conversed with the mirror and produced the biographical portrait and the introspective biography, in the neotechnic phase one poses for the camera, or still more, one acts for the motion picture. The change is from an introspective to a behaviorist psychology, from the fulsome sorrows of Werther to the impassive public mask of an Ernest Hemingway. Facing hunger and death in the midst of a wilderness, a stranded aviator writes in his notes: "I built another raft, and this time took off my clothes to try it. I must have looked good, carrying the big logs on my back in my underwear." Alone, he still thinks of himself as a public character, *being watched:* and to a greater or less degree everyone, from the crone in a remote hamlet to the political dictator on his carefully prepared stage is in the same position. This constant sense of a public world would seem in part, at least, to be the result of the camera and the camera-eye that developed with it. If the eye be absent in reality, one improvises it wryly with a fragment of one's consciousness. The change is significant: not self-examination but self-exposure: not tortured confession but easy open candor: not the proud soul wrapped in his cloak, pacing the lonely beach at midnight, but the matter-of-fact soul, naked, exposed to the sun on the beach at noonday, one of a crowd of naked people. Such reactions are, of course, outside the realm of proof; and even if the influence of the camera were directly demonstrable, there is little reason to think that it is final. Need I stress again that nothing produced by technics is more final than the human needs and interests themselves that have created technics?

Whatever the psychal reactions to the camera and the moving picture and the phonograph may be, there is no doubt, I think, as to their contribution to the economic management of the social heritage. Before they appeared, sound could only be imperfectly represented in the conventions of writing: it is interesting to note that one of the best systems, Bell's Visible Speech, was invented by the father of the man who created the telephone. Other than written and printed documents and paintings on paper, parchment, and canvas, nothing survived of a civilization except its rubbish heaps and its monuments, buildings, sculptures, works of engineering—all bulky, all interfering more or less with the free development of a different life in the same place.

By means of the new devices this vast mass of physical impedimenta could be turned into paper leaves, metallic or rubber discs, or celluloid films which could be far more completely and far more economically preserved. It is no longer necessary to keep vast middens of material in order to have contact, in the mind, with the forms and expressions of the past. These mechanical devices are thus an excellent ally to that other new piece of social apparatus which became common in the nineteenth century: the public museum. They gave modern civilization a direct sense of the past and a more accurate perception of its memorials than any other civilization had, in all probability, had. Not alone did they make the past more immediate: they made the present more historic by narrowing the lapse of time between the actual events themselves and their concrete record. For the first time one might come face to face with the speaking likenesses of dead people and recall in their immediacy forgotten scenes and actions. Faust bartered his soul with Mephistopheles to see Helen of Troy: on much easier terms it will be possible for our descendants to view the Helens of the twentieth century. Thus a new form of immortality was effected; and a late Victorian writer, Samuel Butler, might well speculate upon how completely a man was dead when his words, his image, and his voice were still capable of being resurrected and could have a direct effect upon the spectator and listener.

At first these new recording and reproducing devices have confused the mind and defied selective use: no one can pretend that we have yet employed them, in any sufficient degree, with wisdom or even with ordered efficiency. But they suggest a new relationship between deed and record, between the movement of life and its collective enregistration: above all, they demand a nicer sensitiveness and a higher intelligence. If these inventions have so far made monkeys of us, it is because we are still monkeys.

9: Light and Life

Light shines on every part of the neotechnic world: it filters through solid objects, it penetrates fog, it glances back from the polished surfaces of mirrors and electrodes. And with light, color comes back and the shape of things, once hidden in fog and smoke, becomes sharp as crystal. The glass technics, which had reached its first summit of mechanical perfection in the Venetian mirror, now repeats its triumphs in a hundred different departments: quartz alone is its rival.

In the neotechnic phase the telescope and above all the microscope assume a new importance, for the latter had been left in practical disuse for two centuries, but for the extraordinary work of a Leeuwenhoek and a Spallanzani. To these instruments must be added the spectroscope and the x-ray tube which also utilized light as an instrument of exploration. Clerk-Maxwell's unification of electricity and light is perhaps the outstanding symbol of this new phase. The fine discrimination of color exhibited by Monet and his fellow impressionists, working in the open air and the sunlight was repeated in the laboratory: spectrum analysis and the production of a multitude of aniline dyes derived from coal tar conservation are specifically neotechnic achievements. Now color, hitherto relegated to an unimportant place as a secondary characteristic of matter, becomes an important factor in chemical analysis, with the discovery that each element has its characteristic spectrum. The new dyes, moreover, find their use in the bacteriologist's laboratory for staining specimens: some of them, like gentian violet, have a place as antiseptics, and still others as medicaments in the treatment of certain diseases.

The dark blind world of the machine, the miner's world, began to disappear: heat, light, electricity, and finally matter were all manifestations of energy, and as one pursued the analysis of matter further the old solids became more and more tenuous, until finally they were identified with electric charges: the ultimate building stones of modern physics, as the atom was of the older physical theories. The imperceptible, the ultra-violet and the infra-red series of rays, became commonplace elements in the new physical world at the moment that the dark forces of the unconscious were added to the purely external and rationalized psychology of the human world. Even the unseen was, so to say, illuminated: it was no longer unknown. One might measure and use what one could not see and handle. And while the paleotechnic world had used physical blows and flame to transform matter, the neotechnic was conscious of other forces equally potent under other circumstances: electricity, sound, light, invisible rays and emanations. The mystic's belief in a human aura became as well substantiated by exact science as the alchemist's dream of transmutation was through the Curies' isolation of radium.

The cult of the sun, so dear to Kepler at the beginning of these revolutionary scientific developments, emerged again: the exposure of the naked body to the sun helped, it was found, to prevent rickets and to cure tuberculosis, while direct sunlight sanitated water and reduced the number of pathogenic bacteria in the environment generally. With this new knowledge, founded upon that renewed study of the organism which Pasteur's discoveries promoted, the essentially anti-vital nature of the paleotechnic environment became plain: the darkness and dampness of its typical mines and factories and slum

homes were ideal conditions for breeding bacteria, while its devital-
ized diet resulted in a poor bony structure, defective teeth and weakened resistance to disease. The full effects of these conditions were
amply documented in the examinations for recruits in the British
army toward the end of the century: results which came out with
special clearness because of the predominant urbanization of England. But the Massachusetts mortality tables told the same story: the
farmer's length of life was far greater than the industrial worker's.
Thanks to neotechnic inventions and discoveries the machine became,
for perhaps the first time, a direct ally of life: and in the light of
this new knowledge its previous misdemeanors became more grotesque and incredible.

Mathematical accuracy, physical economy, chemical purity, surgical cleanliness—these are some of the attributes of the new régime.
And mark this: they do not belong to any one department of life.
Mathematical accuracy is necessary in the temperature chart or the
blood count, while cleanliness becomes part of the daily ritual of
neotechnic society with a strictness quite as great as that enforced
by the tabus of the earlier religions like the Jewish or the Moham-
medan. The polished copper of the electric radiator is reflected in
the immaculacy of the operating room: the wide glass windows of
the sanitorium are repeated in the factory, the school, the home.
During the last decade, in the finer communities that have been built
with State aid in Europe the houses themselves are positively helio-
tropic: they are oriented to the sun.

This new technics does not stop short with mechanical inventions:
it begins to call to its aid the biological and psychological sciences,
and the studies of working efficiency and fatigue, for example, establish the fact that to curtail the hours of work may be to increase the
volume of production per unit. The prevention of disease, the substitution of hygiene for belated repair, becomes a characteristic of
neotechnic medicine: a return to Nature, a new confidence in the
organism as a harmonious, self-equilibrating unit. Under the leadership of Osler and his school, the physician relies upon the natural
curative agents: water, diet, sun, air, recreation, massage, change
of scene: in short, upon a balanced and life-enhancing environ-

ment and upon functional readjustment, rather than upon more foreign chemical and mechanical aids without such conditions. Here again the intuition of Hahnemann as to the rôle of minute quantities and the natural therapeutics of his school, anticipated by over a century the new regimen—as Osler himself handsomely acknowledged. The psychological treatment of functional disorders, which made its way into medicine with Freud a generation ago almost completes this new orientation: the social element is alone still largely lacking. As a result of all these advances, one of the major problems for the new technics becomes the removal of the blighted paleotechnic environment, and the re-education of its victims to a more vital regimen of working and living. The dirty crowded houses, the dank airless courts and alleys, the bleak pavements, the sulphurous atmosphere, the over-routinized and dehumanized factory, the drill schools, the second-hand experiences, the starvation of the senses, the remoteness from nature and animal activity—here are the enemies. The living organism demands a life-sustaining environment. So far from seeking to replace this by mechanical substitutes, the neotechnic phase seeks to establish such life-sustaining conditions within the innermost purlieus of technics itself.

The paleotechnic phase was ushered in by a slaughter of the innocents: first in the cradle, and then, if they survived it, in the textile factories and the mines. Child labor remained in the cotton mills in the United States, for example, right down to 1933. As a result of greater care during pregnancy and childbirth, together with a better regimen in infancy, the mortality of children under five years has been enormously decreased—all the more because certain typical children's diseases are, through modern immunology, under better control. This increasing care of life has spread slowly to the occupations of maturity: mark the introduction of safety devices in dangerous industrial operations, such as masks in grinding and spraying, asbestos and mica clothing where the dangers of fire and heat are great, the effort to abolish lead glazes in pottery, to eliminate phosphorous poisoning in the preparation of matches and radium poisoning in the preparation of watch-dials. These negative measures toward health are, of course, but a beginning: the positive fostering of the life-conserving occupations and the discouragement of those forms of industry which decrease the expectation of life without any compensatory intensification of it during production—all this awaits a culture more deeply concerned with life than even the neotechnic one, in which the calculus of energies still takes precedence over the calculus of life.

In surgery likewise neotechnic methods supplement the cruder mechanics of the mid-nineteenth century. There is a large gap between the antiseptic methods of Lister, with his reliance upon that typical coal-tar antiseptic, carbolic acid, and the aseptic technique of modern surgery, first introduced before Lister in operations upon the eye. The use of the x-ray and the tiny electric bulb for exploration, for example, combined with systematic checks provided by the bacteriological laboratory, have increased the possibility of intelligent diagnosis by other means than that offered by the knife.

With prevention rather than cure, and health rather than disease, as the focal points of the new medicine, the psychological side of the mind-body process becomes increasingly the object of scientific investigation. The Descartian notion of a mechanical body presided over by an independent entity called the soul is replaced, as the "matter" of theoretical physics becomes more attenuated, by the notion of the transformation within the organism of mind-states into body-states, and vice-versa. The dualism of the dead mechanical body, belonging to the world of matter, and the vital transcendental soul, belonging to the spiritual realm, disappears before the increasing insight, derived from physiology on one hand and the investigation of neuroses on the other, of a dynamic interpenetration and conversion within the boundaries of organic structures and functions. Now the physical and the psychal become different aspects of the organic process, in much the same way that heat and light are both aspects of energy, differentiated only by the situation to which they refer and by the particular set of receptors upon which they act. This development lays the specialization and isolation of functions, upon which so many mechanical operations are based, open to suspicion. The integral life of the organism is not compatible with extreme isolation of functions: even mechanical efficiency is seriously affected by sexual anxiety and lack of animal health. The fact that simple repetitive operations agree with the psychological constitution of the feeble-minded constitutes a warning as to the limits of sub-divided labor. Mass production under conditions which confirm these limits may exact too high a human price for its cheap products. What is not mechanical enough for a machine to perform may not be human enough for a living man. Efficiency must begin with the utilization of the whole man; and efforts to increase mechanical performance must cease when the balance of the whole man is threatened.

10: The Influence of Biology

In the earlier chapters, we observed that the first step toward mechanism consisted in a counter-movement to life: the substitution of mechanically measured time for duration, of mechanical prime movers for the human body, of drill and regimentation for spontaneous impulses and more cooperative modes of association. During the neotechnic phase this animus was profoundly modified. The investigation of the world of life opened up new possibilities for the machine itself: vital interests, ancient human wishes, influenced the development of new inventions. Flight, telephonic communication, the phonograph, the motion picture all arose out of the more scientific study of living organisms. The studies of the physiologist supplemented those of the physicist.

The belief in mechanical flight grew directly out of the researches of the physiological laboratory. After Leonardo the only scientific study of flight, up to the work of J. B. Pettigrew and E. J. Marey in the eighteen-sixties, was that of the physiologist, Borelli, whose *De Motu Animalium* was published in 1680. Pettigrew, an Edinburgh pathologist, made a detailed study of locomotion in animals, in which he demonstrated that walking, swimming, and flying are in reality only modifications of each other: "the wing," he found, "both when at rest and when in motion, may not inaptly be compared to the blade of an ordinary screw propeller as employed in navigation" . . . while "weight . . . instead of being a barrier to artificial flight, is absolutely necessary to it." From these investigations Pettigrew—and independently Marey—drew the conclusion that human flight was possible.

In this development, flying models, utilizing the new material rubber as motive power, played an important part: Pénaud in Paris, Kress in Vienna, and later Langley in the United States utilized them: but the final touch, necessary for stable flight, came when two bicycle mechanics, Orville and Wilbur Wright, studied the flight of soaring birds, like the gull and the hawk, and discovered the function of warping the tips of the wings to achieve lateral stability. Further improvements in the design of airplanes have been associated, not merely with the mechanical perfection of the wings and the motors, but with the study of the flight of other types of bird, like the duck, and the movement of fish in water.

Similarly, the moving picture was in essence a combination of elements derived from the study of living organisms. The first was the discovery of the basis for the illusion of movement, made by the physiologist Plateau in his investigation of the after-image. Out of this work the succession of paper pictures, passed rapidly before the eye, became a popular child's toy, the phenakistoscope and the zoetrope. The next step was the work of the Frenchman, Marey, in photographing the movements of four-footed animals and of man: a research which was begun in 1870 and finally projected upon a screen in 1889. Meanwhile Edward Muybridge, to decide a bet with Leland Stanford, a horse-lover, undertook to photograph the successive motions of a horse—and later followed this with pictures of an ox, a wild bull, a greyhound, a deer, and birds. In 1887 it occurred to Edison, who was aware of these experiments, to do for the eye what he had already done for the ear, and the invention of the motion picture machine followed, an advance which was in turn dependent upon the invention of the celluloid film in the eighties.

Bell's telephone owes a similar debt to physiology and to human play. Von Kempelen had invented a talking automaton which uttered a few words in 1778. A similar machine, Euphonia, invented by Professor Faber, was exhibited in London; and the elder Bell persuaded Alexander and his brother to make a speaking automaton themselves. Imitating the tongue and the soft parts of the throat with rubber, they made a creditable attempt at a talking machine. Alexander's grandfather had devoted his life to correcting speech defects: his father, A. M. Bell, invented a system of visible speech and was interested in the culture of the voice: he himself was a scientific student of voice production and made great strides in teaching deaf-mutes to talk. Out of this physiological knowledge and these humane interests—aided by Helmholtz's work in physics—grew the telephone: the receiver of which, upon the advice of a Boston surgeon, Dr. C. J. Blake, was directly modeled upon the bones and diaphragm of the human ear.

This interest in living organisms does not stop short with the specific machines that simulate eye or ear. From the organic world came an idea utterly foreign to the paleotechnic mind: the importance of shape.

One can grind a diamond or a piece of quartz to powder: though it has lost its specific crystalline shape, the particles will retain all their chemical properties and most of their physical ones: they will still at least be carbon or silicon dioxide. But the organism that is crushed out of shape is no longer an organism: not merely are its specific properties of growth, renewal, reproduction absent, but the very chemical constitution of its parts undergoes a change. Not even the loosest form of organism, the classic amoeba, can be called a shapeless mass. The technical importance of shape was unappreciated throughout the paleotechnic phase: but for the great mechanical craftsmen, like Maudslay, interest in the esthetic refinement of the machine was non-existent, or, when it came in, it entered as an intrusion, as in the addition of Doric or Gothic ornament, between 1830 and 1860. Except for improvements in specifically eotechnic apparatus, like the clipper sailing ship, shape was looked upon as unimportant. As far back as 1874, for example, the stream-lined locomotive was designed: but the writer in Knight's Dictionary of the Mechanical Arts who described it cited the improvement only to dismiss it. "There is nothing in it," he said with cool contempt. Against possible gains in efficiency by merely altering the shape of a machine, the paleotect put his faith in more power-consumption and greater size.

Only with the development of specifically neotechnic machines, such as the airplane, with the scientific studies of air-resistance that followed close on their heels, did shape begin to play a new rôle in technics. Machines, which had assumed their own characteristic shapes in developing independent of organic forms, were now forced to recognize the superior economy of nature: on actual tests, the blunt heads of many species of fish and the long tapering tail, proved, against naïve intuition, to be the most economic shape of moving through air or water; while, in gliding motion over land, the form of the turtle, developed for walking over a muddy bottom, proved suggestive to the designer. The utilization of aerodynamic curves in the design of the body of the airplane—to say nothing of the wings—increases the lifting power without the addition of a single horsepower: the same principle applied to locomotives and motor cars, eliminating all points of air resistance, lowers the amount of power needed and increases the speed. Indeed, with the knowledge drawn from living forms via the airplane the railroad can now compete once more on even terms with its successor.

In short, the integral esthetic organization of the machine becomes, with the neotechnic economy, the final step in ensuring its efficiency. While the esthetics of the machine is more independent of subjective factors than the esthetics of a painting, there is a point in the background at which they both nevertheless meet: for our emotional responses and our standards of efficiency and beauty are derivable largely in both cases from our reactions to the world of life, where correct adaptations of form have so frequently survived. The eye for form, color, fitness, which the cattle-breeder and horticulturist hitherto had shared with the artist, now made its way into the machine shop and the laboratory: one might judge a machine by some of the criteria one applied to a bull, a bird, an apple. In dentistry the appreciation of the essential physiological function of natural tooth-forms altered the entire technique of tooth-restoration: the crude mechanics and cruder esthetics of an earlier day fell into disrepute. This new interest in form was a direct challenge to the blind ideology of the earlier period. One might reverse Emerson's dictum and say, in the light of the new technology, that the necessary can never divorce itself from the superstructure of the beautiful. I shall return to this fact again when I discuss the assimilation of the machine.

One more phenomenon must be noted, which binds together the machine and the world of life in the neotechnic phase: namely, the respect for minute quantities, unnoticed or invisible before, sometimes below the threshold of consciousness: the part played by the precious alloys in metallurgy, by tiny quantities of energy in radio reception, by the hormones in the body, by the vitamines in the diet, by ultra-violet rays in growth, by the bacteria and filtrable viruses in disease. Not merely is importance in the neotechnic phase no longer symbolized by bulk, but the attention to small quantities leads by habituation to higher standards of refinement in every department of activity. Langley's bolometer can distinguish one one-millionth of a degree centigrade, against the one one-thousandth possible on a mercury thermometer: the Tuckerman strain gauge can read millionths of an inch—the deflection of a brick when bent by the hand—while Bose's high magnification crescograph records the rate of growth as slow as one one-hundred-thousandth of an inch per second. Subtlety, finesse, delicacy, respect for organic complexity and intricacy now characterize the entire range of scientific thought: this has grown in part out of refinements in technical methods, and in turn it has furthered them. The change is recorded in every part of man's experience: from the increased weight placed by psychology upon hitherto unnoticed traumas to the replacement of the pure calory diet, based upon the energy content alone, by the balanced diet which includes even the infinitesimal amounts of iodine and copper that are needed for health. In a word, the quantitative and the mechanical have at last become life-sensitive.

We are still, I must emphasize, probably only at the beginning of this reverse process, whereby technics, instead of benefiting by its abstraction from life, will benefit even more greatly by its integration with it. Already important developments are on the horizon. Two instances must suffice. In 1919 Harvey studied the production of heat during the luminescence of the appropriate substance derived from the crustacean, Cyrpoidina hilgendorfi. He found that the rise of temperature during the luminescent reaction is less than 0.001

degree centigrade, and probably less than 0.0005 degrees. The chemical constituents from which this cold light is made are now known: luciferin and luciferase; and the possibility of synthesizing them and manufacturing them, now theoretically within our grasp, would increase the efficiency of lighting far above anything now possible in the utilization of electricity. The organic production of electricity in certain fishes may likewise furnish a clue to the invention of economic high-powered electric cells—in which case the electric motor, which neither devitalizes nor defiles nor overheats the air would have a new part to play, probably, in all forms of locomotion. Developments like these, which are plainly imminent, point to improvements in technics which will make our present crude utilization of horsepower seem even more wasteful than the practices of paleotechnic engineering do to the designer of a modern power station.

11: From Destruction to Conservation

The paleotechnic period, we have noted, was marked by the reckless waste of resources. Hot in the pursuit of immediate profits, the new exploiters gave no heed to the environment around them, nor to the further consequences of their actions on the morrow. "What had posterity done for them?" In their haste, they over-reached themselves: they threw money into the rivers, let it escape in smoke in the air, handicapped themselves with their own litter and filth, prematurely exhausted the agricultural lands upon which they depended for food and fabrics.

Against all these wastes the neotechnic phase, with its richer chemical and biological knowledge, sets its face. It tends to replace the reckless mining habits of the earlier period with a thrifty and conservative use of the natural environment. Concretely, the conservation and utilization of scrap-metals and scrap-rubber and slag mean a tidying up of the landscape: the end of the paleotechnic middens. Electricity itself aids in this transformation. The smoke pall of paleotechnic industry begins to lift: with electricity the clear sky and the clean waters of the eotechnic phase come back again: the water that runs through the immaculate disks of the turbine, unlike the water filled with the washings of the coal seams or the refuse of the old chemical factories, is just as pure when it emerges. Hydro-electricity, moreover, gives rise to geotechnics: forest cover protection, stream control, the building of reservoirs and power dams.

As early as 1866 George Perkins Marsh, in his classic book on Man and Nature, pointed out the grave dangers of forest destruction and the soil erosion that followed it: here was waste in its primary form—the waste of the precious skin of arable, humus-filled soil with which the more favored regions of the world are covered, a skin that is unreplaceable without centuries of waiting except by transporting new tissue from some other favored region. The skinning of the wheat lands and the cotton lands in order to provide cheap bread and textiles to the manufacturing classes was literally cutting the ground from under their feet. So strongly entrenched were these methods that even in America, no effective steps were taken to combat this wastage until a generation after Marsh's books; indeed, with the invention of the wood-pulp process for making paper, the spoliation of the forest went on more rapidly. Timber-mining and soil-mining proceeded hand in hand.

But during the nineteenth century a series of disastrous experiences began to call attention to the fact that nature could not be ruthlessly invaded and the wild life indiscriminately exterminated by man without bringing upon his head worse evils than he was eliminating. The ecological investigations of Darwin and the later biologists established the concept of the web of life, of that complex interplay of geological formation, climate, soil, plants, animals, protozoa, and bacteria which maintains a harmonious adjustment of species to habitat. To cut down a forest, or to introduce a new species of tree or insect, might be to set in motion a whole chain of remote consequences. In order to maintain the ecological balance of a region, one could no longer exploit and exterminate as recklessly as had been the wont of the pioneer colonist. The region, in short, had some of the characteristics of an individual organism: like the organism, it had various methods of meeting maladjustment and maintaining its balance: but to turn it into a specialized machine for producing a single kind of goods—wheat, trees, coal—and to forget its many-sided potentialities as a habitat for organic life was finally to unsettle and make precarious the single economic function that seemed so important.

With respect to the soil itself, the neotechnic phase produced important conservative changes. One of them was the utilization once more of human excrement for fertilizers, in contrast with the reckless method of befouling stream and tidal water and dissipating the precious nitrogenous compounds. The sewage utilization plants of neotechnic practice, most extensively and systematically introduced perhaps in Germany, not merely avoid the misuse of the environment, but actually enrich it and help bring it to a higher state of cultivation. The presence of such plants is one of the distinguishing characteristics of a neotechnic environment. The second important advance was in the fixation of nitrogen. At the end of the nineteenth century the existence of agriculture seemed threatened by the approaching exhaustion of the Chile nitrate beds. Shortly after this various processes for fixing nitrogen were discovered: the arc process (1903) required cheap electric power: but the synthetic ammonia process, introduced by Haber in 1910, gave a new use to the coke oven. But equally typical of the new technology was the discovering of the nitrogen-forming bacteria at the root-nodules of certain plants like pea and clover and soy bean: some of these plants had been used by the Romans and Chinese for soil regeneration: but now their specific function in restoring nitrogen was definitely established. With this discovery one of the paleotechnic nightmares—that of imminent soil-exhaustion—disappeared. These alternative processes typify another neotechnic fact: namely, that the technical solution it offers for its problems is not confined necessarily to a physical or mechanical means: electro-physics offers one solution, chemistry another, bacteriology and plant physiology still a third.

Plainly, the fixation of nitrogen was a far greater contribution to the efficiency of agriculture than any of the excellent devices that speeded up the processes of ploughing, harrowing, sowing, cultivating, or harvesting. Knowledge of this sort—like the knowledge of the desirable shapes for moving bodies—is characteristic of the neotechnic phase. While on one side neotechnic advances perfect the automatic machine and extend its operations, on the other, they do away with the complications of machinery in provinces where they are not needed. A field of soy beans may, for certain purposes, take the place of a transcontinental railroad, a dock in San Francisco, a port, a railroad, and a mine in Chile, to say nothing of all the labor involved in bringing these machines and pieces of apparatus together. This generalization holds true for other realms than agriculture. One of the first great improvements introduced by Frederick Taylor under the head of scientific management involved only a change in the motion and routine of unskilled laborers carrying ingots. Similarly, a better routine of living and a more adequately planned environment eliminates the need for sun-lamps, mechanical exercisers, constipation remedies, while a knowledge of diet has done away except as a desperate last resort with once fashionable—and highly dangerous—operations upon the stomach.

Whereas the growth and multiplication of machines was a definite characteristic of the paleotechnic period, one may already say pretty confidently that the refinement, the diminution, and the partial elimination of the machine is a characteristic of the emerging neotechnic economy. The shrinkage of the machine to the provinces where its services are unique and indispensable is a necessary consequence of our better understanding of the machine itself and the world in which it functions.

The conservation of the environment has still another neotechnic aspect: that is the building up in agriculture of an appropriate artificial environment. Up to the seventeenth century man's most important artifact was probably the city itself: but during this century the same tactics he had used for his own domestication were applied to agriculture in the building of glass hothouses, and during the nineteenth century, with the increase of glass production and the expanding empirical knowledge of the soils, glass culture became important in the supply of fruits and vegetables. No longer content with taking Nature as it comes, the neotechnic agriculturist seeks to determine the exact conditions of soil, temperature, moisture, insolation that are needed for the specific crop he would grow. Within his cold frames and his hothouses he brings these conditions into existence.

This deliberate and systematic agriculture is seen at its best today, perhaps, in Holland and Belgium, and in dairy farming as carried on in Denmark and Wisconsin. Parallel then with the spread of modern industry throughout the world there is a similar equalization in agriculture. Aided by the cheap production of glass and metal frames, to say nothing of synthetic substitutes for glass which will permit the ultra-violet rays to pass through, there is the prospect of turning part of agriculture into an all-year occupation, thus diminishing the amount of transportation necessary for fresh fruits and vegetables, and even cultivating, under possibly more humane conditions, the tropical fruits and vegetables. In this new phase, the amount of soil available is not nearly of such critical importance as its quality and its manner of use.

The closer inter-planning of rural and urban occupations necessarily follows from the partial industrialization of agriculture. Even without the use of hothouses the widespread distribution of population through the open country is a consequence of neotechnic industry that is actually in the process of realization: this brings with it the possibility of adjusting industrial production to seasonal changes of work enforced by nature in agriculture. And as agriculture becomes more industrialized, not merely will the extreme rustic and the extreme cockney human types tend to diminish, but the rhythms of the two occupations will approach each other and modify each other: if agriculture, freed from the uncertainty of the weather and of insect pests, will become more regular, the organic timing of life processes may modify the beat of industrial organization: a spring rush in mechanical industry, when the fields are beckoning, may be treated not merely as a mark of inefficient planning but as an essential sacrilege. The human gain from this marriage of town and country, of industry and agriculture, was constantly present in the best minds of the nineteenth century, although the state itself seemed an astronomical distance away from them: on this policy the communist Marx, the social tory, Ruskin, and the anarchist Kropotkin were one. It is now one of the obvious objectives of a rationally planned economy.

12: The Planning of Population

Central to the orderly use of resources, the systematic integration of industry, and the planning and development of human regions, is perhaps the most important of all neotechnic innovations: the planning of the growth and distribution of the population.

While births have been controlled from the earliest times by one empirical device or another, from asceticism to abortion, from coitus interruptus to the Athenian method of exposing the newborn infant, the first great improvement in Western Europe came by the sixteenth century via the Arabs. Fallopius, the discoverer of the Fallopian tubes, described the use of both the pessary and the sheath. Like the gardens and palaces of the period, the discovery remained apparently the property of the upper classes in France and Italy: it was only in the early nineteenth century that Francis Place and his disciples attempted to spread the knowledge among the harassed cotton operatives of England. But the rational practice of contraception and the improvement in contraceptive devices awaited not merely the discovery of the exact nature of the germ cell and the process of fertilization: it also awaited improvements in the technological means. Effective general contraception, in other words, post-dates Goodyear and Lister. The first large fall of the English birth rate took place in the decade 1870-1880, the decade we have already marked as that which saw the perfection of the gas engine, the dynamo, the telephone, and the electric filament lamp.

The tabus on sex were so long operative in Christian society that its scientific investigation was delayed long beyond any other function of the body: there are even today textbooks on physiology that skip over the sexual functions with the most hasty allusions: hence a subject of critical importance to the care and nurture of the race is still not altogether out of the hands of empirics and superstition-ridden people, to say nothing of quacks. But the technique of temporary sterilization—so-called birth-control—was perhaps the most important to the human race of all the scientific and technical advances that were carried to completion during the nineteenth century. It was the neotechnic answer to that vast, irresponsible spawning of Western mankind that took place during the paleotechnic phase, partly in response perhaps to the introduction of new staple foods and the extension of new food areas, stimulated and abetted by the fact that copulation was the one art and the one form of recreation which could not be denied to the factory population, however it or they might be brutalized.

The effects of contraception were manifold. As far as the personal life went, it tended to bring about a divorce between the preliminary sexual functions and the parental ones, since sexual intercourse, prudently conducted, no longer brought with it the imminent likelihood of offspring. This tended to prolong the period of romantic love among the newly married: it gave an opportunity for sexual courtship and accomplishment to develop, instead of being reduced and quickly eliminated by early and repeated pregnancies. Contraception likewise naturally gave the opportunity for the exercise of sexual relations before accepting the legal responsibilities of marriage and parenthood: this resulted in a devaluation of mere virginity, while it permitted the erotic life to follow a natural sequence in growth and efflorescence without respect to economic or professional expediency. It therefore lessened to some extent the dangers of arrested sexual and emotional development, with the strains and anxieties that so often attend this arrest, by giving opportunities for sexual intercourse without complete social irresponsibility. Moreover, by permitting intimate sexual knowledge before marriage, it offered a means for avoiding a more or less permanent relationship in the case of two people to whose happy union there might be grave physiological or temperamental obstacles. While contraception, by doing away with the element of finality, perhaps lowered the weight of tragic choices, it tended to stabilize the institution of marriage, by the very fact that it dissociated the social and affectional relation of parenthood from the more capricious incidence of sexual passion.

But important as contraception was to be in sexual life, particularly in the fact that it restored sex with compensatory vigor to a more central rôle in the personality, its wider social effects were equally important.

Whatever the limits of population growth on the planet may be, no one doubts that there are limits. The area of the planet itself is one limit, and the amount of arable soil and fishable water is another. In crowded countries like China and India, the population has in fact pressed close upon the food supply, and security has alternated with famine, despite the immense superiority of Chinese agriculture over most European and American agriculture in the yield it obtains per acre. With the rising pressure of population in European countries from the end of the eighteenth century onward, and with the rate of increase offsetting wars, a high death rate from diseases, and emigration, there was a tidal movement of peoples from the Eastern Hemisphere to the Western, from Russia into Siberia, and from China and Japan into Manchuria. Each sparsely populated area served as a meteorological center of low pressure to attract the cyclonic movement of peoples from areas of high pressure. Had all the population of all countries continued automatically to rise, this movement must in the end have resulted in frantic conflicts—such as that which began in 1932 between China and Japan—with death through starvation and plague as the only alternative to drastic agricultural improvement. Under the stress of blind competition and equally blind fecundity there could be no end to these movements and these mass wars.

With the widespread practice of birth control, however, a vital equilibrium was approached at an early date by France, and is now on the point of being achieved in England and in the United States. This equilibrium reduces the number of variables that must be taken account of in planning, and the size of the population in any area can now theoretically be related to the permanent resources for supporting life that it provides; whilst the waste and wear and dissipation of an uncontrolled birth-rate and a high death-rate is overcome by the lowering of both sides of the ratio at the same time. As yet, birth control has come too tardily into practice to have begun to exercise any measurable control over the affairs of the planet as a whole. Forces which were set in motion in the past may for two or three generations stand in the way of the rational ordering of births, except in the most civilized countries; and the rational re-distribution of the population of the earth into the most desirable habitats awaits the general ebbing of the human tide from the point to which it was whipped up in the nineteenth century.

But the technical means of this change are now for the first time at hand. So strongly do personal and social interests coincide here that it is doubtful if the tabus of religion can withstand them. The very attempts that Catholic physicians have made to discover "safe" periods when conception is unlikely is an earnest of the demand to find a measure which will escape the Church's somewhat capricious ban on artificial methods. Even the religion of nationalism, though stimulated by sadistic exploits, paranoiac delusions of grandeur, and maniacal desires to impose the national will upon other populations —even this religion is not immune to the technological achievement of birth-control, so long as it retains the major elements of modern technology.

Here, then, is another instance of that change from quantitative to qualitative standards that marks the transition from the paleo-technic economy. The first period was marked by an orgy of uncontrolled production and equally uncontrolled reproduction: machine-fodder and cannon-fodder: surplus values and surplus populations. In the neotechnic phase the whole emphasis begins to change: not more births but better births, with greater prospects of survival, with better opportunities for healthy living and healthy parenthood, untainted by ill-health, preventable diseases, and poverty, not spoiled by industrial competitions and national wars. These are the new demands. What rational mind questions their legitimacy? What humane mind would retard their operation?

13: The Present Pseudomorph

So far, in treating the neotechnic phase, I have concerned myself more with description and actuality rather than with prophecy and potentiality. But he who says A in neotechnics has already said B, and it is with the social implications and consequences of the neotechnic economy, rather than with its typical technical instruments, that I purpose to devote the two final chapters of this book.

There is, however, another difficulty in dealing with this phase: namely, we are still in the midst of the transition. The scientific knowledge, the machines and the utilities, the technological methods, the habits of life and the human ends that belong to this economy are far from being dominant in our present civilization. The fact is that in the great industrial areas of Western Europe and America and in the exploitable territories that are under the control of these centers, the paleotechnic phase is still intact and all its essential characteristics are uppermost, even though many of the machines it uses are neotechnic ones or have been made over—as in the electrification of railroad systems—by neotechnic methods. In this persistence of paleotechnic practices the original anti-vital bias of the machine is evident: bellicose, money-centred, life-curbing, we continue to worship the twin deities, Mammon and Moloch, to say nothing of more abysmally savage tribal gods.

Even in the midst of the worldwide economic collapse that began in 1929, the value of what has collapsed was not at first questioned, though the more faint-hearted advocates of the old order have no hope now of reconstituting it. And in the one country, Soviet Russia, that has magnificently attempted to demolish pecuniary standards and interests, even in Soviet Russia, the elements of the neotechnic phase are not clear. For despite Lenin's authentic intuition that "electrification plus socialism equals communism" the worship of size and crude mechanical power, and the introduction of a militarist technique in both government and industry go hand in hand with sane neotechnic achievements in hygiene and education. On one hand the scientific planning of industry: on the other, the mechanistically conceived bonanza farming, in the fashion of America in the seventies: here the great centers of electric power, with a potential decentralization into garden-cities: there the introduction of heavy industries into the already congested and obsolete metropolis of Moscow, and the further waste of energy in the building of costly subways to intensify that congestion. On different lines from non-communist countries, one nevertheless observes in Soviet Russia some of the same confusion and cross-purposes, some of the same baneful survivals, that prevail elsewhere. What is responsible for this miscarriage of the machine?

The answer involves something more complex than a cultural lag or retardation. It is best explained, I think, by a concept put forward by Oswald Spengler in the second volume of the Decline of the West: the concept of the cultural pseudomorph. Spengler points to the common fact in geology that a rock may retain its structure after certain elements have been leached out of it and been replaced by an entirely different kind of material. Since the apparent structure of the old rock remains, the new product is termed a pseudomorph. A similar metamorphosis is possible in culture: new forces, activities, institutions, instead of crystallizing independently into their own appropriate forms, may creep into the structure of an existing civilization. This perhaps is the essential fact of our present situation. As a civilization, we have not yet entered the neotechnic phase; and should a future historian use the present terminology, he would undoubtedly have to characterize the current transition as a mesotechnic period: we are still living, in Matthew Arnold's words, between two worlds, one dead, the other powerless to be born.

For what has been the total result of all these great scientific discoveries and inventions, these more organic interests, these refinements and delicacies of technique? We have merely used our new machines and energies to further processes which were begun under the auspices of capitalist and military enterprise: we have not yet utilized them to conquer these forms of enterprise and subdue them to more vital and humane purposes. The examples of pseudomorphic forms can be drawn from every department. In city growth, for instance, we have utilized electric and gasoline transportation to increase the congestion which was the original result of the capitalistic concentrations of coal and steam power: the new means have been used to extend the area and population of these obsolete and inefficient and humanly defective metropolitan centers. Similarly the steel frame construction in architecture, which permits the fullest use of glass and the most complete utilization of sunlight, has been used in America to increase the overcrowding of buildings and the obliteration of sunlight. The psychological study of human behavior is used to condition people to accept the goods offered by the canny advertisers, despite the fact that science, as applied in the National Bureau of Standards at Washington, gives measurable and rateable levels of performance for commodities whose worth is now putatively established by purely subjective methods. The planning and co-ordination of productive enterprise, in the hands of private bankers rather than public servants, becomes a method of preserving monopoly control for privileged financial groups or privileged countries. Labor saving devices, instead of spreading the total amount of leisure, become means of keeping at a depauperate level an increasing part of the population. The airplane, instead of merely increasing the amount of travel and intercourse between countries, has increased their fear of each other: as an instrument of war, in combination with the latest chemical achievements in poison gas, it promises a ruthlessness of extermination that man has heretofore not been able to apply to either bugs or rats. The neotechnic refinement of the machine, without a coordinate development of higher social purposes, has only magnified the possibilities of depravity and barbarism.

Not alone have the older forms of technics served to constrain the development of the neotechnic economy: but the new inventions and devices have been frequently used to maintain, renew, and stabilize the structure of the old order. There is a political and financial vested interest in obsolete technical equipment: that underlying conflict between business interests and industrial interests, which Veblen analyzed with great acuteness in The Theory of Business Enterprise, is accentuated by the fact that vast amounts of capital are sunk in antiquated machines and burdensome utilities. Financial acquisitiveness which had originally speeded invention now furthers technical inertia. Hence the tardiness in introducing the automatic telephone: hence the continued design of automobiles in terms of superficial fashions, rather than with any readiness to take advantage of aerodynamic principles in building for comfort and speed and economy: hence the continued purchase of patent rights for improvements which are then quietly extirpated by the monopoly holding them.

And this reluctance, this resistance, this inertia have good reason: the old has every cause to fear the superiority of the new. The planned and integrated industry of neotechnic design promises so much greater efficiency than the old that not a single institution appropriate to an economy of parsimony will remain unaltered in an economy of surplus: particularly the institutions limiting ownership and dividends to a small fragment of the population, who thus absorb the purchasing power by excessive re-investment in industrial enterprise and add to its over-expansion. These institutions, indeed, are incompatible with a planned production and distribution of the necessaries of life, for financial values and real goods cannot be equated to the advantage of the whole community on terms that will benefit chiefly the private capitalists by and for whom the original structure of capitalism was created.

One need not wonder that those who affect to control the destinies of industrial society, the bankers, the business men, and the politicians, have steadily put the brakes upon the transition and have sought to limit the neotechnic developments and avoid the drastic changes that must be effected throughout the entire social milieu. The present pseudomorph is, socially and technically, third-rate. It has only a fraction of the efficiency that the neotechnic civilization as a whole may possess, provided it finally produces its own institutional forms and controls and directions and patterns. At present, instead of finding these forms, we have applied our skill and invention in such a manner as to give a fresh lease of life to many of the obsolete capitalist and militarist institutions of the older period. Paleotechnic purposes with neotechnic means: that is the most obvious characteristic of the present order. And that is why a good part of the machines and institutions that boast of being "new" or "advanced" or "progressive" are often so only in the way that a modern battleship is new and advanced: they may in fact be reactionary, and they may stand in the way of the fresh integration of work and art and life that we must seek and create.

CHAPTER VI.   COMPENSATIONS AND
                                      REVERSIONS

1: Summary of Social Reactions

Each of the three phases of machine civilization has left its deposits in society. Each has changed the landscape, altered the physical layout of cities, used certain resources and spurned others, favored certain types of commodity and certain paths of activity, and modified the common technical heritage. It is the sum total of these phases, confused, jumbled, contradictory, cancelling out as well as adding to their forces that constitutes our present mechanical civilization. Some aspects of this civilization are in complete decay; some are alive but neglected in thought; still others are at the earliest stages of development. To call this complicated inheritance the Power Age or the Machine Age is to conceal more facts about it than one reveals. If the machine appears to dominate life today, it is only because society is even more disrupted than it was in the seventeenth century.

But along with the positive transformations of the environment by means of the machine have come the reactions of society against the machine. Despite the long period of cultural preparation, the machine encountered inertia and resistance: in general, the Catholic countries were slower to accept it than were the Protestant countries, and the agricultural regions assimilated it far less completely than the mining districts. Modes of life essentially hostile to the machine have remained in existence: the institutional life of the churches, while often subservient to capitalism, has remained foreign to the naturalistic and mechanistic interests which helped develop the machine. Hence the machine itself has been deflected or metamorphosed to a certain degree by the human reactions which it has set up, or to which, in one manner or another, it has been forced to adapt itself. Many social adjustments have resulted from the machine which were far from the minds of the original philosophers of industrialism. They expected the old social institutions of feudalism to be dissolved by the new order: they did not anticipate that they might be re-crystallized.

It is only in economic textbooks, moreover, that the Economic Man and the Machine Age have ever maintained the purity of their ideal images. Before the paleotechnic period was well under way their images were already tarnished: free competition was curbed from the start by the trade agreements and anti-union collaborations of the very industrialists who shouted most loudly for it. And the retreat from the machine, headed by philosophers and poets and artists, appeared at the very moment that the forces of utilitarianism seemed most coherent and confident. The successes of mechanism only increased the awareness of values not included in a mechanistic ideology—values derived, not from the machine, but from other provinces of life. Any just appreciation of the machine's contribution to civilization must reckon with these resistances and compensations.

2: The Mechanical Routine

Let the reader examine for himself the part played by mechanical routine and mechanical apparatus in his day, from the alarm-clock that wakes him to the radio program that puts him to sleep. Instead of adding to his burden by re-capitulating it, I purpose to summarize the results of his investigations, and analyze the consequences.

The first characteristic of modern machine civilization is its temporal regularity. From the moment of waking, the rhythm of the day is punctuated by the clock. Irrespective of strain or fatigue, despite reluctance or apathy, the household rises close to its set hour. Tardiness in rising is penalized by extra haste in eating breakfast or in walking to catch the train: in the long run, it may even mean the loss of a job or of advancement in business. Breakfast, lunch, dinner, occur at regular hours and are of definitely limited duration: a million people perform these functions within a very narrow band of time, and only minor provisions are made for those who would have food outside this regular schedule. As the scale of industrial organization grows, the punctuality and regularity of the mechanical régime tend to increase with it: the time-clock enters automatically to regulate the entrance and exit of the worker, while an irregular worker—tempted by the trout in spring streams or ducks on salt meadows—finds that these impulses are as unfavorably treated as habitual drunkenness: if he would retain them, he must remain attached to the less routinized provinces of agriculture. "The refractory tempers of work-people accustomed to irregular paroxysms of diligence," of which Ure wrote a century ago with such pious horror, have indeed been tamed.

Under capitalism time-keeping is not merely a means of co-ordinating and inter-relating complicated functions: it is also like money an independent commodity with a value of its own. The school teacher, the lawyer, even the doctor with his schedule of operations conform their functions to a time-table almost as rigorous as that of the locomotive engineer. In the case of child-birth, patience rather than instrumentation is one of the chief requirements for a successful normal delivery and one of the major safeguards against infection in a difficult one. Here the mechanical interference of the obstetrician, eager to resume his rounds, has apparently been largely responsible for the current discreditable record of American physicians, utilizing the most sanitary hospital equipment, in comparison with midwives who do not attempt brusquely to hasten the processes of nature. While regularity in certain physical functions, like eating and eliminating, may in fact assist in maintaining health, in other matters, like play, sexual intercourse, and other forms of recreation the strength of the impulse itself is pulsating rather than evenly recurrent: here habits fostered by the clock or the calendar may lead to dullness and decay.

Hence the existence of a machine civilization, completely timed and scheduled and regulated, does not necessarily guarantee maximum efficiency in any sense. Time-keeping establishes a useful point of reference, and is invaluable for co-ordinating diverse groups and functions which lack any other common frame of activity. In the practice of an individual's vocation such regularity may greatly assist concentration and economize effort. But to make it arbitrarily rule over human functions is to reduce existence itself to mere time-serving and to spread the shades of the prison-house over too large an area of human conduct. The regularity that produces apathy and atrophy—that *acedia* which was the bane of monastic existence, as it is likewise of the army—is as wasteful as the irregularity that produces disorder and confusion. To utilize the accidental, the unpredictable, the fitful is as necessary, even in terms of economy, as to utilize the regular: activities which exclude the operations of chance impulses forfeit some of the advantages of regularity.

In short: mechanical time is not an absolute. And a population trained to keep to a mechanical time routine at whatever sacrifice to health, convenience, and organic felicity may well suffer from the strain of that discipline and find life impossible without the most strenuous compensations. The fact that sexual intercourse in a modern city is limited, for workers in all grades and departments, to the fatigued hours of the day may add to the efficiency of the working life only by a too-heavy sacrifice in personal and organic relations. Not the least of the blessings promised by the shortening of working hours is the opportunity to carry into bodily play the vigor that has hitherto been exhausted in the service of machines.

Next to mechanical regularity, one notes the fact that a good part of the mechanical elements in the day are attempts to counteract the effects of lengthening time and space distances. The refrigeration of eggs, for example, is an effort to space their distribution more uniformly than the hen herself is capable of doing: the pasteurization of milk is an attempt to counteract the effect of the time consumed in completing the chain between the cow and the remote consumer. The accompanying pieces of mechanical apparatus do nothing to improve the product itself: refrigeration merely halts the process of decomposition, while pasteurization actually robs the milk of some of its value as nutriment. Where it is possible to distribute the population closer to the rural centers where milk and butter and green vegetables are grown, the elaborate mechanical apparatus for counteracting time and space distances may to a large degree be diminished.

One might multiply such examples from many departments; they

point to a fact about the machine that has not been generally recognized by those quaint apologists for machine-capitalism who look
upon every extra expenditure of horsepower and every fresh piece
of mechanical apparatus as an automatic net gain in efficiency. In
The Instinct of Workmanship Veblen has indeed wondered whether
the typewriter, the telephone, and the automobile, though creditable
technological achievements "have not wasted more effort and substance than they have saved," whether they are not to be credited
with an appreciable economic loss, because they have increased the
pace and the volume of correspondence and communication and travel
out of all proportion to the real need. And Mr. Bertrand Russell has
noted that each improvement in locomotion has increased the area
over which people are compelled to move: so that a person who
would have had to spend half an hour to walk to work a century
ago must still spend half an hour to reach his destination, because
the contrivance that would have enabled him to save time had he
remained in his original situation now—by driving him to a more
distant residential area—effectually cancels out the gain.

One further effect of our closer time co-ordination and our instantaneous communication must be noted here: broken time and broken
attention. The difficulties of transport and communication before
1850 automatically acted as a selective screen, which permitted
no more stimuli to reach a person than he could handle: a certain
urgency was necessary before one received a call from a long distance or was compelled to make a journey oneself: this condition
of slow physical locomotion kept intercourse down to a human scale,
and under definite control. Nowadays this screen has vanished: the
remote is as close as the near: the ephemeral is as emphatic as the
durable. While the tempo of the day has been quickened by instantaneous communication the rhythm of the day has been broken: the
radio, the telephone, the daily newspaper clamor for attention, and
amid the host of stimuli to which people are subjected, it becomes
more and more difficult to absorb and cope with any one part of
the environment, to say nothing of dealing with it as a whole. The
common man is as subject to these interruptions as the scholar or
the man of affairs, and even the weekly period of cessation from

familiar tasks and contemplative reverie, which was one of the great contributions of Western religion to the discipline of the personal life, has become an ever remoter possibility. These mechanical aids to efficiency and cooperation and intelligence have been mercilessly exploited, through commercial and political pressure: but so far—since unregulated and undisciplined—they have been obstacles to the very ends they affect to further. We have multiplied the mechanical demands without multiplying in any degree our human capacities for registering and reacting intelligently to them. With the successive demands of the outside world so frequent and so imperative, without any respect to their real importance, the inner world becomes progressively meager and formless: instead of active selection there is passive absorption ending in the state happily described by Victor Branford as "addled subjectivity."

3: Purposeless Materialism: Superfluous Power

Growing out of its preoccupation with quantity production is the machine's tendency to center effort exclusively upon the production of material goods. There is a disproportionate emphasis on the physical means of living: people sacrifice time and present enjoyments in order that they acquire a greater abundance of physical means; for there is supposed to be a close relation between well-being and the number of bathtubs, motor cars, and similar machine-made products that one may possess. This tendency, not to satisfy the physical needs of life, but to expand toward an indefinite limit the amount of physical equipment that is applied to living is not exclusively characteristic of the machine, because it has existed as a natural accompaniment of other phases of capitalism in other civilizations. What is typical of the machine is the fact that these ideals, instead of being confined to a class, have been vulgarized and spread—at least as an ideal—in every section of society.

One may define this aspect of the machine as "purposeless materialism." Its particular defect is that it casts a shadow of reproach upon all the non-material interests and occupations of mankind: in particular, it condemns liberal esthetic and intellectual interests because "they serve no useful purpose." One of the blessings of inven-

tion, among the naïve advocates of the machine, is that it does away
with the need for the imagination: instead of holding a conversation
with one's distant friend in reverie, one may pick up a telephone
and substitute his voice for one's fantasy. If stirred by an emotion,
instead of singing a song or writing a poem, one may turn on a
phonograph record. It is no disparagement of either the phonograph
or the telephone to suggest that their special functions do not take
the place of a dynamic imaginative life, nor does an extra bathroom, however admirably instrumental, take the place of a picture
or a flower-garden. The brute fact of the matter is that our civilization is now weighted in favor of the use of mechanical instruments,
because the opportunities for commercial production and for the
exercise of power lie there: while all the direct human reactions or
the personal arts which require a minimum of mechanical parapher-
nalia are treated as negligible. The habit of producing goods whether
they are needed or not, of utilizing inventions whether they are
useful or not, of applying power whether it is effective or not pervades almost every department of our present civilization. The result
is that whole areas of the personality have. been slighted: the telic,
rather than the merely adaptive, spheres of conduct exist on suffer-
ance. This pervasive instrumentalism places a handicap upon vital
reactions which cannot be closely tied to the machine, and it magni-
fies the importance of physical goods as symbols—symbols of intelligence and ability and far-sightedness—even as it tends to characterize
their absence as a sign of stupidity and failure. And to the extent
that this materialism is purposeless, it becomes final: the means are
presently converted into an end. If material goods need any other
justification, they have it in the fact that the effort to consume them
keeps the machines running.

These space-contracting, time-saving, goods-enhancing devices are
likewise manifestations of modern power production: and the same
paradox holds of power and power-machinery: its economies have
been partly cancelled out by increasing the opportunity, indeed the
very necessity, for consumption. The situation was put very neatly
a long time ago by Babbage, the English mathematician. He relates
an experiment performed by a Frenchman, M. Redelet, in which a

block of squared stone was taken as the subject for measuring the effort required to move it. It weighed 1080 pounds. In order to drag the stone, roughly chiseled, along the floor of the quarry, it required a force equal to 758 pounds. The same stone dragged over a floor of planks required 652 pounds; on a platform of wood, drawn over a floor of planks, it required 606 pounds. After soaping the two surfaces of wood which slid over each other it required 182 pounds. The same stone was now placed upon rollers three inches in diameter, when it required to put it in motion along the floor of the quarry only 34 pounds, while to drag it by these rollers over a wooden floor it needed but 22 pounds.

This is a simple illustration of the two ways open in applying power to modern production. One is to increase the expenditure of power; the other is to economize in the application of it. Many of our so-called gains in efficiency have consisted, in effect, of using power-machines to apply 758 pounds to work which could be just as efficiently accomplished by careful planning and preparation with an expenditure of 22 pounds: our illusion of superiority is based on the fact that we have had 736 pounds to waste. This fact explains some of the grotesque miscalculations and misappraisals that have been made in comparing the working efficiency of past ages with the present. Some of our technologists have committed the blunder of confusing the increased load of equipment and the increased expenditure of energy with the quantity of effective work done. But the billions of horsepower available in modern production must be balanced off against losses which are even greater than those for which Stuart Chase has made a tangible estimate in his excellent study of The Tragedy of Waste. While a net gain can probably be shown for modern civilization, it is not nearly so great as we have imagined through our habit of looking only at one side of the balance sheet.

The fact is that an elaborate mechanical organization is often a temporary and expensive substitute for an effective social organization or for a sound biological adaptation. The secret of analyzing motions, of harnessing energies, of designing machines was discovered before we began an orderly analysis of modern society and attempted to control the unconscious drift of technic and economic forces. Just as the ingenious mechanical restorations of teeth begun in the nineteenth century anticipated our advance in physiology and nutrition, which will reduce the need for mechanical repair, so many of our other mechanical triumphs are merely stopgaps, to serve society whilst it learns to direct its social institutions, its biological conditions, and its personal aims more effectively. In other words, much of our mechanical apparatus is useful in the same way that a crutch is useful when a leg is injured. Inferior to the normal functioning leg, the crutch assists its user to walk about whilst bone and tissue are being repaired. The common mistake is that of fancying that a society in which everyone is equipped with crutches is thereby more efficient than one in which the majority of people walk on two legs.

We have with considerable cleverness devised mechanical apparatus to counteract the effect of lengthening time and space distances, to increase the amount of power available for performing unnecessary work, and to increase the waste of time attendant upon irrelevant and superficial intercourse. But our success in doing these things has blinded us to the fact that such devices are not by themselves marks of efficiency or of intelligent social effort. Canning and refrigeration as a means of distributing a limited food supply over the year, or of making it available in areas distant from the place originally grown, represent a real gain. The use of canned goods, on the other hand, in country districts when fresh fruits and vegetables are available comes to a vital and social loss. The very fact that mechanization lends itself to large-scale industrial and financial organization, and marches in step with the whole distributing mechanism of capitalist society frequently gives an advantage to such indirect and ultimately more inefficient methods. There is, however, no virtue whatever in eating foods that are years old or that have been transported thousands of miles, when equally good foods are available without going out of the locality. It is a lack of rational distribution that permits this process to go on in our society. Power machines have given a sort of licence to social inefficiency. This licence was tolerated all the more easily because what the community as a whole lost through these misapplied energies enterprising individuals gained in profits.

The point is that efficiency is currently confused with adaptability to large-scale factory production and marketing: that is to say, with fitness for the present methods of commercial exploitation. But in terms of social life, many of the most extravagant advances of the machine have proved to rest on the invention of intricate means of doing things which can be performed at a minor cost by very simple ones. Those complicated pieces of apparatus, first devised by American cartoonists, and later carried onto the stage by comedians like Mr. Joe Cook, in which a whole series of mechanisms and involved motions are created in order to burst a paper bag or lick a postage stamp are not wild products of the American imagination: they are merely transpositions into the realm of the comic of processes which can be witnessed at a hundred different points in actual life. Elaborate antiseptics are offered in expensive mechanically wrapped packages, made tempting by lithographs and printed advertisements, to take the place which common scientific knowledge indicates is amply filled by one of the most common minerals, sodium chloride. Vacuum pumps driven by electric motors are forced into American households for the purpose of cleaning an obsolete form of floor covering, the carpet or the rug, whose appropriateness for use in interiors, if it did not disappear with the caravans where it originated, certainly passed out of existence with rubber heels and steam-heated houses. To count such pathetic examples of waste to the credit of the machine is like counting the rise in the number of constipation remedies a proof of the benefits of leisure.

The third important characteristic of the machine process and machine environment is uniformity, standardization, replaceability. Whereas handicraft, by the very nature of human work, exhibits constant variations and adaptations, and boasts of the fact that no two products are alike, machine work has just the opposite characteristic: it prides itself on the fact that the millionth motor car built to a specific pattern is exactly like the first. Speaking generally, the machine has replaced an unlimited series of variables with a limited number of constants: if the range of possibility is lessened, the area of prediction and control is increased.

And while the uniformity of performance in human beings, pushed beyond a certain point, deadens initiative and lowers the whole tone of the organism, uniformity of performance in machines and standardization of the product works in the opposite direction. The dangers of standardized products have in fact been over-rated by people who have applied the same criterion to machines as they would to the behavior of living beings. This danger has been further over-stressed by those who look upon uniformity as in itself bad, and upon variation as in itself good: whereas monotony (uniformity) and variety are in reality polar characteristics, neither of which can or should be eliminated in the conduct of life. Standardization and repetition have in fact the part in our social economy that habit has in the human organism: by pushing below the level of consciousness certain recurrent elements in our experience, they free attention for the non-mechanical, the unexpected, the personal. (I shall deal with the social and esthetic importance of this fact when I discuss the assimilation of our machine culture.)

4: Co-operation *versus* Slavery

One of the by-products of the development of mechanical devices and mechanical standards has been the nullification of skill: what has taken place here within the factory has also taken place in the final utilization of its products. The safety razor, for example, has changed the operation of shaving from a hazardous one, best left to a trained barber, to a rapid commonplace of the day which even the most inept males can perform. The automobile has transformed engine-driving from the specialized task of the locomotive engineer to the occupation of millions of amateurs. The camera has in part transformed the artful reproductions of the wood engraver to a relatively simple photo-chemical process in which anyone can acquire at least the rudiments. As in manufacture the human function first becomes specialized, then mechanized, and finally automatic or at least semi-automatic.

When the last stage is reached, the function again takes on some of its original non-specialized character: photography helps recultivate the eye, the telephone the voice, the radio the ear, just as the motor car has restored some of the manual and operative skills that the machine was banishing from other departments of existence at the same time that it has given to the driver a sense of power and autonomous direction—a feeling of firm command in the midst of potentially constant danger—that had been taken away from him in other departments of life by the machine. So, too, mechanization, by lessening the need for domestic service, has increased the amount of personal autonomy and personal participation in the household. In short, mechanization creates new occasions for human effort; and on the whole the effects are more educative than were the semi-automatic services of slaves and menials in the older civilizations. For the mechanical nullification of skill can take place only up to a certain point. It is only when one has completely lost the power of discrimination that a standardized canned soup can, without further preparation, take the place of a home-cooked one, or when one has lost prudence completely that a four-wheel brake can serve instead of a good driver. Inventions like these increase the province and multiply the interests of the amateur. When automatism becomes general and the benefits of mechanization are socialized, men will be back once more in the Edenlike state in which they have existed in regions of natural increment, like the South Seas: the ritual of leisure will replace the ritual of work, and work itself will become a kind of game. That is, in fact, the ideal goal of a completely mechanized and automatized system of power production: the elimination of work: the universal achievement of leisure. In his discussion of slavery Aristotle said that when the shuttle wove by itself and the plectrum played by itself chief workmen would not need helpers nor masters slaves. At the time he wrote, he believed that he was establishing the eternal validity of slavery; but for us today he was in reality justifying the existence of the machine. Work, it is true, is the constant form of man's interaction with his environment, if by work one means the sum total of exertions necessary to maintain life; and lack of work usually means an impairment of function and a breakdown in organic relationship that leads to substitute forms of work, such as invalidism and neurosis. But work in the form of unwilling drudgery or of that sedentary routine which, as Mr. Alfred Zimmern reminds us, the Athenians so properly despised—work in these degrading forms is the true province of machines. Instead of reducing human beings to work-mechanisms, we can now transfer the main part of burden to automatic machines. This potentiality, still so far from effective achievement for mankind at large, is perhaps the largest justification of the mechanical developments of the last thousand years.

From the social standpoint, one final characterization of the machine, perhaps the most important of all, must be noted: the machine imposes the necessity for collective effort and widens its range. To the extent that men have escaped the control of nature they must submit to the control of society. As in a serial operation every part must function smoothly and be geared to the right speed in order to ensure the effective working of the process as a whole, so in society at large there must be a close articulation between all its elements. Individual self-sufficiency is another way of saying technological crudeness: as our technics becomes more refined it becomes impossible to work the machine without large-scale collective cooperation, and in the long run a high technics is possible only on a basis of worldwide trade and intellectual intercourse. The machine has broken down the relative isolation—never complete even in the most primitive societies—of the handicraft period: it has intensified the need for collective effort and collective order. The efforts to achieve collective participation have been fumbling and empirical: so for the most part, people are conscious of the necessity in the form of limitations upon personal freedom and initiative—limitations like the automatic traffic signals of a congested center, or like the red-tape in a large commercial organization. The collective nature of the machine process demands a special enlargement of the imagination and a special education in order to keep the collective demand itself from becoming an act of external regimentation. To the extent that the collective discipline becomes effective and the various groups in society are worked into a nicely interlocking organization, special provisions must be made for isolated and anarchic elements that are not included in such a wide-reaching collectivism—elements that cannot without danger be ignored or repressed. But to abandon the social collectivism imposed by modern technics means to return to nature and be at the mercy of natural forces.

The regularization of time, the increase in mechanical power, the multiplication of goods, the contraction of time and space, the standardization of performance and product, the transfer of skill to automata, and the increase of collective interdependence—these, then, are the chief characteristics of our machine civilization. They are the basis of the particular forms of life and modes of expression that distinguish Western Civilization, at least in degree, from the various earlier civilizations that preceded it.

In the translation of technical improvements into social processes, however, the machine has undergone a perversion: instead of being utilized as an instrument of life, it has tended to become an absolute. Power and social control, once exercised chiefly by military groups who had conquered and seized the land, have gone since the seventeenth century to those who have organized and controlled and owned the machine. The machine has been valued because—it increased the employment of machines. And such employment was the source of profits, power, and wealth to the new ruling classes, benefits which had hitherto gone to traders or to those who monopolized the land. Jungles and tropical islands were invaded during the nineteenth century for the purpose of making new converts to the machine: explorers like Stanley endured incredible tortures and hardships in order to bring the benefits of the machine to inaccessible regions tapt by the Congo: insulated countries like Japan were entered forcibly at the point of the gun in order to make way for the trader: natives in Africa and the Americas were saddled with false debts or malicious taxes in order to give them an incentive to work and to consume in the machine fashion—and thus to supply an outlet for the goods of America and Europe, or to ensure the regular gathering of rubber and lac.

The injunction to use machines was so imperative, from the standpoint of those who owned them and whose means and place in society depended upon them, that it placed upon the worker a special burden, the duty to consume machine-products, while it placed upon the manufacturer and the engineer the duty of inventing products weak enough and shoddy enough—like the safety razor blade or the common run of American woolens—to lend themselves to rapid replacement. The great heresy to the machine was to believe in an institution or a habit of action or a system of ideas that would lessen this service to the machines: for under capitalist direction the aim of mechanism is not to save labor but to eliminate all labor except that which can be channeled at a profit through the factory.

At the beginning, the machine was an attempt to substitute quantity for value in the calculus of life. Between the conception of the machine and its utilization, as Krannhals points out, a necessary psychological and social process was skipped: the stage of evaluation. Thus a steam turbine may contribute thousands of horsepower, and a speedboat may achieve speed: but these facts, which perhaps satisfy the engineer, do not necessarily integrate them in society. Railroads may be quicker than canalboats, and a gas-lamp may be brighter than a candle: but it is only in terms of human purpose and in relation to a human and social scheme of values that speed or brightness have any meaning. If one wishes to absorb the scenery, the slow motion of a canalboat may be preferable to the fast motion of a motor car; and if one wishes to appreciate the mysterious darkness and the strange forms of a natural cave, it is better to penetrate it with uncertain steps, with the aid of a torch or a lantern, than to descend into it by means of an elevator, as in the famous caves of Virginia, and to have the mystery entirely erased by a grand display of electric lights—a commercialized perversion that puts the whole spectacle upon the low dramatic level of a cockney amusement park.

Because the process of social evaluation was largely absent among the people who developed the machine in the eighteenth and nineteenth centuries the machine raced like an engine without a governor, tending to overheat its own bearings and lower its efficiency without any compensatory gain. This left the process of evaluation to groups who remained outside the machine milieu, and who unfortunately often lacked the knowledge and the understanding that would have made their criticisms more pertinent.

The important thing to bear in mind is that the failure to evaluate the machine and to integrate it in society as a whole was not due simply to defects in distributing income, to errors of management, to the greed and narrow-mindedness of the industrial leaders: it was also due to a weakness of the entire philosophy upon which the new techniques and inventions were grounded. The leaders and enterprisers of the period believed that they had avoided the necessity for introducing values, except those which were automatically recorded in profits and prices. They believed that the problem of justly distributing goods could be sidetracked by creating an abundance of them: that the problem of applying one's energies wisely could be cancelled out simply by multiplying them: in short, that most of the difficulties that had hitherto vexed mankind had a mathematical or mechanical—that is a quantitative—solution. The belief that values could be dispensed with constituted the new system of values. Values, divorced from the current processes of life, remained the concern of those who reacted against the machine. Meanwhile, the current processes justified themselves solely in terms of quantity production and cash results. When the machine as a whole overspeeded and purchasing power failed to keep pace with dishonest overcapitalization and exorbitant profits—then the whole machine went suddenly into reverse, stripped its gears, and came to a standstill: a humiliating failure, a dire social loss.

One is confronted, then, by the fact that the machine is ambivalent. It is both an instrument of liberation and one of repression. It has economized human energy and it has misdirected it. It has created a wide framework of order and it has produced muddle and chaos. It has nobly served human purposes and it has distorted and denied them. Before I attempt to discuss in greater detail those aspects of the machine that have been effectively assimilated and that have worked well, I purpose to discuss the resistances and compensations created by the machine. For neither this new type of civilization nor its ideal has gone unchallenged: the human spirit has not bowed to the machine in complete submission. In every phase of existence the machine has stirred up antipathies, dissents, reactions, some weak, hysterical, unjustified, others that are in their nature so inevitable, so sound, that one cannot touch the future of the machine without taking them into account. Similarly the compensations that have arisen to overcome or mitigate the effects of the new routine of life and work call attention to dangers in the partial integration that now exists.

5: Direct Attack on the Machine

The conquest of Western Civilization by the machine was not accomplished without stubborn resistance on the part of institutions and habits and impulses which did not lend themselves to mechanical organization. From the very beginning the machine provoked compensatory or hostile reactions. In the world of ideas, romanticism and utilitarianism go side by side: Shakespeare with his cult of the individual hero and his emphasis of nationalism appeared at the same time as the pragmatic Bacon, and the emotional fervor of Wesley's Methodism spread like fire in dry grass through the very depressed classes that were subject to the new factory régime. The direct reaction of the machine was to make people materialistic and rational: its indirect action was often to make them hyper-emotional and irrational. The tendency to ignore the second set of reactions because they did not logically coincide with the claims of the machine has unfortunately been common in many critics of the new industrial order: even Veblen was not free from it.

Resistance to mechanical improvements took a wide variety of forms. The most direct and simple form was to smash the offending machine itself or to murder its inventor.

The destruction of machines and the prohibition of invention, which so beneficently transformed the society of Butler's Erewhon, might have been accomplished by the working classes of Europe but for two facts. First: the direct war against the machine was an unevenly matched struggle; for the financial and military powers were on the side of the classes that were bent on exploiting the machine, and in a pinch the soldiery, armed with their new machines, could demolish the resistance of the handworkers with a volley of musketry. As long as invention took place sporadically, the introduction of a single machine could well be retarded by direct attack: once it operated on a wide and united front no mere local rebellion could more than temporarily hold up its advance: a successful challenge would have needed a degree of organization which in the very nature of the case the working classes did not have—indeed lack even today.

The second point was equally important: life and energy and adventure were at first on the side of the machine: handicraft was associated with the fixed, the sessile, the superannuated, the dying: it manifestly shrank away from the new movements in thought and from the ordeal of the new reality. The machine meant fresh revelations, new possibilities of action: it brought with it a revolutionary élan. Youth was on its side. Seeking only the persistence of old ways, the enemies of the machine were fighting a rear-guard retreat, and they were on the side of the dead even when they espoused the organic against the mechanical.

As soon as the machine came to predominate in actual life, the only place where it could be successfully attacked or resisted was in the attitudes and interests of those who worked it. The extent to which unmechanical ideologies and programs have flourished since the seventeenth century, despite the persistent habituation of the machine, is in part a measure of the amount of resistance that the machine has, directly or indirectly, occasioned.

6: Romantic and Utilitarian

The broadest general split in ideas occasioned by the machine was that between the Romantic and the Utilitarian. Carried along by the industrial and commercial ideals of his age, the utilitarian was at one with its purposes. He believed in science and inventions, in profits and power, in machinery and progress, in money and comfort, and he believed in spreading these ideals to other societies by means of free trade, and in allowing some of the benefits to filter down from the possessing classes to the exploited—or as they are now euphemistically called, the "underprivileged"—provided that this was done prudently enough to keep the lower classes diligently at work in a state of somnolent and respectful submission.

The newness of the mechanical products was, from the utilitarian standpoint, a guarantee of their worth. The utilitarian wished to put as much distance as possible between his own society of unfettered money-making individuals and the ideals of a feudal and corporate life. These ideals, with their traditions, loyalties, sentiments, constituted a brake upon the introduction of changes and mechanical improvements: the sentiments that clustered around an old house might stand in the way of opening a mine that ran underneath it, even as the affection that often entered into the relation of master and servant under the more patriarchal older régime might stand in the way of that enlightened self-interest which would lead to the dismissal of the worker as soon as the market was slack. What most obviously prevented a clean victory of capitalistic and mechanical ideals was the tissue of ancient institutions and habits of thought: the belief that honor might be more important than money or that friendly affection and comradeship might be as powerful a motive in life as profit making: or that present animal health might be more precious than future material acquisitions—in short, that the whole man might be worth preserving at the expense of the utmost success and power of the Economic Man. Indeed, some of the sharpest criticism of the new mechanical creed came from the tory aristocrats in England, France, and in the Southern States of the United States.

Romanticism in all its manifestations, from Shakespeare to William Morris, from Goethe and the Brothers Grimm to Nietzsche, from Rousseau and Chateaubriand to Hugo, was an attempt to restore the essential activities of human life to a central place in the new scheme, instead of accepting the machine as a center, and holding all its values to be final and absolute.

In its animus, romanticism was right; for it represented those vital and historic and organic attributes that had been deliberately eliminated from the concepts of science and from the methods of the earlier technics, and it provided necessary channels of compensation. Vital organs of life, which have been amputated through historic accident, must be restored at least in fantasy, as preliminary to their actual rebuilding in fact: a psychosis is sometimes the only possible alternative to complete disruption and death. Unfortunately, in its comprehension of the forces that were at work in society the romantic movement was weak: overcome by the callous destruction that attended the introduction of the machine, it did not distinguish between the forces that were hostile to life and those that served it, but tended to lump them all in the same compartment, and to turn its back upon them. In its effort to find remedies for the dire weakness and perversions of industrial society, romanticism avoided the very energies by which alone it could hope to create a more sufficient pattern of existence—namely, the energies that were focussed in science and technics and in the mass of new machine-workers themselves. The romantic movement was retrospective, walled-in, sentimental: in a word, regressive. It lessened the shock of the new order, but it was, for the greater part, a movement of escape.

But to confess this is not to say that the romantic movement was unimportant or unjustified. On the contrary, one cannot comprehend the typical dilemmas of the new civilization unless one understands the reason and the rationale of the romantic reaction against it, and sees how necessary it is to import the positive elements in the romantic attitude into the new social synthesis. Romanticism as an *alternative* to the machine is dead: indeed it never was alive: but the forces and ideas once archaically represented by romanticism are necessary ingredients in the new civilization, and the need today is to translate them into direct social modes of expression, instead of continuing them in the old form of an unconscious or deliberate regression into a past that can be retrieved only in phantasy.

The romantic reaction took many forms: and I shall consider only the three dominant ones: the cult of history and nationalism, the cult of nature, and the cult of the primitive. The same period saw likewise the cult of the isolated individual, and the revival of old theologies and theosophies and supernaturalisms, which owed their existence and much of their strength no doubt to the same denials and emptinesses that prompted the more specially romantic revivals: but it is next to impossible to distinguish clearly between the continued interests of religion and their modern revivals; so I shall confine this analysis to the romantic reaction proper; for this plainly accompanied and probably grew out of the new situation.

7: The Cult of the Past

The cult of the past did not immediately develop in response to the machine; it was, in Italy, an attempt to resume the ideas and forms of classic civilization, and during the Renascence the cult was, in fact, a sort of secret ally to the machine. Did it not, like the machine, challenge the validity of the existing traditions in both philosophy and daily life? Did it not give more authority to the manuscripts of ancient authors, to Hero of Alexandria in physics, to Vitruvius in architecture, to Columella in farming, than it did to the existing body of tradition and the practices of contemporary masters? Did it not, by breaking with the immediate past, encourage the future to break with the present?

The recovery of the classic past during the Renascence caused a break in the historic continuity of Western Europe; and this gap, which opened in education and the formal arts, made a breach of which the machine promptly took advantage. By the eighteenth century the Renascence culture itself was sterilized, pedanticized, formalized: it gave itself over to the recovery and reproduction of dead forms; and though a Poussin or a Piranesi could revitalize these forms with a little of the flair and confidence that the men of the late fifteenth century had felt, the neo-classic and the mechanical played into each other's hands: in the sense of being divorced from life, the first was even more mechanical than mechanism itself. It is not perhaps altogether an accident that at a distance the palaces of Versailles and St. Petersburg have the aspect of modern factory buildings. When the cult of the past revived again, it was directed against both the arid humanism of the eighteenth century and the equally arid dehumanism of the mechanical age. William Blake, with his usual true instinct for fundamental differences, attacked with equal vehemence Sir Joshua Reynolds and Sir Isaac Newton.

In the eighteenth century a cultured man was one who knew his Greek and Latin classics; an enlightened man was one who regarded any part of the globe as suitable for human habitation, provided that its laws were just and their administration impartial; a man of taste was one who knew that standards of proportion and beauty in architecture and sculpture and painting had been fixed forever by classic precedent. The living tissue of customs and traditions, the vernacular architecture, the folkways and the folk-tales, the vulgar languages and dialects that were spoken outside Paris and London— all these things were looked upon by the eighteenth century gentleman as a mass of follies and barbarisms. Enlightenment and progress meant the spreading of London, Paris, Vienna, Berlin, Madrid, and St. Petersburg over wider and wider areas.

Thanks to the dominance of the machine, to books and bayonets, to printed calicos and missionary pocket-handkerchiefs, to brummagem jewelry and cutlery and beads, a layer of this civilization began to spread like a film of oil over the planet at large: machine textiles supplanted hand-woven ones, aniline dyes eventually took the place of vegetable dyes locally made, and even in distant Polynesia calico dresses and stove-pipe hats and shame covered up the proud bodies of the natives, while syphilis and rum, introduced at the same time as the Bible, added a special physical horror to their degradation. Wherever this film of oil spread, the living fish were poisoned and their bloated bodies rose to the surface of the water, adding their own decay to the stench of the oil itself. The new mechanical civilization respected neither place nor past. In the reaction that it provoked place and past were the two aspects of existence that were over-stressed.

This reaction appeared definitely in the eighteenth century, just at the moment that the paleotechnic revolution was getting under way. It began as an attempt to take up the old threads of life at the point where the Renascence had dropped them: it was thus a return to the Middle Ages and a re-reading of their significance, absurdly by Walpole, coldly by Robert Adam, graphically by Scott, faithfully by von Scheffel, esthetically by Goethe and Blake, piously by Pugin and the members of the Oxford movement, moralistically by Carlyle and Ruskin, imaginatively by Victor Hugo. These poets and architects and critics disclosed once more the wealth and interest of the old local life in Europe: they showed how much engineering had lost by deserting gothic forms for the simpler post and lintel construction of classic architecture, and how much literature had forfeited by its extravagant interest in classic forms and themes and its snobbish parade of classic allusions, while the most poignant emotions were embodied in the local ballads that still lingered on in the countryside.

By this "gothic" revival a slight check was placed upon the centralizing, exploitative, and de-regionalizing processes of the machine civilization. Local folk lore and local fairy tales were collected by scholars like the Brothers Grimm and historically minded novelists like Scott; local monuments of archaeology were preserved, and the glorious stained glass and wall paintings of the medieval and early Renascence churches were saved here and there from the glazier and plasterer, still erasing these remnants of "gothic barbarity" in the name of progress and good taste. Local legends were collected: indeed, one of the most remarkable poems of the romantic movement, Tam O'Shanter, was written merely to serve as letterpress for a picture of Alloway's auld haunted kirk. Most potent of all, local languages and dialects were pounced upon, in the very act of dying, and restored to life by turning them to literary uses.

The nationalist movement took advantage of these new cultural interests and attempted to use them for the purpose of fortifying the political power of the unified nationalist state, that mighty engine for preserving the economic *status quo* and for carrying out imperialistic policies of aggression among the weaker races. In this manner, amorphous entities like Germany and Italy became self-conscious and realized a certain degree of political self-sufficiency. But the new interests and revivals struck much deeper than political nationalism, and were more concentrated in their sphere of action: moreover, they touched aspects of life to which a mere power politics was as indifferent as was a power economics. The creation of nationalist states was essentially a movement of protest against alien political powers, wielded without the consent and participation of the governed: a protest against the largely arbitrary political groupings of the dynastic period. But the nations, once they achieved independent nationality, speedily began with the introduction of coal-industrialism to go through the same process of de-regionalization as those that had had no separate national existence; and it was only with the growth of a more intensive and self-conscious regionalism that the process began to work in the opposite direction.

The revival of place interests and language interests, focussed in the new appreciation of regional history, is one of the definite characteristics of nineteenth century culture. Because it was in direct conflict with the cosmopolitan free-trade imperialism of the leading economic thought of the period—and political economy had a hallowed status among the social sciences during this period, because of its useful mythological character—this new regionalism was never carefully appraised or sufficiently appreciated in the early days of its existence. Even now it is still often looked upon as a queer aberration: for plainly it does not fit in altogether with the doctrines of industrial world-conquest or with those of "progress." The movement did not in fact crystallize, despite the valuable preliminary work of the romantics, until the middle of the nineteenth century; and instead of disappearing with the more universal triumph of the machine it went on after that with accelerating speed and intensity. First France: then Denmark: now every part of the world has felt at least a tremor of the countering shock of regionalism, sometimes a definite upheaval.

At the beginning, the main impulse came from the historic regions whose existence was threatened by the mechanical and political unifications of the nineteenth century. The movement had indeed a definite beginning in time, namely 1854; in that year occurred the first meeting of the Félibrigistes, who gathered together for the purpose of restoring the language and the autonomous cultural life of Provence. The Provençal language had all but been destroyed by the Albigensian crusades: Provence had been, so to say, a conquered province of the Church, which had decimated it by a strenuous use of the secular arm; and although an attempt had been made by the Seven Poets of Toulouse, in 1324, to revive the language, the movement had not succeeded: the speech of Ronsard and Racine had finally prevailed. In their consciousness of the part played by language as a means of establishing and helping to build up their identity with their region, a group of literary men, headed by Frédéric Mistral, started to institute the regionalist movement.

This movement has gone through a similar set of stages in every country where it has taken place: in Denmark, in Norway, in Ireland, in Catalonia, in Brittany, in Wales, in Scotland, in Palestine, and similar signs are already visible in various regions in North America. There is, as M. Jourdanne has put it, at first a poetic cycle: this leads to the recovery of the language and literature of the folk, and the attempt to use it as a vehicle for contemporary expression on the basis of largely traditional forms. The second is the cycle of prose, in which the interest in the language leads to an interest in the totality of a community's life and history, and so brings the movement directly onto the contemporary stage. And finally there is the cycle of action, in which regionalism forms for itself fresh objectives, political, economic, civic, cultural, on the basis, not of a servile restoration of the past, but of a growing integration of the new forces that have attached themselves to the main trunk of tradition. The only places where regionalism has not been militantly self-conscious are places like the cities and provinces of Germany in which—until the recent centralization of power by the Totalitarian State—an autonomous and effective local life had never entirely disappeared.

The besetting weakness of regionalism lies in the fact that it is in part a blind reaction against outward circumstances and disruptions, an attempt to find refuge within an old shell against the turbulent invasions of the outside world, armed with its new engines: in short, an aversion from what is, rather than an impulse toward what may be. For the merely sentimental regionalist, the past was an absolute. His impulse was to fix some definite moment in the past, and to keep on living it over and over again, holding the "original" regional costumes, which were in fact merely the fashion of a certain century, maintaining the regional forms of architecture, which were merely the most convenient and comely constructions at a certain moment of cultural and technical development; and he sought, more or less, to keep these "original" customs and habits and interests fixed forever in the same mould: a neurotic retreat. In that sense regionalism, it seems plain, was anti-historical and anti-organic: for it denied both the fact of change and the possibility that anything of value could come out of it.

While it would be dishonest to gloss over this weakness, one must understand it in terms of the circumstances that conspired to produce it. It was a flat reaction against the equally exaggerated neglect of the traditions and historic monuments of a community's life, fostered by the abstractly progressive minds of the nineteenth century. For the new industrialist, "history was bunk." Is it any wonder that the new regionalist overcompensated for that contempt and ignorance by holding that even the dustiest relics of the past were sacred? What was mistaken was not the interest but the tactics. Vis-à-vis the machine, the regionalist was in the position of a swimmer facing a strong incoming tide: if he attempts to stand up against the high waves he is knocked down: if he seeks safety by retreating unaided to the shore, he is caught in the undertow of the receding wave and can neither reach land nor keep his footing: his welfare depends upon his confidence in meeting the wave and plunging along with it at the moment it is about to break, thus utilizing the energy of the very force he is attempting to escape. These were the tactics of Bishop Grundtvig of Denmark, who not merely revived the old ballads but founded the cooperative agricultural movement: they are the basis of a dynamic regionalism.

The fact is, at all events, that the development of local languages and regional cultures, though springing immediately perhaps out of a reactionary impulse, was not limited to negations, neither was it hopelessly remote from those currents of modern life which strengthen the bonds between regions and universalize the common benefits of Western Civilization: it was rather complementary to them. A world that is united physically by the airplane, the radio, the cable, must eventually, if cooperation is to increase, devise a common language to take care of all its practical matters—its news despatches, its business communications, its international broadcasts, and the relatively simple needs and curiosities of travellers. Precisely as the boundaries of mechanical intercourse widen and become worldwide, a universal language must supplant the tongue of even the most influential national aggregation. From this point of view, one of the worst blows to internationalism was that struck by the pedants of the Renascence when in their worship of the classics they abandoned scholastic Latin, the universal language of the learned classes.

But along with this pragmatic development of a common tongue a more intimate language is needed for the deeper sort of cooperation and communication. Languages equipped for this special cultural purpose have been spontaneously growing up or reviving all over the Western World from the middle of the nineteenth century onwards. Welsh, Gaelic, Hebrew, Catalan, Flemish, Czech, Norwegian, Landsmåal, Africaans are some of the languages that are either new, or have been renovated and popularized recently for combined vernacular and literary use. While the growth of travel and communication will doubtless lead to a consolidation of dialects, reducing, say, the three hundred odd languages of India to a handful of major languages, it is already being counteracted by the opposite process of re-differentiation: the gap between English and American is much wider now than it was when Noah Webster codified the slightly more archaic American forms and pronunciations.

There is no reason to think that any single national language can now dominate the world, as the French and the English people have by turns dreamed: for unless an international language can be made relatively fixed and lifeless, it will go through a babel-like differentiation in precisely the same fashion as Latin did. It is much more likely that bi-lingualism will become universal—that is, an arranged and purely artificial world-language for pragmatic and scientific uses, and a cultural language for local communication.

The revival of these cultural languages and literatures and the stimulation of local life that has resulted from their use, must be counted as one of the most effective measures society has taken for protection against the automatic processes of machine civilization. Against the dream of universal and complete standardization, the dream of the universal cockney, and of one long street, called the Tottenham Court Road or Broadway threading over the globe, and of one language spoken everywhere and on all occasions—against this now archaic dream one must place the fact of cultural re-individua·

tion. While the reaction has often been blind and arbitrary, it has been no more so than the equally "forward-looking" movements it was attempting to halt. Behind it lies the human need to control the machine, if not at the point of origin, then at the point of application.

8: The Return to Nature

The historical revival of regionalism was re-enforced by another movement: the Return to Nature.

The cultivation of nature for its own sake, and the pursuit of rural modes of living and the appreciation of the rural environment became in the eighteenth century one of the chief means of escaping the counting house and the machine. So long as the country was uppermost, the cult of nature could have no meaning: being a part of life, there was no need to make it a special object of thought. It was only when the townsman found himself closed in by his methodical urban routine and deprived in his new urban environment of the sight of sky and grass and trees, that the value of the country manifested itself clearly to him. Before this, an occasional rare adventurer would seek the solitude of the mountains to cultivate his soul: but in the eighteenth century Jean-Jacques Rousseau, preaching the wisdom of the peasant and the sanity of the simple rural occupations, led a whole succession of generations outside the gates of their cities: they botanized, they climbed mountains, they sang peasant songs, they swam in the moonlight, they helped in the harvest field; and those who could afford to built themselves rural retreats. This impulse to recapture nature had a powerful influence upon the cultivation of the environment as a whole and upon the development of cities: but I reserve this for discussion in another book.

The important thing is to realize that at the very moment life was becoming more constricted and routinized, a great safety valve for the aboriginal human impulses had been found—the raw, unexplored, and relatively uncultivated regions of America and Africa, and even the less formidable islands of the South Seas: above all, the most steadfast of primitive environments, the ocean, had been thrown open to the discontented and the adventurous. Failing to accept the destiny that the inventors and the industrialists were creating, failing to welcome the comforts and the conveniences of civilized existence and accept the high value placed upon them by the reigning bourgeoisie, those who possessed hardier virtues and a quicker sense of values could escape from the machine. In the forests and grasslands of the new worlds they could wring a living from the soil, and on the sea they could face the elemental forces of wind and water. Here, likewise, those too weak to face the machine could find temporary refuge.

This solution was perhaps almost a too perfect one: for the new settlers and pioneers not merely satisfied their own spiritual needs by colonizing the less inhabited areas of the globe, but in the act of so doing they provided raw materials for the new industries, they likewise afforded a market for their manufactured goods, and they paved the way for the eventual introduction of the machine. Rarely have the inner impulses of different parts of society balanced so neatly with the outer conditions of its success: rarely has there been a social situation which was satisfactory to so many different types of personality and so many varieties of human effort. For a brief hundred years—roughly from 1790 to 1890 in North America, and perhaps a little earlier and a little later for South America and Africa—the land pioneer and the industrial pioneer were in close partnership. The thrifty, aggressive, routinized men built their factories and regimented their workers: the tough, sanguine, spirited, non-mechanical men fought the aborigines, cleared the land, scoured the forests for game and clove the virgin soils with their plows. If the new agricultural opportunities were still too tame and respectable, even though old customs and solidarities were disregarded and old precedents flouted, there were horses to be roped on the pampas, petroleum to be tapt in Pennsylvania, gold was to be found in California and Australia, rubber and tea to be planted in the East, and virgin lands in the steaming heart of Africa or in the coldest north could be trodden for the first time by white men, seeking food or knowledge or adventure or psychal remoteness from their own kind.

Not until the new lands were completely occupied and exploited did the machine come in, to claim its special form of dominion over those who had shown neither courage nor luck nor cunning in exploiting Nature. For millions of men and women, the new lands staved off the moment of submission. By accepting the shackles of nature they could evade for a brief while the complicated interdependence of the machine civilization. The more humane or fanatic types, in the company of their fellows, could even make an equally brief effort to realize their dream of the perfect society or the Heavenly City: from the Shaker colonies in New England to the Mormons of Utah there stretched a weak faint line of perfectionists, seeking to circumvent both the aimless brutality of nature and the more purposeful brutality of man.

Movements as vast and complex as the migration of peoples from the seventeenth to the twentieth century cannot of course be accounted for by a single cause or a single set of circumstances. The pressure of population-growth by itself is not sufficient to explain it, for not merely did the movement precede the growth, but the fact is that this pressure was considerably eased in Europe by the introduction of the potato, the improvement of the winter cattle fodder crops and the overthrow of the three-field system, at the very moment that the exodus to the new world was greatly accelerated. Nor can it be explained on purely political terms as an attempt to escape obsolete ecclesiastical and political institutions, or a result of the desire to breathe the free unpolluted air of republican institutions. Nor again was it merely a practical working out of the desire to return to Nature, although Rousseau had plainly influenced people who talked Rousseau and acted Rousseau without ever perhaps having heard his name. But all these motives were in existence: the desire to be free from social compulsion, the desire for economic security, the desire to return to nature; and they played into each other's hands. They provided both the excuse and the motive power for escaping from the new mechanical civilization that was closing in upon the Western World. To shoot, to trap, to chop trees, to hold a plow, to prospect, to face a seam—all these primitive occupations, out of which technics had originally sprung, all these occupations that had been closed and stabilized by the very advances of technics, were now open to the pioneer: he might be hunter, fisher, miner, woodman, and farmer by turn, and by engaging in these occupations people could restore their plain animal vigor as men and women, temporarily freed from the duties of a more orderly and servile existence.

Within a short century this savage idyll practically came to an end. The industrial pioneer caught up with the land pioneer and the latter could only rehearse in play what his forefathers had done out of sheer necessity. But as long as the opportunities were open in the unsettled countries, people took advantage of them in numbers that would be astounding if the blessings of an orderly, acquisitive, mechanized civilization were as great as the advocates of Progress believed and preached. Millions of people chose a lifetime of danger, heroic toil, deprivation and hardships, battling with the forces of Nature, rather than accept life on the terms that it was offered alike to the victorious and the vanquished in the new hives of industry. The movement was in part the reverse of that great organizing effort of the eleventh and twelfth centuries which cleared the forests and marshes and erected cities from one end of Europe to the other: it was rather a tendency to disperse, to escape from a close, systematic, cultivated life into an open and relatively barbarous existence.

With the occupation of the remaining open lands, this modern movement of population tapered off, and our mechanical civilization lost one of its main safety valves. The most simple human reaction that fear of the machine could provoke—running away from it—had ceased to be possible without undermining the basis of livelihood. So complete has the victory of the machine been during the last generation that in the periodic exodus from the machine which takes place on holidays in America the would-be exiles escape in motor cars and carry into the wilderness a phonograph or a radio set. And ultimately, then, the reaction of the pioneer was far less effective, though it so soon found practical channels, than the romanticism of the poets and architects and painters who merely created in the mind the ideal image of a more humane life.

Yet the lure of more primitive conditions of life, as an alternative to the machine, remains. Some of those who shrink from the degree of social control necessary to operate the machine rationally, are now busy with plans for scrapping the machine and returning to a bare subsistence level in little island utopias devoted to sub-agriculture and sub-manufacture. The advocates of these measures for returning to the primitive forget only one fact: what they are proposing is not an adventure but a bedraggled retreat, not a release but a confession of complete failure. They propose to return to the physical conditions of pioneer existence without the positive spiritual impulse that made the original conditions tolerable and the original efforts possible. If such defeatism becomes widespread it would mean something more than the collapse of the machine: it would mean the end of the present cycle of Western Civilization.

9: Organic and Mechanical Polarities

During the century and a half that followed Rousseau the cult of the primitive took many forms. Joining up with historical romanticism, which had other roots, it expressed itself on the imaginative level as an interest in the folk arts and in the products of primitive people, no longer dismissed as crude and barbarous, but valued precisely for these qualities, which were often conspicuously lacking in more highly developed communities. Not by accident was the interest in the art of the African negroes, one of the manifestations of this cult in our century, the product of the same group of Parisian painters who accepted with utmost heartiness the new forms of the machine: Congo maintained the balance against the motor works and the subway.

But on the wider platform of personal behavior, the primitive disclosed itself during the twentieth century in the insurgence of sex. The erotic dances of the Polynesians, the erotic music of the African negro tribes, these captured the imagination and presided over the recreation of the mechanically disciplined urban masses of Western Civilization, reaching their swiftest development in the United States, the country that had most insistently fostered mechanical gadgets and mechanical routines. To the once dominantly masculine relaxation of drunkenness was added the hetero-sexual relaxation of the dance and the erotic embrace, two phases of the sexual act that were now performed in public. The reaction grew in proportion to the external restraint imposed by the day's grind; but instead of enriching the erotic life and providing deep organic satisfactions, these com-

pensatory measures tended to keep sex at a constant pitch of stimula.
tion and ultimately of irritation: for the ritual of sexual excitation
pervaded not merely recreation but business: it appeared in the office
and the advertisement, to remind and to tantalize without providing
sufficient occasions for active release.

The distinction between sexual expression as one of the modes of
life and sex as a compensating element in a monotonous and restricted existence must not be lost, even though it be difficult to define.
For sex, I need hardly say, manifested itself in both forms during
this period, and with the positive side of this development and its
many fruitful and far-reaching consequences, I purpose to deal at
length in another place. But in its extreme forms, the compensatory
element could easily be detected: for it was marked by an abstract-
ness and a remoteness, derived from the very environment that the
populace was desperately trying to escape. The weakness of these
primitive compensations disclosed itself in the usually synthetic
obscenities of the popular joke, the remote glamor of the embraces
of moving picture stars, the voluptuous contortions of dancers on the
stage and of experiences taken in at second or third hand through the
bawdy mimicry of the popular song or, a little closer to reality,
snatched hastily and furtively at the end of an automobile ride or a
fatiguing day in the office or the factory. Those who escaped the
anxiety and frustration of such embraces did so only by deadening
their higher nerve-centers by means of alcohol or by the chemistry
of some form of psychal anesthesia which took the outward form of
coarseness and debasement.

In brief, most of the sexual compensations were little above the
level of abject fantasy; whereas when sex is accepted as an important
mode of life, lovers reject these weak and secondary substitutes for
it, and devote their minds and energies to courtship and expression
themselves: necessary steps to those enlargements and enrichments
and sublimations of sex that alike maintain the species and energize
the entire cultural heritage. It was a miner's son, D. H. Lawrence,
who distinguished most sharply between the degradation of sex which
occurs when it is merely a means of getting away from the sordid
environment and oppressive dullness of a low-grade industrial town,

and the exhilaration that arises when sex is genuinely respected and celebrated in its own right.

The weakness of the sexual relapse into the primitive was not indeed unlike that which overtook the more general cultivation of the body through sport. The impulse that excited it was genuine and justified; but the form it took did not lead to a transformation of the original condition: rather, it became the mechanism by means of which the original condition was remedied sufficiently to continue in existence. Sex had a larger part of life to claim than it filched for itself in the instinctive reaction against the machine.

As the machine tended toward the pole of regularity and complete automatism, it became severed, finally, from the umbilical cord that bound it to the bodies of men and women: it became an absolute. That was the danger Samuel Butler jestingly prophesied in Erewhon, the danger that the human being might become a means whereby the machine perpetuated itself and extended its dominion. The recoil from the absolute of mechanism was into an equally sterile absolute of the organic: the raw primitive. The organic processes, reduced to shadows by the machine, made a violent effort to retrieve their position. The machine, which acerbically denied the flesh, was offset by the flesh, which denied the rational, the intelligent, the orderly processes of behavior that have entered into all man's cultural developments—even those developments that most closely derive from the organic. The spurious notion that mechanism had naught to learn from life was supplanted by the equally false notion that life had nothing to learn from mechanism. On one side is the gigantic printing press, a miracle of fine articulation, which turns out the tabloid newspaper: on the other side are the contents of the tabloid itself, symbolically recording the most crude and elementary states of emotion, feeling, barely vestigial thought. Here the impersonal and the cooperative and the objective: over against it the limited, the subjective, the recalcitrant, violent ego, full of hatreds, fears, blind frenzies, crude impulses toward destruction. Mechanical instruments, potentially a vehicle of rational human purposes, are scarcely a blessing when they enable the gossip of the village idiot and the deeds of the thug to be broadcast to a million people each day.

The effect of this return to the absolute primitive, like so many other neurotic adaptations that temporarily bridge the chasm, develops stresses of its own which tend to push the two sides of existence still further apart. That hiatus limits the efficiency of the compensatory reaction: ultimately it spells ruin for the civilization that seeks to maintain the raw mechanical by weighting it with the raw primitive. For in its broadest reaches, including all those cultural interests and sentiments and admirations which sustain the work of the scientist, the technician, the artist, the philosopher, even when they do not appear directly in the particular work itself—in its broadest reaches this civilization cannot be run by barbarians. A hairy ape in the stokehold is a grave danger signal: a hairy ape on the bridge means speedy shipwreck. The appearance of such apes, in the forms of those political dictators who attempt to accomplish by calculated brutality and aggression what they lack the intelligence and magnanimity to consummate by more humane direction, indicates on what an infirm and treacherous foundation the machine at present rests. For, more disastrous than any mere physical destruction of machines by the barbarian is his threat to turn off or divert the human motive power, discouraging the cooperative processes of thought and the disinterested research which are responsible for our major technical achievements.

Toward the end of his life Herbert Spencer viewed with proper alarm the regression into imperialism, militarism, servility that he saw all around him at the beginning of the present century; and in truth he had every reason for his forebodings. But the point is that these forces were not merely archaic survivals that had failed to be extirpated by the machine: they were rather underlying human elements awakened into stertorous activity by the very victory of the machine as an absolute and non-conditioned force in human life. The machine, by failing as yet—despite neotechnic advances—to allow sufficient play in social existence to the organic, has opened the way for its return in the narrow and inimical form of the primitive. Western society is relapsing at critical points into pre-civilized modes of thought, feeling, and action because it has acquiesced too easily in the dehumanization of society through capitalist exploitation and military conquest. The retreat into the primitive is, in sum, a maudlin effort to avoid the more basic and infinitely more difficult transformation which our thinkers and leaders and doers have lacked the candor to face, the intelligence to contrive, and the will to effect— the transition beyond the historic forms of capitalism and the equally limited original forms of the machine to a life-centered economy.

10: Sport and the "Bitch-goddess"

The romantic movements were important as a corrective to the machine because they called attention to essential elements in life that were left out of the mechanical world-picture: they themselves prepared some of the materials for a richer synthesis. But there is within modern civilization a whole series of compensatory functions that, so far from making better integration possible, only serve to stabilize the existing state—and finally they themselves become part of the very regimentation they exist to combat. The chief of these institutions is perhaps mass-sports. One may define these sports as those forms of organized play in which the spectator is more important than the player, and in which a good part of the meaning is lost when the game is played for itself. Mass-sport is primarily a spectacle.

Unlike play, mass-sport usually requires an element of mortal chance or hazard as one of its main ingredients: but instead of the chance's occurring spontaneously, as in mountain climbing, it must take place in accordance with the rules of the game and must be increased when the spectacle begins to bore the spectators. Play in one form or another is found in every human society and among a great many animal species: but sport in the sense of a mass-spectacle, with death to add to the underlying excitement, comes into existence when a population has been drilled and regimented and depressed to such an extent that it needs at least a vicarious participation in difficult feats of strength or skill or heroism in order to sustain its waning life-sense. The demand for circuses, and when the milder spectacles are still insufficiently life-arousing, the demand for sadistic exploits and finally for blood is characteristic of civilizations that are losing their grip: Rome under the Caesars, Mexico at the time of Montezuma, Germany under the Nazis. These forms of surrogate manliness and bravado are the surest signs of a collective impotence and a pervasive death wish. The dangerous symptoms of that ultimate decay one finds everywhere today in machine civilization under the guise of mass-sport.

The invention of new forms of sport and the conversion of play into sport were two of the distinctive marks of the last century: baseball is an example of the first, and the transformation of tennis and golf into tournament spectacles, within our own day, is an example of the second. Unlike play, sport has an existence in our mechanical civilization even in its most abstract possible manifestation: the crowd that does not witness the ball game will huddle around the scoreboard in the metropolis to watch the change of counters. If it does not see the aviator finish a record flight around the world, it will listen over the radio to the report of his landing and hear the frantic shouts of the mob on the field: should the hero attempt to avoid a public reception and parade, he would be regarded as cheating. At times, as in horse-racing, the elements may be reduced to names and betting odds: participation need go no further than the newspaper and the betting booth, provided that the element of chance be there. Since the principal aim of our mechanical routine in industry is to reduce the domain of chance, it is in the glorification of chance and the unexpected, which sport provides, that the element extruded by the machine returns, with an accumulated emotional charge, to life in general. In the latest forms of mass-sport, like air races and motor races, the thrill of the spectacle is intensified by the promise of immediate death or fatal injury. The cry of horror that escapes from the crowd when the motor car overturns or the airplane crashes is not one of surprise but of fulfilled expectation: is it not fundamentally for the sake of exciting just such bloodlust that the competition itself is held and widely attended? By means of the talking picture that spectacle and that thrill are repeated in a thousand theatres throughout the world as a mere incident in the presentation of the week's news: so that a steady habituation to blood-letting and exhibitionistic murder and suicide accompanies the spread of the machine and, becoming stale by repetition in its milder forms, encourages the demand for more massive and desperate exhibitions of brutality.

Sport presents three main elements: the spectacle, the competition, and the personalities of the gladiators. The spectacle itself introduces the esthetic element, so often lacking in the paleotechnic industrial environment itself. The race is run or the game is played within a frame of spectators, tightly massed: the movements of this mass, their cries, their songs, their cheers, are a constant accompaniment of the spectacle: they play, in effect, the part of the Greek chorus in the new machine-drama, announcing what is about to occur and underlining the events of the contest. Through his place in the chorus, the spectator finds his special release: usually cut off from close physical associations by his impersonal routine, he is now at one with a primitive undifferentiated group. His muscles contract or relax with the progress of the game, his breath comes quick or slow, his shouts heighten the excitement of the moment and increase his internal sense of the drama: in moments of frenzy he pounds his neighbor's back or embraces him. The spectator feels himself contributing by his presence to the victory of his side, and sometimes, more by hostility to the enemy than encouragement to the friend, he does perhaps exercize a visible effect on the contest. It is a relief from the passive rôle of taking orders and automatically filling them, of conforming by means of a reduced "I" to a magnified "It," for in the sports arena the spectator has the illusion of being completely mobilized and utilized. Moreover, the spectacle itself is one of the richest satisfactions for the esthetic sense that the machine civilization offers to those that have no key to any other form of culture: the spectator knows the style of his favorite contestants in the way that the painter knows the characteristic line or palette of his master, and he reacts to the bowler, the pitcher, the punter, the server, the air ace, with a view, not only to his success in scoring, but to the esthetic spectacle itself. This point has been stressed in bull-fighting; but of course it applies to every form of sport. There remains, nevertheless, a conflict between the desire for a skilled exhibition and the desire for a brutal outcome: the maceration or death of one or more of the contestants.

Now in the competition two elements are in conflict: chance and record-making. Chance is the sauce that stimulates the excitement of the spectator and increases his zest for gambling: whippet-racing and horse-racing are as effective in this relation as games where a greater degree of human skill is involved. But the habits of the mechanical régime are as difficult to combat in sport as in the realm of sexual behavior: hence one of the most significant elements in modern sport is the fact that an abstract interest in record-making has become one of its main preoccupations. To cut the fifth of a second off the time of running a race, to swim the English channel twenty minutes faster than another swimmer, to stay up in the air an hour longer than one's rival did—these interests come into the competition and turn it from a purely human contest to one in which the real opponent is the previous record: time takes the place of a visible rival. Sometimes, as in dance marathons or flag-pole squat-tings, the record goes to feats of inane endurance: the blankest and dreariest of sub-human spectacles. With the increase in professional-ized skill that accompanies this change, the element of chance is further reduced: the sport, which was originally a drama, becomes an exhibition. As soon as specialism reaches this point, the whole performance is arranged as far as possible for the end of making possible the victory of the popular favorite: the other contestants are, so to say, thrown to the lions. Instead of "Fair Play" the rule now becomes "Success at Any Price."

Finally, in addition to the spectacle and the competition, there comes onto the stage, further to differentiate sport from play, the new type of popular hero, the professional player or sportsman. He is as specialized for the vocation as a soldier or an opera singer: he represents virility, courage, gameness, those talents in exercizing and commanding the body which have so small a part in the new mechanical regimen itself: if the hero is a girl, her qualities must be Amazonian in character. The sports hero represents the masculine virtues, the Mars complex, as the popular motion picture actress or the bathing beauty contestant represents Venus. He exhibits that com-plete skill to which the amateur vainly aspires. Instead of being looked upon as a servile and ignoble being, because of the very perfection of his physical efforts, as the Athenians in Socrates' time looked upon the professional athletes and dancers, this new hero represents the summit of the amateur's effort, not at pleasure but at efficiency. The hero is handsomely paid for his efforts, as well as being rewarded by praise and publicity, and he thus further restores to sport its connection with the very commercialized existence from which it is supposed to provide relief—restores it and thereby sanctifies it. The few heroes who resist this vulgarization—notably Lindbergh—fall into popular or at least into journalistic disfavor, for they are only playing the less important part of the game. The really successful sports hero, to satisfy the mass-demand, must be midway between a pander and a prostitute.

Sport, then, in this mechanized society, is no longer a mere game empty of any reward other than the playing: it is a profitable business: millions are invested in arenas, equipment, and players, and the maintenance of sport becomes as important as the maintenance of any other form of profit-making mechanism. And the technique of mass-sport infects other activities: scientific expeditions and geographic explorations are conducted in the manner of a speed stunt or a prizefight—*and for the same reason.* Business or recreation or mass spectacle, sport is always a means: even when it is reduced to athletic and military exercizes held with great pomp within the sports arenas, the aim is to gather a record-breaking crowd of performers and spectators, and thus testify to the success or importance of the movement that is represented. Thus sport, which began originally, perhaps, as a spontaneous reaction against the machine, has become one of the mass-duties of the machine age. It is a part of that universal regimentation of life—for the sake of private profits or nationalistic exploit—from which its excitement provides a temporary and only a superficial release. Sport has turned out, in short, to be one of the least effective reactions against the machine. There is only one other reaction less effective in its final result: the most ambitious as well as the most disastrous. I mean war.

11: The Cult of Death

Conflict, of which war is a specialized institutional drama, is a recurrent fact in human societies: it is indeed inevitable when society has reached any degree of differentiation, because the absence of conflict would presume a unanimity that exists only in placentals between embryos and their female parents. The desire to achieve that kind of unity is one of the most patently regressive characteristics of totalitarian states and other similar attempts at tyranny in smaller groups.

But war is that special form of conflict in which the aim is not to resolve the points of difference but to annihilate physically the defenders of opposing points or reduce them by force to submission. And whereas conflict is an inevitable incident in any active system of cooperation, to be welcomed just because of the salutary variations and modifications it introduces, war is plainly a specialized perversion of conflict, bequeathed perhaps by the more predatory hunting groups; and it is no more an eternal and necessary phenomenon in group life than is cannibalism or infanticide.

War differs in scale, in intention, in deadliness, and in frequency with the type of society: it ranges all the way from the predominantly ritualistic warfare of many primitive societies to the ferocious slaughters instituted from time to time by barbarian conquerors like Ghengis Khan and the systematic combats between entire nations that now occupy so much of the time and energy and attention of "advanced" and "peaceful" industrial countries. The impulses toward destruction have plainly not decreased with progress in the means: indeed there is some reason to think that our original collecting and food-gathering ancestors, before they had invented weapons to aid them in hunting, were more peaceful in habit than their more civilized descendants. As war has increased in destructiveness, the sporting element has grown smaller. Legend tells of an ancient conqueror who spurned to capture a town by surprise at night because it would be too easy and would take away the glory: today a well-organized army attempts to exterminate the enemy by artillery fire before it advances to capture the position.

In almost all its manifestations, however, war indicates a throwback to an infantile psychal pattern on the part of people who can no longer stand the exacting strain of life in groups, with all the necessities for compromise, give-and-take, live-and-let-live, understanding and sympathy that such life demands, and with all the complexities of adjustment involved. They seek by the knife and the gun to unravel the social knot. But whereas national wars today are essentially collective competitions in which the battlefield takes the place of the market, the ability of war to command the loyalty and interests of the entire underlying population rests partly upon its peculiar psychological reactions: it provides an outlet and an emotional release. "Art degraded, imagination denied," as Blake says, "war governed the nations."

For war is the supreme drama of a completely mechanized society; and it has an element of advantage that puts it high above all the other preparatory forms of mass-sport in which the attitudes of war are mimicked: war is real, while in all the other mass-sports there is an element of make-believe: apart from the excitements of the game and the gains or losses from gambling, it does not really matter who is victorious. In war, there is no doubt as to the reality: success may bring the reward of death just as surely as failure, and it may bring it to the remotest spectator as well as to the gladiators in the center of the vast arena of the nations.

But war, for those actually engaged in combat, likewise brings a release from the sordid motives of profit-making and self-seeking that govern the prevailing forms of business enterprise, including sport: the action has the significance of high drama. And while warfare is one of the principal sources of mechanism, and its drill and regimentation are the very pattern of old-style industrial effort, it provides, far more than the sport-field, the necessary compensations to this routine. The preparation of the soldier, the parade, the smartness and polish of the equipment and uniform, the precise movement of large bodies of men, the blare of bugles, the punctuation of drums, the rhythm of the march, and then, in actual battle itself, the final explosion of effort in the bombardment and the charge, lend an esthetic and moral grandeur to the whole performance. The death or maiming of the body gives the drama the element of a tragic sacrifice, like that which underlies so many primitive religious rituals: the effort is sanctified and intensified by the scale of the holocaust. For peoples that have lost the values of culture and can no longer respond with interest or understanding to the symbols of culture, the abandonment of the whole process and the reversion to crude faiths and non-rational dogmas, is powerfully abetted by the processes of war. If no enemy really existed, it would be necessary to create him, in order to further this development.

Thus war breaks the tedium of a mechanized society and relieves it from the pettiness and prudence of its daily efforts, by concentrating to their last degree both the mechanization of the means of production and the countering vigor of desperate vital outbursts. War sanctions the utmost exhibition of the primitive at the same time that it deifies the mechanical. In modern war, the raw primitive and the clockwork mechanical are one.

In view of its end products—the dead, the crippled, the insane, the devastated regions, the shattered resources, the moral corruption, the anti-social hates and hoodlumisms—war is the most disastrous outlet for the repressed impulses of society that has been devised. The evil consequences have increased in magnitude and in human distress in proportion as the actual elements of fighting have become more mechanized: the threat of chemical warfare against the civilian population as well as the military arm places in the hands of the armies of the world instruments of ruthlessness of which only the most savage conquerors in the past would have taken advantage. The difference between the Athenians with their swords and shields fighting on the fields of Marathon, and the soldiers who faced each other with tanks, guns, flame-throwers, poison gases, and hand-grenades on the Western Front, is the difference between the ritual of the dance and the routine of the slaughter house. One is an exhibition of skill and courage with the chance of death present, the other is an exhibition of the arts of death, with the almost accidental by-product of skill and courage. But it is in death that these repressed and regimented populations have their first glimpse of effective life; and the cult of death is a sign of their throwback to the corrupt primitive.

As a back-fire against mechanism, war, even more than mass-sport, has increased the area of the conflagration without stemming its advance. Still, as long as the machine remains an absolute, war will represent for this society the sum of its values and compensations: for war brings people back to the earth, makes them face the battle with the elements, unleashes the brute forces of their own nature, releases the normal restraints of social life, and sanctions a return to the primitive in thought and feeling, even as it further sanctions infantility in the blind personal obedience it exacts, like that of the archetypal father with the archetypal son, which divests the latter of the need of behaving like a responsible and autonomous personality. Savagery, which we have associated with the not-yet-civilized, is equally a reversionary mode that arises with the mechanically over-civilized. Sometimes the mechanism against which reaction takes place is a compulsive morality or social regimentation: in the case of Western peoples it is the too-closely regimented environment we associate with the machine. War, like a neurosis, is the destructive solution of an unbearable tension and conflict between organic impulses and the code and circumstances that keep one from satisfying them.

This destructive union of the mechanized and the savage primitive is the alternative to a mature, humanized culture capable of directing the machine to the enhancement of communal and personal life. If our life were an organic whole this split and this perversion would not be possible, for the order we have embodied in machines would be more completely exemplified in our personal life, and the primitive impulses, which we have diverted or repressed by excessive preoccupation with mechanical devices, would have natural outlets in their appropriate cultural forms. Until we begin to achieve this culture, however, war will probably remain the constant shadow of the machine: the wars of national armies, the wars of gangs, the wars of classes: beneath all, the incessant preparation by drill and propaganda towards these wars. A society that has lost its life values will tend to make a religion of death and build up a cult around its worship—a religion not less grateful because it satisfies the mounting number of paranoiacs and sadists such a disrupted society necessarily produces.

12: The Minor Shock-Absorbers

From all the forms of resistance and compensation we have been examining it is plain that the introduction of the machine was not smooth, nor were its characteristic habits of life undisputed. The

reactions would probably have been more numerous and more decisive had it not been for the fact that old habits of thought and old
ways of life continued in existence: this bridged the gap between the
old and the new, and kept the machine from dominating life as a
whole as much as it controlled the routine of industrial activity.
In part, these existing institutions, while they stabilized society, prevented it from absorbing and reacting upon the cultural elements
derived from the machine: so that they lessened the valuable offices
of the machine in the act of mitigating its defects.

In addition to the stabilizing inertia of society as a whole, and to
the many-sided attempts to combat the machine by the force of ideas
and institutional contrivances, there were still other reactions that
served, as it were, as cushions and shock-absorbers. So far from
stopping the machine or undermining the purely mechanical program, they perhaps decreased the tensions that the machine produced.
Thus the tendency to destroy the memorials of older cultures, exhibited by the utilitarians in their first vigor of self-confidence and
creative effort, was met in part among the very classes that were most
active in this attack, by the cult of antiquarianism.

This cult lacked the passionate conviction that one period or another of the past was of supreme value: it merely held that almost
anything old was *ipso facto* valuable or beautiful, whether it was
a fragment of Roman statuary, a wooden image of a fifteenth century
saint, or an iron door knocker. The exponents of this cult attempted
to create private environments from which every hint of the machine
was absent: they burned wooden logs in the open fireplaces of imitation Norman manor houses, which were in reality heated by steam,
designed with the help of a camera and measured drawings, and
supported, where the architect was a little uncertain of his skill
or materials, with concealed steel beams. When handicraft articles
could not be filched from the decayed buildings of the past, they
were copied with vast effort by belated handworkers: when the demand for such copies filtered down through the middle classes, they
were then reproduced by means of power machinery in a fashion
capable of deceiving only the blind and ignorant: a double prevari-
cation.

Oppressed by a mechanical environment they had neither mastered nor humanized nor succeeded esthetically in appreciating, the ruling classes and their imitators among the lesser bourgeoisie retreated from the factory or the office into a fake non-mechanical environment, in which the past was modified by the addition of physical comforts, such as tropical temperature in the winter, and springs and padding on sofas, lounges, beds. Each successful individual produced his own special antiquarian environment: a private world.

This private world, as lived in Suburbia or in the more palatial country houses, is not to be differentiated by any objective standard from the world in which the lunatic attempts to live out the drama in which he appears to himself to be Lorenzo the Magnificent or Louis XIV. In each case the difficulty of maintaining an equilibrium in relation to a difficult or hostile external world is solved by withdrawal, permanent or temporary, into a private retreat, untainted by most of the conditions that public life and effort lay down. These antiquarian stage-settings, which characterized for the most part the domestic equipment of the more successful members of the bourgeoisie from the eighteenth century onward—with a minor interlude of self-confident ugliness during the high paleotechnic period— these stage-settings were, on a strict psychological interpretation, cells: indeed, the addition of "comforts" made them padded cells. Those who lived in them were stable, "normal," "adjusted" people. In relation to the entire environment in which they worked and thought and lived, they merely behaved *as if* they were in a state of neurotic collapse, *as if* there were a deep conflict between their inner drive and the mechanical environment they had helped to create, *as if* they had been unable to resolve their divided activities into a single consistent pattern.

The other side of this conservatism of taste and this refusal to recognize natural change was the tendency to take refuge in change for its own sake, and to hasten the very process that was introduced by the machine. Changing the style of an object, altering its superficial shape or color, without effecting any real improvement, became part of the routine of modern society just because the natural variations and breaks in life were absent: the answer to excessive regimentation came in through an equally heightened and over-stimulated demand for novelties. In the long run, unceasing change is as monotonous as unceasing sameness: real refreshment implies both uncertainty and choice, and to have to abandon choice merely because for external reasons a style has changed is to forfeit what real gain has been made. Here again change and novelty are no more sacred, no more inimical, than stability and monotony: but the purposeless materialism and imbecile regimentation of production resulted in the aimless change and the absence of real stimuli and effective adjustments in consumption; and so far from resolving the difficulty the resistance only increased it. The itch for change: the itch for movement: the itch for novelty infected the entire system of production and consumption and severed them from the real standards and norms which it was highly important to devise. When people's work and days were varied they were content to remain in the same place; when their lives were ironed out into a blank routine they found it necessary to move; and the more rapidly they moved the more standardized became the environment in which they moved: there was no getting away from it. So it went in every department of life.

Where the physical means of withdrawal were inadequate, pure fantasy flourished without any other external means than the word or the picture. But these external means were put upon a mechanized collective basis during the nineteenth century, as a result of the cheapened processes of production made possible by the rotary press, the camera, photo-engraving, and the motion picture. With the spread of literacy, literature of all grades and levels formed a semi-public world into which the unsatisfied individual might withdraw, to live a life of adventure following the travellers and explorers in their memoirs, to live a life of dangerous action and keen observation by participating in the crimes and investigations of a Dupin or a Sherlock Holmes, or to live a life of romantic fulfillment in the love stories and erotic romances that became everyone's property from the eighteenth century onward. Most of these varieties of day-dream and private fantasy had of course existed in the past: now they became part of a gigantic collective apparatus of escape. So important was the function of popular literature as escape that many modern psychologists have treated literature as a whole as a mere vehicle of withdrawal from the harsh realities of existence: forgetful of the fact that literature of the first order, so far from being a mere pleasure-device, is a supreme attempt to face and encompass reality—an attempt beside which a busy working life involves a shrinkage and represents a partial retreat.

During the nineteenth century vulgar literature to a large extent replaced the mythological constructions of religion: the austere cosmical sweep and the careful moral codes of the more sacred religions were, alas! a little too much akin to the machine itself, from which people were trying to escape. This withdrawal into fantasy was immensely re-enforced from 1910 on, by the motion-picture, which came into existence just when the pressure from the machine was beginning to bear down more and more inexorably. Public daydreams of wealth, magnificence, adventure, irregularity and spontaneous action—identification with the criminal defying the forces of order—identification with the courtesan practicing openly the allurements of sex—these scarcely adolescent fantasies, created and projected with the aid of the machine, made the machine-ritual tolerable to the vast urban or urbanized populations of the world. But these dreams were no longer private, and what is more they were no longer spontaneous and free: they were promptly capitalized on a vast scale as the "amusement business," and established as a vested interest. To create a more liberal life that might do without such anodynes was to threaten the safety of investments, built on the certainty of continued dullness, boredom and defeat.

Too dull to think, people might read: too tired to read, they might look at the moving pictures: unable to visit the picture theater they might turn on the radio: in any case, they might avoid the call to action: surrogate lovers, surrogate heroes and heroines, surrogate wealth filled their debilitated and impoverished lives and carried the perfume of unreality into their dwellings. And as the machine itself became, as it were, more active and human, reproducing the organic properties of eye and ear, the human beings who employed the machine as a mode of escape have tended to become more passive and mechanical. Unsure of their own voices, unable to hold a tune, they carry a phonograph or a radio set with them even on a picnic: afraid to be alone with their own thoughts, afraid to confront the blankness and inertia of their own minds, they turn on the radio and eat and talk and sleep to the accompaniment of a continuous stimulus from the outside world: now a band, now a bit of propaganda, now a piece of public gossip called news. Even such autonomy as the poorest drudge once had, left like Cinderella to her dreams of Prince Charming when her sisters went off to the ball, is gone in this mechanical environment: whatever compensations her present-day counterpart may have, it must come through the machine. Using the machine alone to escape from the machine, our mechanized populations have jumped from a hot frying pan into a hotter fire. The shock-absorbers are of the same order as the environment itself. The moving picture deliberately glorifies the cold brutality and homicidal lusts of gangsterdom: the newsreel prepares for battles to come by exhibiting each week the latest devices of armed combat, accompanied by a few persuasive bars from the national anthem. In the act of relieving psychological strain these various devices only increase the final tension and support more disastrous forms of release. After one has lived through a thousand callous deaths on the screen one is ready for a rape, a lynching, a murder, or a war in actual life: when the surrogate excitements of the film and the radio begin to pall, a taste of real blood becomes necessary. In short: the shock-absorber prepares one for a fresh shock.

13: Resistance and Adjustment

In all these efforts to attack, to resist, or to retreat from the machine the observer may be tempted to see nothing more than the phenomenon that Professor W. F. Ogburn has described as the "cultural lag." The failure of "adjustment" may be looked upon as a failure of art and morals and religion to change with the same degree of rapidity as the machine and to change in the same direction.

This seems to me an essentially superficial interpretation. For one thing, change in a direction *opposite* to the machine may be as important in ensuring adjustment as change in the same direction, when it happens that the machine is taking a course that would, unless compensated, lead to human deterioration and collapse. For another thing, this interpretation regards the machine as an independent structure, and it holds the direction and rate of change assumed by the machine as a norm, to which all the other aspects of human life must conform. In truth, interactions between organisms and their environments take place in both directions, and it is just as correct to regard the machinery of warfare as retarded in relation to the morality of Confucius as to take the opposite position. In his The Instinct of Workmanship Thorstein Veblen carefully avoided the one-sided notion of adjustment: but later economists and sociologists have not always been so unparochial, and they have treated the machine as if it were final and as if it were something other than the projection of one particular side of the human personality.

All the arts and institutions of man derive their authority from the nature of human life as such. This applies as fully to technics as to painting. A particular economic or technical régime may deny this nature, as some particular social custom, like that of binding the feet of women or enforcing virginity, may deny the patent facts of physiology and anatomy: but such erroneous views and usages do not eliminate the fact they deny. At all events, the mere bulk of technology, its mere power and ubiquitousness, give no proof whatever of its relative human value or its place in the economy of an intelligent human society. The very fact that one encounters resistances, reversions, archaicisms at the moment of the greatest technical achievement—even among those classes who have, from the standpoint of wealth and power, benefited most by the victory of the machine—makes one doubt both the effectiveness and the sufficiency of the whole scheme of life the machine has so far brought into existence. And who is so innocent today as to think that maladjustment to the machine can be solved by the simple process of introducing greater quantities of machinery?

Plainly, if human life consisted solely in adjustment to the dominant physical and social environment, man would have left the world as he found it, as most of his biological companions have done: the machine itself would not have been invented. Man's singular ability consists in the fact that he creates standards and ends of his own, not given directly in the external scheme of things, and in fulfilling his own nature in cooperation with the environment, he creates a third realm, the realm of the arts, in which the two are harmonized and ordered and made significant. Man is that part of nature in which causality may, under appropriate circumstances, give place to finality: in which the ends condition the means. Sometimes man's standards are grotesque and arbitrary: untempered by positive knowledge and a just sense of his limitations, man is capable of deforming the human anatomy in pursuit of a barbarous dream of beauty, or, to objectify his fears and his tortured desires, he may resort to horrible human sacrifices. But even in these perversions there is an acknowledgment that man himself in part creates the conditions under which he lives, and is not merely the impotent prisoner of circumstances.

If this has been man's attitude toward Nature, why should he assume a more craven posture in confronting the machine, whose physical laws he discovered, whose body he created, whose rhythms he anticipated by external feats of regimentation in his own life? It is absurd to hold that we must continue to accept the bourgeoisie's overwhelming concern for power, practical success, above all for comfort, or that we must passively absorb, without discrimination and selection—which implies, where necessary, rejection—all the new products of the machine. It is equally foolish to believe that we must conform our living and thinking to the antiquated ideological system which helped create the numerous brilliant short cuts that attended the early development of the machine. The real question before us lies here: do these instruments further life and enhance its values, or not? Some of the results, as I shall show in the next chapter, are admirable, far more admirable even than the inventor and the industrialist and the utilitarian permitted himself to imagine. Other aspects of the machine are on the contrary trifling, and still others, like modern mechanized warfare, are deliberately antagonistic to every ideal of humanity—even to the old-fashioned ideal of the soldier who once risked his life in equal combat. In these latter cases, our problem is to eliminate or subjugate the machine, unless we ourselves wish to be eliminated. For it is not automatism and standardization and order that are dangerous: what is dangerous is the restriction of life that has so often attended their untutored acceptance. By what inept logic must we bow to our creation if it be a machine, and spurn it as "unreal" if it happens to be a painting or a poem? The machine is just as much a creature of thought as the poem: the poem is as much a fact of reality as the machine. Those who use the machine when they need to react to life directly or employ the humane arts, are as completely lacking in efficiency as if they studied metaphysics in order to learn how to bake bread. The question in each case is: what is the appropriate life-reaction? How far does this or that instrument further the biological purposes or the ideal ends of life?

Every form of life, as Patrick Geddes has expressed it, is marked not merely by adjustment to the environment, but by insurgence against the environment: it is both creature and creator, both the victim of fate and the master of destiny: it lives no less by domination than by acceptance. In man this insurgence reaches its apex, and manifests itself most completely, perhaps, in the arts, where dream and actuality, the imagination and its limiting conditions, the ideal and the means, are fused together in the dynamic act of expression and in the resultant body that is expressed. As a being with a social heritage, man belongs to a world that includes the past and the future, in which he can by his selective efforts create passages and ends not derived from the immediate situation, and alter the blind direction of the senseless forces that surround him.

To recognize these facts is perhaps the first step toward dealing rationally with the machine. We must abandon our futile and lamentable dodges for resisting the machine by stultifying relapses into savagery, by recourse to anesthetics and shock-absorbers. Though they temporarily may relieve the strain, in the end they do more harm than they avoid. On the other hand, the most objective advocates of the machine must recognize the underlying human validity of the Romantic protest against the machine: the elements originally embodied in literature and art in the romantic movement are essential parts of the human heritage that can not be neglected or flouted: they point to a synthesis more comprehensive than that developed through the organs of the machine itself. Failing to create this synthesis, failing to incorporate it in our personal and communal life, the machine will be able to continue only with the aid of shock-absorbers which confirm its worst characteristics, or with the compensatory adjustment of vicious and barbaric elements which will, in all probability, ruin the entire structure of our civilization.

CHAPTER VII.

ASSIMILATION OF
THE MACHINE

1: New Cultural Values

The tools and utensils used during the greater part of man's history were, in the main, extensions of his own organism: they did not have—what is more important they did not *seem* to have—an independent existence. But though they were an intimate part of the worker, they reacted upon his capacities, sharpening his eye, refining his skill, teaching him to respect the nature of the material with which he was dealing. The tool brought man into closer harmony with his environment, not merely because it enabled him to re-shape it, but because it made him recognize the limits of his capacities. In dream, he was all powerful: in reality he had to recognize the weight of stone and cut stones no bigger than he could transport. In the book of wisdom the carpenter, the smith, the potter, the peasant wrote, if they did not sign, their several pages. And in this sense, technics has been, in every age, a constant instrument of discipline and education. A surviving primitive might, here and there, vent his anger on a cart that got stuck in the mud by breaking up its wheels, in the same fashion that he would beat a donkey that refused to move: but the mass of mankind learned, at least during the period of the written record, that certain parts of the environment can neither be intimidated nor cajoled. To control them, one must learn the laws of their behavior, instead of petulantly imposing one's own wishes. Thus the lore and tradition of technics, however empirical, tended to create the picture of an objective reality. Something of this fact remained in the Victorian definition of science as "organized common sense."

Because of their independent source of power, and their semi-automatic operation even in their cruder forms, machines have seemed to have a reality and an independent existence apart from the user. Whereas the educational values of handicraft were mainly in the process, those of the machine were largely in the preparatory design: hence the process itself was understood only by the machinists and technicians responsible for the design and operation of the actual machinery. As production became more mechanized and the discipline of the factory became more impersonal and the work itself became less rewarding, apart from such slight opportunities for social intercourse as it furthered, attention was centered more and more upon the product: people valued the machine for its external achievements, for the number of yards of cloth it wove, for the number of miles it carried them. The machine thus appeared purely as an external instrument for the conquest of the environment: the actual forms of the products, the actual collaboration and intelligence manifested in creating them, the educational possibilities of this impersonal cooperation itself—all these elements were neglected. We assimilated the objects rather than the spirit that produced them, and so far from respecting that spirit, we again and again attempted to make the objects themselves seem to be something other than a product of the machine. We did not expect beauty through the machine any more than we expected a higher standard of morality from the laboratory: yet the fact remains that if we seek an authentic sample of a new esthetic or a higher ethic during the nineteenth century it is in technics and science that we will perhaps most easily find them.

The practical men themselves were the very persons who stood in the way of our recognizing that the significance of the machine was not limited to its practical achievements. For, on the terms that the inventors and industrialists considered the machine, it did not carry over from the factory and the marketplace into any other department of human life, except as a means. The possibility that technics had become a creative force, carried on by its own momentum, that it was rapidly ordering a new kind of environment and was producing a third estate midway between nature and the humane arts, that it was not merely a quicker way of achieving old ends but an effective way of expressing new ends—the possibility in short that the machine furthered a new mode of *living* was far from the minds of those who actively promoted it. The industrialists and engineers themselves did not believe in the qualitative and cultural aspects of the machine. In their indifference to these aspects, they were just as far from appreciating the nature of the machine as were the Romantics: only what the Romantics, judging the machine from the standpoint of life, regarded as a defect the utilitarian boasted of as a virtue: for the latter the absence of art was an assurance of practicality.

If the machine had really lacked cultural values, the Romantics would have been right, and their desire to seek these values, if need be, in a dead past would have been justified by the very desperateness of the case. But the interests in the factual and the practical, which the industrialist made the sole key to intelligence, were only two in a whole series of new values that had been called into existence by the development of the new technics. Matters of fact and practice had usually in previous civilizations been treated with snobbish contempt by the leisured classes: as if the logical ordering of propositions were any nobler a technical feat than the articulation of machines. The interest in the practical was symptomatic of that wider and more intelligible world in which people had begun to live, a world in which the taboos of class and caste could no longer be considered as definitive in dealing with events and experiences. Capitalism and technics had both acted as a solvent of these clots of prejudice and intellectual confusion; and they were thus at first important liberators of life.

From the beginning, indeed, the most durable conquests of the machine lay not in the instruments themselves, which quickly became outmoded, nor in the goods produced, which quickly were consumed, but in the modes of life made possible via the machine and in the machine: the cranky mechanical slave was also a pedagogue. While the machine increased the servitude of servile personalities, it also promised the further liberation of released personalities: it challenged thought and effort as no previous system of technics had done. No part of the environment, no social conventions, could be taken for granted, once the machine had shown how far order and system and intelligence might prevail over the raw nature of things.

What remains as the permanent contribution of the machine, carried over from one generation to another, is the technique of cooperative thought and action it has fostered, the esthetic excellence of the machine forms, the delicate logic of materials and forces, which has added a new canon—the machine canon—to the arts: above all, perhaps, the more objective personality that has come into existence through a more sensitive and understanding intercourse with these new social instruments and through their deliberate cultural assimilation. *In projecting one side of the human personality into the concrete forms of the machine, we have created an independent environment that has reacted upon every other side of the personality.*

In the past, the irrational and demonic aspects of life had invaded spheres where they did not belong. It was a step in advance to discover that bacteria, not brownies, were responsible for curdling milk, and that an air-cooled motor was more effective than a witch's broomstick for rapid long distance transportation. This triumph of order was pervasive: it gave a confidence to human purposes akin to that which a well-drilled regiment has when it marches in step. Creating the illusion of invincibility, the machine actually added to the amount of power man can exercize. Science and technics stiffened our morale: by their very austerities and abnegations they enhanced the value of the human personality that submitted to their discipline: they cast contempt on childish fears, childish guesses, equally childish assertions. By means of the machine man gave a concrete and external and impersonal form to his desire for order: and in a subtle way he thus set a new standard for his personal life and his more organic attitudes. Unless he was better than the machine he would only find himself reduced to its level: dumb, servile, abject, a creature of immediate reflexes and passive unselective responses.

While many of the boasted achievements of industrialism are merely rubbish, and while many of the goods produced by the machine are fraudulent and evanescent, its esthetic, its logic, and its factual technique remain a durable contribution: they are among man's supreme conquests. The practical results may be admirable or dubious: but the method that underlies them has a permanent importance to the race, apart from its immediate consequences. For the machine has added a whole series of arts to those produced by simple tools and handicraft methods and it has added a new realm to the environment in which the cultured man works and feels and thinks. Similarly, it has extended the power and range of human organs and has disclosed new esthetic spectacles, new worlds. The exact arts produced with the aid of the machine have their proper standards and give their own peculiar satisfactions to the human spirit. Differing in technique from the arts of the past, they spring nevertheless from the same source: for the machine itself, I must stress for the tenth time, is a human product, and its very abstractions make it more definitely human in one sense than those humane arts which on occasion realistically counterfeit nature.

Here, beyond what appears at the moment of realization, is the vital contribution of the machine. What matters the fact that the ordinary workman has the equivalent of 240 slaves to help him, if the master himself remains an imbecile, devouring the spurious news, the false suggestions, the intellectual prejudices that play upon him in the press and the school, giving vent in turn to tribal assertions and primitive lusts under the impression that he is the final token of progress and civilization. One does not make a child powerful by placing a stick of dynamite in his hands: one only adds to the dangers of his irresponsibility. Were mankind to remain children, they would exercize more effective power by being reduced to using a lump of clay and an old-fashioned modelling tool. But if the machine is one of the aids man has created toward achieving further intellectual growth and attaining maturity, if he treats this powerful automaton of his as a challenge to his own development, if the exact arts fostered by the machine have their own contribution to make to the mind, and are aids in the orderly crystallization of experience, then these contributions are vital ones indeed. The machine, which reached such overwhelming dimensions in Western Civilization partly because it sprang out of a disrupted and one-sided culture, nevertheless may help in enlarging the provinces of culture itself and thereby in building a greater synthesis: in that case, it will carry an antidote to its own poison. So let us consider the machine more closely as an instrument of culture and examine the ways in which we have begun, during the last century, to assimilate it.

2: The Neutrality of Order

Before the machine pervaded life, order was the boast of the gods and absolute monarchs. Both the deity and his representatives on earth had, however, the misfortune to be inscrutable in their judgment and frequently capricious and cruel in their assertion of mastery. On the human level, their order was represented by slavery: complete determination from above: complete subservience without question or understanding below. Behind the gods and the absolute monarchs stood brute nature itself, filled with demons, djinns, trolls, giants, contesting the reign of the gods. Chance and the accidental malice of the universe cut across the purposes of men and the observable regularities of nature. Even as a symbol the absolute monarch was weak as an exponent of order: his troops might obey with absolute precision, but he might be undone, as Hans Andersen pointed out in one of his fairy tales, by the small torture of a gnat.

With the development of the sciences and with the articulation of the machine in practical life, the realm of order was transferred from the absolute rulers, exercizing a personal control, to the universe of impersonal nature and to the particular group of artifacts and customs we call the machine. The royal formula of purpose—"I will"—was translated into the causal terms of science—"It must." By partly supplanting the crude desire for personal dominion by an impersonal curiosity and by the desire to understand, science prepared the way for a more effective conquest of the external environment and ultimately for a more effective control of the agent, man, himself. That a part of the order of the universe was a contribution by man himself, that the limitations imposed upon scientific research by human instruments and interests tend to produce an orderly and mathematically analyzable result, does not lessen the wonder and the beauty of the system: it rather gives to the conception of the universe itself some of the character of a work of art. To acknowl-edge the limitations imposed by science, to subordinate the wish to the fact, and to look for order as an emergent in observed relations, rather than as an extraneous scheme imposed upon these relations— these were the great contributions of the new outlook on life. Expressing regularities and recurrent series, science widened the area of certainty, prediction, and control.

By deliberately cutting off certain phases of man's personality, the warm life of private sensation and private feelings and private perceptions, the sciences assisted in building up a more public world which gained in accessibility what it lost in depth. To measure a weight, a distance, a charge of electricity, by reference to pointer readings established within a mechanical system, deliberately constructed for this purpose, was to limit the possibility of errors of interpretation, and cancel out the differences of individual experience and private history. And the greater the degree of abstraction and limitation, the greater was the accuracy of reference. By isolating simple systems and simple causal sequences the sciences created confidence in the possibility of finding a similar type of order in every aspect of experience: it was, indeed, by the success of science in the realm of the inorganic that we have acquired whatever belief we may legitimately entertain in the possibility of achieving similar understanding and control in the vastly more complex domain of life.

The first steps in the physical sciences did not go very far. Compared to organic behavior, in which any one of a given set of stimuli may create the same reaction, or in which a single stimulus may under different conditions create a number of different reactions, in which the organism as a whole responds and changes at the same time as the isolated part one seeks to investigate, compared to behavior within this frame the most complicated physical reaction is gratifyingly simple. But the point is that by means of the analyses and instruments developed in the physical sciences and embodied in technics, some of the necessary preliminary instruments for biological and social exploration have been created. All measurement involves the reference of certain parts of a complex phenomenon to a simpler one whose characteristics are relatively independent and fixed and determinable. The whole personality was a useless instrument for investigating limited mechanical phenomena. In its uncritical state, it was likewise useless for investigating organic systems, whether they were animal organisms or social groups. By a process of dismemberment science created a more useful type of order: an order external to the self. In the long run that special limitation fortified the ego as perhaps no other achievement in thought had done.

Although the most intense applications of the scientific method were in technology, the interests that it satisfied and re-excited, the desire for order that it expressed, translated themselves in other spheres. More and more factual research, the document, the exact calculation became a preliminary to expression. Indeed, the respect for quantities became a new condition of what had hitherto been crude qualitative judgments. Good and bad, beauty and ugliness, are determined, not merely by their respective natures but by the quantity one may assign to them in any particular situation. To think closely with respect to quantities is to think more accurately about the essential nature and the actual functions of things: arsenic is a tonic in grains and a poison in ounces: the quantity, the local composition, and the environmental relation of a quality are as important, so to say, as its original sign *as quality*. It is for this reason that a whole series of ethical distinctions, based upon the notion of pure and absolute qualities without relation to their amounts, has been instinctively discarded by a considerable part of mankind: while Samuel Butler's dictum, that every virtue should be mixed with a little of its opposite, implying as it does that qualities are altered by their quantitative relations, seems much closer to the heart of the matter. This respect for quantity has been grossly caricatured by dull pedantic minds who have sought by mathematical means to eliminate the qualitative aspects of complicated social and esthetic situations: but one need not be led by their mistake into failing to recognize the peculiar contribution that our quantitative technique has made in departments apparently remote from the machine.

One must distinguish between the cult of Nature as a standard and a criterion of human expression and the general influence of the scientific spirit. As for the first, one may say that though Ruskin, an esthetic disciple of science, rejected the Greek fret in decoration because it had no counterpart among flowers, minerals, or animals, for us today nature is no longer an absolute: or rather, we no longer regard nature as if man himself were not implicated in her, and as if his modifications of nature were not themselves a part of the natural order to which he is born. Even when emphasizing the impersonality of the machine one must not forget the busy humanizing that goes on before man even half completes his picture of an objective and indifferent nature. All the tools man uses, his eyes with their limited field of vision and their insensitiveness to ultra-violet and infra-red rays, his hands which can hold and manipulate only a limited number of objects at one time, his mind which tends to create categories of twos and threes because, without intensive training, to hold as many ideas together as a musician can hold notes of the piano puts an excessive strain upon his intelligence—still more his microscopes and balances, all bear the imprint of his own character as well as the general characteristics imposed by the physical environment. It has only been by a process of reasoning and inference—itself not free from the taint of his origin—that man has established the neutral realm of nature. Man may arbitrarily define nature as that part of his experience which is neutral to his desires and interests: but he with his desires and interests, to say nothing of his chemical constitution, has been formed by nature and inescapably is part of the system of nature. Once he has picked and chosen from this realm, as he does in science, the result is a work of art—*his* art: certainly it is no longer in a state of nature.

In so far as the cult of nature has made men draw upon a wider experience, to discover themselves in hitherto unexplored environments, and to contrive new isolations in the laboratory which will enable them to make further discoveries, it has been a good influence: man should be at home among the stars as well as at his own fireside. But although the new canon of order has a deep esthetic as well as an intellectual status, external nature has no finally independent authority: it exists, as a result of man's collective experience, and as a subject for his further improvisations by means of science, technics, and the humane arts.

The merit of the new order was to give man by projection an outer world which helped him to make over the hot spontaneous

world of desire he carried within. But the new order, the new impersonality, was but a fragment transplanted from the personality as a
whole: it had existed as part of man before he cut it off and gave
it an independent milieu and an independent root system. The comprehension and transformation of this impersonal "external" world
of technics was one of the great revelations of the painters and artists
and poets of the last three centuries. Art is the re-enactment of reality,
of a reality purified, freed from constraints and irrelevant accidents,
unfettered to the material circumstances that confuse the essence.
The passage of the machine into art was in itself a signal of release—
a sign that the hard necessities of practice, the preoccupation with the
immediate battle was over—a sign that the mind was free once more
to see, to contemplate, and so to enlarge and deepen all the practical
benefits of the machine.

Science had something other to contribute to the arts than the
notion that the machine was an absolute. It contributed, through
its effects upon invention and mechanization, a new type of order to
the environment: an order in which power, economy, objectivity,
the collective will play a more decisive part than they had played
before even in such absolute forms of dominion as in the royal
priesthood—and engineers—of Egypt or Babylon. The sensitive
apprehension of this new environment, its translation into terms
which involve human affections and feelings, and that bring into
play once more the full personality, became part of the mission of
the artist: and the great spirits of the nineteenth century, who first
fully greeted this altered environment, were not indifferent to it.
Turner and Tennyson, Emily Dickinson and Thoreau, Whitman and
Emerson, all saluted with admiration the locomotive, that symbol
of the new order in Western Society. They were conscious of the fact
that new instruments were changing the dimensions and to some
extent therefore the very qualities of experience; these facts were
just as clear to Thoreau as to Samuel Smiles; to Kipling as to H. G.
Wells. The telegraph wire, the locomotive, the ocean steamship, the
very shafts and pistons and switches that conveyed and canalized or
controlled the new power, could awaken emotion as well as the harp

and the war-horse: the hand at the throttle or the switch was no less regal than the hand that had once held a scepter.

The second contribution of the scientific attitude was a limiting one: it tended to destroy the lingering mythologies of Greek goddesses and Christian heroes and saints; or rather, it prevented a naive and repetitious use of these symbols. But at the same time, it disclosed new universal symbols, and widened the very domain of the symbol itself. This process took place in all the arts: it affected poetry as well as architecture. The pursuit of science, however, suggested new myths. The transformation of the medieval folk-legend of Dr. Faustus from Marlowe to Goethe, with Faust ending up as a builder of canals and a drainer of swamps and finding the meaning of life in sheer activity, the transformation of the Prometheus myth in Melville's Moby Dick, testify not to the destruction of myths by positive knowledge but to their more pregnant application. I can only repeat here what I have said in another place: "What the scientific spirit has actually done has been to exercise the imagination in finer ways than the autistic wish—the wish of the infant possessed of the illusions of power and domination—was able to express. Faraday's ability to conceive the lines of force in a magnetic field was quite as great a triumph as the ability to conceive of fairies dancing in a ring: and, Mr. A. N. Whitehead has shown, the poets who sympathized with this new sort of imagination, poets like Shelley, Wordsworth, Whitman, Melville, did not feel themselves robbed of their specific powers, but rather found them enlarged and refreshed.

"One of the finest love poems in the nineteenth century, Whitman's Out of the Cradle Endlessly Rocking, is expressed in such an image as Darwin or Audubon might have used, were the scientist as capable of expressing his inner feelings as of noting 'external' events: the poet haunting the seashore and observing the mating of the birds, day after day following their life, could scarcely have existed before the nineteenth century. In the early seventeenth century such a poet would have remained in the garden and written about a literary ghost, Philomel, and not about an actual pair of birds; in Pope's time the poet would have remained in the library and written about the birds on a lady's fan. Almost all the important works of the nineteenth century were cast in this mode and expressed the new imaginative range: they respect the fact: they are replete with observation: they project an ideal realm in and through, not transcendentally over, the landscape of actuality. Notre Dame might have been written by an historian, War and Peace by a sociologist, The Idiot might have been created by a psychiatrist, and Salammbô might have been the work of an archaeologist. I do not say that these books were scientific by intention, or that they might be replaced by a work of science without grave loss; far from it. I merely point out that they were conceived in the same spirit; that they belong to a similar plane of consciousness."

Once the symbol was focussed, the task of the practical arts became more purposive. Science gave the artist and the technician new objectives: it demanded that he respond to the nature of the machine's functions and refrain from seeking to express his personality by irrelevant and surreptitious means upon the objective material. The woodiness of wood, the glassiness of glass, the metallic quality of steel, the movement of motion—these attributes had been analyzed out by chemical and physical means, and to respect them was to understand and work with the new environment. Ornament, conceived apart from function, was as barbarous as the tattooing of the human body: the naked object, whatever it was, had its own beauty, whose revealment made it more human, and more close to the new personality than could any amount of artful decoration. While the Dutch gardeners of the seventeenth century had often, for example, turned the privet and the box into the shapes of animals and arbitrary figures, a new type of gardening appeared in the twentieth century which respected the natural ecological partnerships, and which not merely permitted plants to grow in their natural shapes but sought simply to clarify their natural relationships: scientific knowledge was one of the facts that indirectly contributed to the esthetic pleasure. That change symbolizes what has been steadily happening, sometimes slowly, sometimes rapidly, in all the arts. For finally, if nature itself is not an absolute, and if the facts of external nature are not the artist's sole materials, nor its literal imitation his guarantee of esthetic success, science nevertheless gives him the assurance of a partly independent realm which defines the limits of his own working powers. In ·creating his union of the inner world and the outer, of his passions and affections with the thing that exists, the artist need not remain the passive victim of his neurotic caprices and hallucinations: hence even when he departs from some external objective form or some tried convention, he still has a common measure of the extent of his deviation. While the determinism of the object—if one may coin a phrase—is more emphatic in the mechanical arts than in the humane ones, a binding thread runs through both realms.

Co-ordinate with the intellectual assimilation of the machine by the technician and the artist, which came partly through habit, partly through workaday experience, and partly through the extension of systematic training in science, came the esthetic and emotional apprehension of the new environment. Let us consider this in detail.

3: The Esthetic Experience of the Machine

The developed environment of the machine in the twentieth century has its kinship with primitive approximations to this order in the castles and fortifications and bridges from the eleventh to the thirteenth centuries, and even later: the bridge at Tournay or the brickwork and vaults of the Marienkirche at Lübeck: these earliest touches of the practical have the same fine characteristics that the latest grain elevators or steel cranes have. But the new characteristics now touch almost every department of experience. Observe the derricks, the ropes, the stanchions and ladders of a modern steamship, close at hand in the night, when the hard shadows mingle obliquely with the hard white shapes. Here is a new fact of esthetic experience; and it must be transposed in the same hard way: to look for gradation and atmosphere here is to miss a fresh quality that has emerged through the use of mechanical forms and mechanical modes of lighting. Or stand on a deserted subway platform and contemplate the low cavity becoming a black disc into which, as the train rumbles toward the station, two green circles appear as pin-points widening into plates. Or follow the spidery repetition of boundary lines, defining unoccupied cubes, which make the skeleton of a modern skyscraper: an effect not given even in wood before machine-sawed beams were possible. Or pass along the waterfront in Hamburg, say, and review the line of gigantic steel birds with spread legs that preside over the filling and emptying of the vessels in the basin: that span of legs, that long neck, the play of movement in this vast mechanism, the peculiar pleasure derived from the apparent lightness combined with enormous strength in its working, never existed on this scale in any other environment: compared to these cranes the pyramids of Egypt belong to the order of mud-pies. Or put your eye at the eyepiece of a microscope, and focus the high-powered lens on a thread, a hair, a section of leaf, a drop of blood: here is a world with forms and colors as varied and mysterious as those one finds in the depths of the sea. Or stand in a warehouse and observe a row of bathtubs, a row of siphons, a row of bottles, each of identical size, shape, color, stretching away for a quarter of a mile: the special visual effect of a repeating pattern, exhibited once in great temples or massed armies, is now a commonplace of the mechanical environment. There is an esthetic of units and series, as well as an esthetic of the unique and the non-repeatable.

Absent from such experiences, for the most part, is the play of surfaces, the dance of subtle lights and shadows, the nuances of color, tones, atmosphere, the intricate harmonies that human bodies and specifically organic settings display—all the qualities that belong to the traditional levels of experience and to the unordered world of nature. But face to face with these new machines and instruments, with their hard surfaces, their rigid volumes, their stark shapes, a fresh kind of perception and pleasure emerges: to interpret this order becomes one of the new tasks of the arts. While these new qualities existed as facts of mechanical industry, they were not generally recognized as values until they were interpreted by the painter and the sculptor; and so they existed in an indifferent anonymity for more than a century. The new forms were sometimes appreciated, perhaps, as symbols of Progress: but art, as such, is valued for what it is, not for what it indicates, and the sort of attention needed for the appreciation of art was largely lacking in the industrial environment of the nineteenth century, and except for the work of an occasional engineer of great talent, like Eiffel, was looked upon with deep suspicion.

At the very moment when the praise of industrialism was loudest and most confident, the environment of the machine was regarded as inherently ugly: so ugly that it mattered not how much additional ugliness was created by litter, refuse, slag-piles, scrap metal, or removable dirt. Just as Watt's contemporaries demanded more noise in the steam engine, as a proclamation of power, so did the paleotechnic mind glory, for the most part, in the anti-esthetic quality of the machine.

The Cubists were perhaps the first school to overcome this association of the ugly and the mechanical: they not merely held that beauty could be produced through the machine: they even pointed to the fact that it had been produced. The first expression of Cubism indeed dates back to the seventeenth century: Jean Baptiste Bracelle, in 1624, did a series of Bizarreries which depicted mechanical men, thoroughly cubist in conception. This anticipated in art, as Glanvill did in science, our later interests and inventions. What did the modern Cubists do? They extracted from the organic environment just those elements that could be stated in abstract geometrical symbols: they transposed and readjusted the contents of vision as freely as the inventor readjusted organic functions: they even created on canvas or in metal mechanical equivalents of organic objects: Léger painted human figures that looked as if they had been turned in a lathe, and Duchamp-Villon modeled a horse as if it were a machine. This whole process of rational experiment in abstract mechanical forms was pushed further by the constructivists. Artists like Grabo and Moholy-Nagy put together pieces of abstract sculpture, composed of glass, metal plates, spiral springs, wood, which were the non-utilitarian equivalents of the apparatus that the physical scientist was using in his laboratory. They created in form the semblance of the mathematical equations and physical formulae that had produced our new environment, seeking in this new sculpture to observe the physical laws of equipose or to evolve dynamic equivalents for the solid sculpture of the past by rotating a part of the object through space.

The ultimate worth of such efforts did not perhaps lie in the art itself: for the original machines and instruments were often just as stimulating as their equivalents, and the new pieces of sculpture were just as limited as the machines. No: the worth of these efforts lay in the increased sensitiveness to the mechanical environment that was produced in those who understood and appreciated this art. The esthetic experiment occupied a place comparable to the scientific experiment: it was an attempt to use a certain kind of physical apparatus for the purpose of isolating a phenomenon in experience and for determining the values of certain relations: the experiment was a guide to thought and an approach to action. Like the abstract paintings of Braque, Picasso, Léger, Kandinsky, these constructivist experiments sharpened the response to the machine as an esthetic object. By analyzing, with the aid of simple constructions, the effects produced, they showed what to look for and what values to expect. Calculation, invention, mathematical organization played a special rôle in the new visual effects produced by the machine, while the constant lighting of the sculpture and the canvas, made possible by electricity, profoundly altered the visual relationship. By a process of abstraction the new paintings finally, in some of the painters like Mondrian, approached a purely geometrical formula, with a mere residue of visual content.

Perhaps the most complete as well as the most brilliant interpretations of the capacities of the machine was in the sculpture of Brancusi: for he exhibited both form, method, and symbol. In Brancusi's work one notes first of all the importance of the material, with its specific weight, shape, texture, color, finish: when he models in wood he still endeavors to keep the organic shape of the tree, emphasizing rather than reducing the part given by nature, whereas when he models in marble he brings out to the full the smooth satiny texture, in the smoothest and most egg-like of forms. The respect for material extends further into the conception of the subject treated: the individual is submerged, as in science, into the class: instead of representing in marble the counterfeit head of a mother and child, he lays two blocks of marble side by side with only the faintest depression of surface to indicate the features of the face: it is by relations of volume that he presents the generic idea of mother and child: the idea in its most tenuous form. Again, in his famous bird, he treats the object itself, in the brass model, as if it were the piston of an engine: the tapering is as delicate, the polish is as high, as if it were to be fitted into the most intricate piece of machinery, in which so much as a few specks of dust would interfere with its perfect action: looking at the bird, one thinks of the shell of a torpedo. As for the bird itself, it is no longer any particular bird, but a generic bird in its most birdlike aspect, the function of flight. So, too, with his metallic or marble fish, looking like experimental forms developed in an aviation laboratory, floating on the flawless surface of a mirror. Here is the equivalent in art of the mechanical world that lies about us on every hand: with this further perfection of the symbol, that in the highly polished metallic forms the world as a whole and the spectator himself, are likewise mirrored: so that the old separation between subject and object is now figuratively closed. The obtuse United States customs officer who wished to classify Brancusi's sculpture as machinery or plumbing was in fact paying it a compliment. In Brancusi's sculpture the idea of the machine is objectified and assimilated in equivalent works of art.

In this perception of the machine as a source of art, the new painters and sculptors clarified the whole issue and delivered art from the romantic prejudice against the machine as necessarily hostile to the world of feeling. At the same time, they began to interpret intuitively the new conceptions of time and space that distinguish the present age from the Renascence. The course of this development can perhaps be followed best in the photograph and the motion picture: the specific arts of the machine.

4: Photography as Means and Symbol

The history of the camera, and of its product, the photograph, illustrates the typical dilemmas that have arisen in the development of the machine process and its application to objects of esthetic value. Both the special feats of the machine and its possible perversions are equally manifest.

At first, the limitations of the camera were a safeguard to its intelligent use. The photographer, still occupied with difficult photo-chemical and optical problems, did not attempt to extract from the photograph any other values than those rendered immediately by the technique itself; and as a result, the grave portraiture of some of the early photographers, particularly that of David Octavius Hill of Edinburgh, reached a high pitch of excellence: indeed it has not often been surpassed by any of the later work. As the technical problems were solved one by one, through the use of better lenses, more sensitive emulsions, new textures of paper to replace the shiny surface of the daguerreotype, the photographer became more conscious of the esthetic arrangements of the subjects before him: instead of carrying the esthetic of the light-picture further, he returned timidly to the canons of painting, and endeavored to make his pictures fit certain preconceptions of beauty as achieved by the classical painters. Far from glorying in minute and tangled representation of life, as the mechanical eye confronts it, the photographer from the eighties onward sought by means of soft lenses a foggy impressionism, or by care of arrangement and theatrical lighting he attempted to imitate the postures and sometimes the costumes of Holbein and Gainsborough. Some experimenters even went so far as to imitate in the photographic print the smudgy effect of charcoal or the crisp lines of the etching. This relapse from clean mechanical processes to an artful imitativeness worked ruin in photography for a full generation: it was like that relapse in the technique of furniture making which used modern machinery to imitate the dead forms of antique handicraft. In back of it was the failure to understand the intrinsic esthetic importance of the new mechanical device in terms of its own peculiar possibilities.

Every photograph, no matter how painstaking the observation of the photographer or how long the actual exposure, is essentially a snapshot: it is an attempt to penetrate and capture the unique esthetic moment that singles itself out of the thousand of chance compositions, uncrystallized and insignificant, that occur in the course of a day. The photographer cannot rearrange his material on his own terms. He must take the world as he finds it: at most his rearrangement is limited to a change in position or an alteration of the direction and intensity of the light or in the length of the focus. He must respect and understand sunlight, atmosphere, the time of day, the season of the year, the capabilities of the machine, the processes of chemical development; for the mechanical device does not function automatically, and the results depend upon the exact correlation of the esthetic moment itself with the appropriate physical means. But whereas an underlying technique conditions both painting and photography—for the painter, too, must respect the chemical composition of his colors and the physical conditions which will give them permanence and visibility—photography differs from the other graphic arts in that the process is determined at every state by the external conditions that present themselves: his inner impulse, instead of spreading itself in subjective fantasy, must always be in key with outer circumstances. As for the various kinds of *montage* photography, they are in reality not photography at all but a kind of painting, in which the photograph is used—as patches of textiles are used in crazy-quilts—to form a mosaic. Whatever value the montage may have derives from the painting rather than the camera.

Rare though painting of the first order is, photography of the first rank is perhaps even rarer. The gamut of emotion and significance represented in photography by the work of Alfred Stieglitz in America is one that the photographer rarely spans. Half the merit of Stieglitz' work is due to his rigorous respect for the limitations of the machine and to the subtlety with which he effects the combination of image and paper. He plays no tricks, he has no affectations, not even the affectation of being hard-boiled, for life and the object have their soft moments and their tender aspects. The mission of the photograph is to clarify the object. This objectification, this clarification, are important developments in the mind itself: it is perhaps the prime psychological fact that emerges with our rational assimilation of the machine. To see as they are, as if for the first time, a boatload of immigrants, a tree in Madison Square Park, a woman's breast, a cloud lowering over a black mountain—that requires patience and understanding. Ordinarily we skip over and schematize these objects, relate them to some practical need, or subordinate them to some immediate wish: photography gives us the ability to recognize them in the independent form created by light and shade and shadow. Good photography, then, is one of the best educations toward a rounded sense of reality. Restoring to the eye, otherwise so preoccupied with the abstractions of print, the stimulus of things roundly seen as things, shapes, colors, textures, demanding for its enjoyment a previous experience of light and shade, this machine process in itself counteracts some of the worst defects of our mechanical environment. It is the hopeful antithesis to an emasculated and segregated esthetic sensibility, the cult of pure form, which endeavors to hide away from the world that ultimately gives shape and significance to its remotest symbols.

If photography has become popular again in our own day, after its first great but somewhat sentimental outburst in the eighties, it is perhaps because, like an invalid returning to health, we are finding a new delight in being, seeing, touching, feeling; because in a rural or a neotechnic environment the sunlight and pure air that make it possible are present; because, too, we have at least learned Whitman's lesson and behold with a new respect the miracle of our finger joints or the reality of a blade of grass: photography is not least effective when it is dealing with such ultimate simplicities. To disdain photography because it cannot achieve what El Greco or Rembrandt or Tintoretto achieved is like dismissing science because its view of the world is not comparable to the visions of Plotinus or the mythologies of Hinduism. Its virtue lies precisely in the fact that it has conquered another and quite different department of reality. For photography, finally, gives the effect of permanence to the transient and the ephemeral: photography—and perhaps photography alone—is capable of coping with and adequately presenting the complicated, inter-related aspects of our modern environment. As histories of the human comedy of our times, the photographs of Atget in Paris and of Stieglitz in New York are unique both as drama and as document: not merely do they convey to us the very shape and touch of this environment, but by the angle of vision and the moment of observation throw an oblique light upon our inner lives, our hopes, our values, our humours. And this art, of all our arts, is perhaps the most widely used and the most fully enjoyed: the amateur, the specialist, the news-photographer, and the common man have all participated in this eye-opening experience, and in this discovery of that esthetic moment which is the common property of all experience, at all its various levels from ungoverned dream to brute action and rational idea.

What has been said of the photograph applies even more, perhaps, to the motion picture. In its first exploitation the motion picture emphasized its unique quality: the possibility of abstracting and reproducing objects in motion: the simple races and chases of the early pictures pointed the art in the right direction. But in its subsequent commercial development it was degraded a little by the attempt to make it the vehicle of a short-story or a novel or a drama: a mere imitation in vision of entirely different arts. So one must distinguish between the motion picture as an indifferent reproductive device, less satisfactory in most ways than direct production on the stage, and the motion picture as an art in its own right. The great achievements of the motion picture have been in the presentation of history or natural history, the sequences of actuality, or in their interpretation of the inner realm of fantasy, as in the pure comedies of Charlie Chaplin and René Clair and Walt Disney. Unlike the photograph, the extremes of subjectivism and of factualism meet in the motion picture. Nanook of the North, Chang, the S.S. Potemkin—these pictures got their dramatic effect through their interpretation of an immediate experience and through a heightened delight in actuality. Their exoticism was entirely accidental: an equally good eye would abstract the same order of significant events from the day's routine of a subway guard or a factory-hand: indeed, the most consistently interesting pictures have been those of the newsreel—despite the insufferable banality of the announcers who too often accompany them.

Not plot in the old dramatic sense, but historic and geographic sequences is the key to the arrangement of these new kinetic compositions: the passage of objects, organisms, dream images through time and space. It is an unfortunate social accident—as has happened in so many departments of technics—that this art should have been grossly diverted from its proper function by the commercial necessity for creating sentimental shows for an emotionally empty metropolitanized population, living vicariously on the kisses and cocktails and crimes and orgies and murders of their shadow-idols. For the motion picture symbolizes and expresses, better than do any of the traditional arts, our modern world picture and the essential conceptions of time and space which are already part of the unformulated experience of millions of people, to whom Einstein or Bohr or Bergson or Alexander are scarcely even names.

In Gothic painting one may recall time and space were successive and unrelated: the immediate and the eternal, the near and the far, were confused: the faithful time ordering of the medieval chroniclers is marred by the jumble of events presented and by the impossibility of distinguishing hearsay from observation and fact from conjecture. In the Renascence space and time were co-ordinated within a single system: but the axis of these events remained fixed, so to say, within a single frame established at a set distance from the observer, whose existence with reference to the system was innocently taken for granted. Today, in the motion picture, which symbolizes our actual perceptions and feelings, time and space are not merely co-ordinated on their own axis, but in relation to an observer who himself, by his position, partly determines the picture, and who is no longer fixed but is likewise capable of motion. The moving picture, with its close-ups and its synoptic views, with its shifting events and its ever-present camera eye, with its spatial forms always shown through time, with its capacity for representing objects that interpenetrate, and for placing distant environments in immediate juxtaposition—as happens in instantaneous communication—with its ability, finally, to represent subjective elements, distortions, hallucinations, it is today the only art that can represent with any degree of concreteness the emergent world-view that differentiates our culture from every preceding one.

Even with weak and trivial subjects, the art focusses interests and captures values that the traditional arts leave untouched. Music alone heretofore has represented movement through time: but the motion picture synthesizes movement through both time and space, and in the very fact that it can co-ordinate visual images with sound and release both of these elements from the boundaries of apparent space and a fixed location, it contributes something to our picture of the world not given completely in direct experience. Utilizing our daily experience of motion in the railroad train and the motor car, the motion picture re-creates in symbolic form a world that is otherwise beyond our direct perception or grasp. Without any conscious notion of its destination, the motion picture presents us with a world of interpenetrating, counter-influencing organisms: and it enables us to think about that world with a greater degree of concreteness. This is no small triumph in cultural assimilation. Though it has been so stupidly misused, the motion picture nevertheless announces itself as a major art of the neotechnic phase. Through the machine, we have new possibilities of understanding the world we have helped to create.

But in the arts, it is plain that the machine is an instrument with manifold and conflicting possibilities. It may be used as a passive substitute for experience; it may be used to counterfeit older forms of art; it may also be used, in its own right, to concentrate and intensify and express new forms of experience. As substitutes for primary experience, the machine is worthless: indeed it is actually debilitating. Just as the microscope is useless unless the eye itself is keen, so all our mechanical apparatus in the arts depends for its success upon the due cultivation of the organic, physiological, and spiritual aptitudes that lie behind its use. The machine cannot be used as a shortcut to escape the necessity for organic experience. Mr. Waldo Frank has put the matter well: "Art," he says, "cannot become a language, hence an experience, unless it is practiced. To the man who plays, a mechanical reproduction of music may mean much, since he already has the experience to assimilate. But where reproduction becomes the norm, the few music makers will grow more isolate and sterile, and the ability to experience music will disappear. The same is true with the cinema, dance, and even sport."

Whereas in industry the machine may properly replace the human being when he has been reduced to an automaton, in the arts the machine can only extend and deepen man's original functions and intuitions. In so far as the phonograph and the radio do away with the impulse to sing, in so far as the camera does away with the impulse to see, in so far as the automobile does away with the impulse to walk, the machine leads to a lapse of function which is but one step away from paralysis. But in the application of mechanical instruments to the arts it is not the machine itself that we must fear. The chief danger lies in the failure to integrate the arts themselves with the totality of our life-experience: the perverse triumph of the machine follows automatically from the abdication of the spirit. Consciously to assimilate the machine is one means of reducing its omnipotence. We cannot, as Karl Buecher wisely said, "give up the hope that it will be possible to unite technics and art in a higher rhythmical unity, which will restore to the spirit the fortunate serenity and to the body the harmonious cultivation that manifest themselves at their best among primitive peoples." The machine has not destroyed that promise. On the contrary, through the more conscious cultivation of the machine arts and through greater selectivity in their use, one sees the pledge of its wider fulfillment throughout civilization. For at the bottom of that cultivation there must be the direct and immediate experience of living itself: we must directly see, feel, touch, manipulate, sing, dance, communicate before we can extract from the machine any further sustenance for life. If we are empty to begin with, the machine will only leave us emptier; if we are passive and powerless to begin with, the machine will only leave us more feeble.

5: The Growth of Functionalism

But modern technics, even apart from the special arts that it fostered, had a cultural contribution to make in its own right. Just as science underlined the respect for fact, so technics emphasized the importance of function: in this domain, as Emerson pointed out, the beautiful rests on the foundations of the necessary. The nature of this contribution can best be shown, perhaps, by describing the way in which the problem of machine design was first faced, then evaded, and finally solved.

One of the first products of the machine was the machine itself. As in the organization of the first factories the narrowly practical considerations were uppermost, and all the other needs of the personality were firmly shoved to one side. The machine was a direct expression of its own functions: the first cannon, the first crossbows, the first steam engines were all nakedly built for action. But once the primary problems of organization and operation had been solved, the human factor, which had been left out of the picture, needed somehow to be re-incorporated. The only precedent for this fuller integration of form came naturally from handicraft: hence over the incomplete, only partly realized forms of the early cannon, the early bridges, the early machines, a meretricious touch of decoration was added: a mere relic of the happy, semi-magical fantasies that painting and carving had once added to every handicraft object. Because perhaps the energies of the eotechnic period were so completely engrossed in the technical problems, it was, from the standpoint of design, amazingly clean and direct: ornament flourished in the utilities of life, flourished often perversely and extravagantly, but one looks for it in vain among the machines pictured by Agricola or Besson or the Italian engineers: they are as direct and factual as was architecture from the tenth to the thirteenth century.

The worst sinners—that is the most obvious sentimentalists—were the engineers of the paleotechnic period. In the act of recklessly deflowering the environment at large, they sought to expiate their failures by adding a few sprigs or posies to the new engines they were creating: they embellished their steam engines with Doric columns or partly concealed them behind Gothic tracery: they decorated the frames of their presses and their automatic machines with cast-iron arabesque, they punched ornamental holes in the iron framework of their new structures, from the trusses of the old wing of the Metropolitan Museum to the base of the Eiffel tower in Paris. Everywhere similar habits prevailed: the homage of hypocrisy to art. One notes identical efforts on the original steam radiators, in the floral decorations that once graced typewriters, in the nondescript ornament that still lingers quaintly on shotguns and sewing machines, even if it has at length disappeared from cash registers and Pullman cars—as long before, in the first uncertainties of the new technics, the same division had appeared in armor and in crossbows.

The second stage in machine design was a compromise. The object was divided into two parts. One of them was to be precisely designed for mechanical efficiency. The other was to be designed for looks. While the utilitarian claimed the working parts of the structure the esthete was, so to speak, permitted slightly to modify the surfaces with his unimportant patterns, his plutonic flowers, his aimless filigree, provided he did not seriously weaken the structure or condemn the function to inefficiency. Mechanically utilizing the machine, this type of design shamefully attempted to conceal the origins that were still felt as low and mean. The engineer had the uneasiness of a parvenu, and the same impulse to imitate the most archaic patterns of his betters.

Naturally the next stage was soon reached: the utilitarian and the esthete withdrew again to their respective fields. The esthete, insisting with justice that the structure was integral with the decoration and that art was something more fundamental than the icing the pastrycook put on the cake, sought to make the old decoration real by altering the nature of the structure. Taking his place as workman, he began to revive the purely handicraft methods of the weaver, the cabinet maker, the printer, arts that had survived for the most part only in the more backward parts of the world, untouched by the tourist and the commercial traveller. The old workshops and ateliers were languishing and dying out in the nineteenth century, especially in progressive England and in America, when new ones, like those devoted to glass under William de Morgan in England, and John La Farge in America, and Lalique in France, or to a miscellany of handicrafts, such as that of William Morris in England, sprang into existence, to prove by their example that the arts of the past could survive. The industrial manufacturer, isolated from this movement yet affected by it, contemptuous but half-convinced, made an effort to retrieve his position by attempting to copy mechanically the dead forms of art he found in the museum. So far from gaining from the handicrafts movement by this procedure he lost what little virtue his untutored designs possessed, issuing as they sometimes did out of an intimate knowledge of the processes and the materials.

The weakness of the original handicrafts movement was that it assumed that the only important change in industry had been the intrusion of the soulless machine. Whereas the fact was that everything had changed, and all the shapes and patterns employed by technics were therefore bound to change, too. The world men carried in their heads, their idolum, was entirely different from that which set the medieval mason to carving the history of creation or the lives of the saints above the portals of the cathedral, or a jolly image of some sort above his own doorway. An art based like handicraft upon a certain stratification of the classes and the social differentiation of the arts could not survive in a world where men had seen the French Revolution and had been promised some rough share of equality. Modern handicraft, which sought to rescue the worker from the slavery of shoddy machine production, merely enabled the well-to-do to enjoy new objects that were as completely divorced from the dominant social milieu as the palaces and monasteries that the antiquarian art dealer and collector had begun to loot. The *educational aim* of the arts and crafts movement was admirable; and, in so far as it gave courage and understanding to the amateur, it was a success. If this movement did not add a sufficient amount of good handicraft it at least took away a great deal of false art. William Morris's dictum, that one should not possess anything one did not believe to be beautiful or know to be useful was, in the shallow showy bourgeois world he addressed, a revolutionary dictum.

But the social outcome of the arts and crafts movement was not commensurate with the needs of the new situation; as Mr. Frank Lloyd Wright pointed out in his memorable speech at Hull House in 1908, the machine itself was as much an instrument of art, in the hands of an artist, as were the simple tools and utensils. To erect a social barrier between machines and tools was really to accept the false notion of the new industrialist who, bent on exploiting the machine, which they owned, and jealous of the tool, which might still be owned by the independent worker, bestowed on the machine an exclusive sanctity and grace it did not merit. Lacking the courage to use the machine as an instrument of creative purpose, and being unable to attune themselves to new objectives and new standards, the esthetes were logically compelled to restore a medieval ideology in order to provide a social backing for their anti-machine bias. In a word, the arts and crafts movement did not grasp the fact that the new technics, by expanding the rôle of the machine, had altered the entire relation of handwork to production, and that the exact processes of the machine were not necessarily hostile to handicraft and fine workmanship. In its modern form handicraft could no longer serve as in the past when it had worked under the form of an intensive caste-specialization. To survive, handicraft would have to adapt itself to the amateur, and it was bound to call into existence, even in pure handwork, those forms of economy and simplicity which the machine was claiming for its own, and to which it was adapting mind and hand and eye. In this process of re-integration certain "eternal" forms would be recovered: there are handicraft forms dating back to a distant past which so completely fulfill their functions that no amount of further calculation or experiment will alter them for the better. These type-forms appear and reappear from civilization to civilization; and if they had not been discovered by handicraft, the machine would have had to invent them.

The new handicraft was in fact to receive presently a powerful lesson from the machine. For the forms created by the machine, when they no longer sought to imitate old superficial patterns of hand-work, were closer to those that could be produced by the amateur than were, for example, the intricacies of special joints, fine inlays, matched woods, beads and carvings, complicated forms of metallic ornament, the boast of handicraft in the past. While in the factory the machine was often reduced to producing fake handicraft, in the workshop of the amateur the reverse process could take place with a real gain: he was liberated by the very simplicities of good machine forms. Machine technique as a means to achieving a simplified and purified form relieved the amateur from the need of respecting and imitating the perversely complicated patterns of the past—patterns whose complications were partly the result of conspicuous waste, partly the outcome of technical virtuosity, and partly the result of a different state of feeling. But before handicraft could thus be restored as an admirable form of play and an efficacious relief from a physically untutored life, it was necessary to dispose of the machine itself as a social and esthetic instrument. So the major contribution to art was made, after all, by the industrialist who remained on the job and saw it through.

With the third stage in machine design an alteration takes place. The imagination is not applied to the mechanical object after the practical design has been completed: it is infused into it at every stage in development. The mind works through the medium of the machine directly, respects the conditions imposed upon it, and—not content with a crude quantitative approximation—seeks out a more positive esthetic fulfillment. This must not be confused with the dogma, so often current, that any mechanical contraption that works necessarily is esthetically interesting. The source of this fallacy is plain. In many cases, indeed, our eyes have been trained to recognize beauty in nature, and with certain kinds of animals and birds we have an especial sympathy. When an airplane becomes like a gull it has the advantage of this long association and we properly couple the beauty with the mechanical adequacy, since the poise and swoop of a gull's flight casts in addition a reflective beauty on its animal structure. Having no such association with a milkweed seed, we do not feel the same beauty in the autogyro, which is kept aloft by a similar principle. While genuine beauty in a thing of use must always be joined to mechanical adequacy and therefore involves a certain amount of intellectual recognition and appraisal, the relation is not a simple one: it points to a common source rather than an identity.

In the conception of a machine or of a product of the machine there is a point where one may leave off for parsimonious reasons without having reached esthetic perfection: at this point perhaps every mechanical factor is accounted for, and the sense of incompleteness is due to the failure to recognize the claims of the human agent. Esthetics carries with it the implication of alternatives between a number of mechanical solutions of equal validity: and unless this awareness is present at every stage of the process, in smaller matters of finish, fineness, trimness, it is not likely to come out with any success in the final stage of design. Form follows function, underlining it, crystallizing it, clarifying it, making it real to the eye. Makeshifts and approximations express themselves in incomplete forms: forms like the absurdly cumbrous and ill-adjusted telephone apparatus of the past, like the old-fashioned airplane, full of struts, wires, extra supports, all testifying to an anxiety to cover innumerable unknown or uncertain factors; forms like the old automobile in which part after part had been added to the effective mechanism without having been absorbed into the body of the design as a whole; forms like our oversized steel-work which were due to our carelessness in using cheap materials and our desire to avoid the extra expense of calculating them finely and expending the necessary labor to work them up. The impulse that creates a complete mechanical object is akin to that which creates an esthetically finished object; and the fusion of the two at every stage in the process will necessarily be effected by the environment at large: who can gauge how much the slatternliness and disorder of the paleotechnic environment undermined good design, or how much the order and beauty of our neotechnic plants—like that of the Van Nelle factory in Rotterdam—will eventually aid it? Esthetic interests can not suddenly be introduced from without: they must be constantly operative, constantly visible.

Expression through the machine implies the recognition of relatively new esthetic terms: precision, calculation, flawlessness, simplicity, economy. Feeling attaches itself in these new forms to different qualities than those that made handicraft so entertaining. Success here consists in the elimination of the non-essential, rather than, as in handicraft decoration, in the willing production of superfluity, contributed by the worker out of his own delight in the work. The elegance of a mathematical equation, the inevitability of a series of physical inter-relations, the naked quality of the material itself, the tight logic of the whole—these are the ingredients that go into the design of machines: and they go equally into products that have been properly designed for machine production. In handicraft it is the worker who is represented: in machine design it is the work. In handicraft, the personal touch is emphasized, and the imprint of the worker and his tool are both inevitable: in machine work the impersonal prevails, and if the worker leaves any tell-tale evidence of his part in the operation, it is a defect or a flaw. Hence the burden of machine design is in the making of the original pattern: it is here that trials are made, that errors are discovered and buried, that the creative process as a whole is concentrated. Once the master-pattern is set, the rest is routine: beyond the designing room and the laboratory there is—for goods produced on a serial basis for a mass market—no opportunity for choice and personal achievement. Hence apart from those commodities that can be produced automatically, the effort of sound industrial production must be to increase the province of the designing room and the laboratory, reducing the scale of the production, and making possible an easier passage back and forth between the designing and the operative sections of the plant.

Who discovered these new canons of machine design? Many an engineer and many a machine worker must have mutely sensed them and reached toward them: indeed, one sees the beginning of them in very early mechanical instruments. But only after centuries of more or less blind and unformulated effort were these canons finally demonstrated with a certain degree of completeness in the work of the great engineers toward the end of the nineteenth century—particularly the Roeblings in America and Eiffel in France—and formulated after that by theoreticians like Riedler and Meyer in Germany. The popularization of the new esthetic awaited, as I have pointed out, the post-impressionist painters. They contributed by breaking away from the values of purely associative art and by abolishing an undue concern for natural objects as the basis of the painter's interest: if on one side this led to completer subjectivism, on the other it tended toward a recognition of the machine as both form and symbol. In the same direction Marcel Duchamp, for example, who was one of the leaders of this movement, made a collection of cheap, ready-made articles, produced by the machine, and called attention to their esthetic soundness and sufficiency. In many cases, the finest designs had been achieved before any conscious recognition of the esthetic had taken place. With the coming of a commercialized designer, seeking to add "art" to a product which *was* art, the design has more often than not been trifled with and spoiled. The studious botching of the kodak, the bathroom fixture, and the steam radiator under such stylicizing is a current commonplace.

The key to this fresh appreciation of the machine as a source of new esthetic forms has come through a formulation of its chief esthetic principle: the principle of economy. This principle is of course not unknown in other phases of art: but the point is that in mechanical forms it is at all times a controlling one, and it has for its aid the more exact calculations and measurements that are now possible. The aim of sound design is to remove from the object, be it an automobile or a set of china or a room, every detail, every moulding, every variation of the surface, every extra part except that which conduces to its effective functioning. Toward the working out of this principle, our mechanical habits and our unconscious impulses have been tending steadily. In departments where esthetic choices are not consciously uppermost our taste has often been excellent and sure. Le Corbusier has been very ingenious in picking out manifold objects, buried from observation by their very ubiquity, in which this mechanical excellence of form has manifested itself without pretence or fumbling. Take the smoking pipe: it is no longer carved to look like a human head nor does it bear, except among college students, any heraldic emblems: it has become exquisitely anonymous, being nothing more than an apparatus for supplying drafts of smoke to the human mouth from a slow-burning mass of vegetation. Take the ordinary drinking glass in a cheap restaurant: it is no longer cut or cast or engraved with special designs: at most it may have a slight bulge near the top to keep one glass from sticking to another in stacking: it is as clean, as functional, as a high tension insulator. Or take the present watch and its case and compare it with the forms that handicraft ingenuity and taste and association created in the sixteenth or seventeenth centuries. In all the commoner objects of our environment the machine canons are instinctively accepted: even the most sentimental manufacturer of motor cars has not been tempted to paint his coach work to resemble a sedan chair in the style of Watteau, although he may live in a house in which the furniture and decoration are treated in that perverse fashion.

This stripping down to essentials has gone on in every department of machine work and has touched every aspect of life. It is a first step toward that completer integration of the machine with human needs and desires which is the mark of the neotechnic phase, and will be even more the mark of the biotechnic period, already visible over the edge of the horizon. As in the social transition from the paleotechnic to the neotechnic order, the chief obstacle to the fuller development of the machine lies in the association of taste and fashion with waste and commercial profiteering. For the rational development of genuine technical standards, based on function and performance, can come about only by a wholesale devaluation of the scheme of bourgeois civilization upon which our present system of production is based.

Capitalism, which along with war played such a stimulating part in the development of technics, now remains with war the chief obstacle toward its further improvement. The reason should be plain. The machine devaluates rarity: instead of producing a single unique object, it is capable of producing a million others just as good as the master model from which the rest are made. The machine devaluates age: for age is another token of rarity, and the machine, by placing its emphasis upon fitness and adaptation, prides itself on the brand-new rather than on the antique: instead of feeling comfortably authentic in the midst of rust, dust, cobwebs, shaky parts, it prides itself on the opposite qualities—slickness, smoothness, gloss, cleanness. The machine devaluates archaic taste: for taste in the bourgeois sense is merely another name for pecuniary reputability, and against that standard the machine sets up the standards of function and fitness. The newest, the cheapest, the commonest objects may, from the standpoint of pure esthetics, be immensely superior to the rarest, the most expensive, and the most antique. To say all this is merely to emphasize that the modern technics, by its own essential nature, imposes a great purification of esthetics: that is, it strips off from the object all the barnacles of association, all the sentimental and pecuniary values which have nothing whatever to do with esthetic form, and it focusses attention upon the object itself.

The social devaluation of caste, enforced by the proper use and appreciation of the machine, is as important as the stripping down of essential forms in the process itself. One of the happiest signs of this during the last decade was the use of cheap and common materials in jewelry, first introduced, I believe, by Lalique: for this implied a recognition of the fact that an esthetically appropriate form, even in the adornment of the body, has nothing to do with rarity or expense, but is a matter of color, shape, line, texture, fitness, symbol. The use of cheap cottons in dress by Chanel and her imitators, which was another post-war phenomenon, was an equally happy recognition of the essential values in our new economy: it at last put our civilization, if only momentarily, on the level of those primitive cultures which gladly bartered their furs and ivory for the white man's colored glass beads, by the adroit use of which the savage artist often proved to any disinterested observer that they—contrary to the white man's fatuous conceit—had gotten the better of the bargain. Because of the fact that woman's dress has a peculiarly compensatory rôle to play in our megalopolitan society, so that it more readily indicates what is absent than calls attention to what is present in it, the victory for genuine esthetics could only be a temporary one. But these forms of dress and jewelry pointed to the goal of machine production: the goal at which each object would be valued in terms of its direct mechanical and vital and social function, apart from its pecuniary status, the snobberies of caste, or the dead sentiments of historical emulation.

This warfare between a sound machine esthetic and what Veblen has called the "requirements of pecuniary reputability" has still another side. Our modern technology has, in its inner organization, produced a collective economy and its typical products are collective products. Whatever the politics of a country may be, the machine is a communist: hence the deep contradictions and conflicts that have kept on developing in machine industry since the end of the eighteenth century. At every stage in technics, the work represents a collaboration of innumerable workers, themselves utilizing a large and ramifying technological heritage: the most ingenious inventor, the most brilliant individual scientist, the most skilled designer contributes but a moiety to the final result. And the product itself necessarily bears the same impersonal imprint: it either functions or it does not function on quite impersonal lines. There can be no qualitative difference between a poor man's electric bulb of a given candlepower and a rich man's, to indicate their differing pecuniary status in society, although there was an enormous difference between the rush or stinking tallow of the peasant and the wax candles or sperm oil used by the upper classes before the coming of gas and electricity.

In so far as pecuniary differences are permitted to count in the machine economy, they can alter only the scale of things—not, in terms of present production, the kind. What applies to electric light bulbs applies to automobiles: what applies there applies equally to every manner of apparatus or utility. The frantic attempts that have been made in America by advertising agencies and "designers" to stylicize machine-made objects have been, for the most part, attempts to pervert the machine process in the interests of caste and pecuniary distinction. In money-ridden societies, where men play with poker chips instead of with economic and esthetic realities, every attempt is made to disguise the fact that the machine has achieved potentially a new collective economy, in which the possession of goods is a meaningless distinction, since the machine can produce all our essential goods in unparalleled qualities, falling on the just and the unjust, the foolish and the wise, like the rain itself.

The conclusion is obvious: we cannot intelligently accept the practical benefits of the machine without accepting its moral imperatives and its esthetic forms. Otherwise both ourselves and our society will be the victims of a shattering disunity, and one set of purposes, that which created the order of the machine, will be constantly at war with trivial and inferior personal impulses bent on working out in covert ways our psychological weaknesses. Lacking on the whole this rational acceptance, we have lost a good part of the practical benefits of the machine and have achieved esthetic expression only in a spotty, indecisive way. The real social distinction of modern technics, however, is that it tends to eliminate social distinctions. Its immediate goal is effective work. Its means are standardization: the emphasis of the generic and the typical: in short, conspicuous economy. Its ultimate aim is leisure—that is, the release of other organic capacities.

The powerful esthetic side of this social process has been obscured by speciously pragmatic and pecuniary interests that have inserted themselves into our technology and have imposed themselves upon its legitimate aims. But in spite of this deflection of effort, we have at last begun to realize these new values, these new forms, these new modes of expression. Here is a new environment—man's extension of nature in terms discovered by the close observation and analysis and abstraction of nature. The elements of this environment are hard and crisp and clear: the steel bridge, the concrete road, the turbine and the alternator, the glass wall. Behind the façade are rows and rows of machines, weaving cotton, transporting coal, assembling food, printing books, machines with steel fingers and lean muscular arms, with perfect reflexes, sometimes even with electric eyes. Alongside them are the new utilities—the coke oven, the transformer, the dye vats—chemically cooperating with these mechanical processes, assembling new qualities in chemical compounds and materials. Every effective part in this whole environment represents an effort of the collective mind to widen the province of order and control and provision. And here, finally, the perfected forms begin to hold human interest even apart from their practical performances: they tend to produce that inner composure and equilibrium, that sense of balance between the inner impulse and the outer environment, which is one of the marks of a work of art. The machines, even when they are not works of art, underlie our art—that is, our organized perceptions and feelings—in the way that Nature underlies them, extending the basis upon which we operate and confirming our own impulse to order. The economic: the objective: the collective: and finally the integration of these principles in a new conception of the organic—these are the marks, already discernible, of our assimilation of the machine not merely as an instrument of practical action but as a valuable mode of life.

6: The Simplification of the Environment

As a practical instrument, the machine has enormously complicated the environment. When one compares the shell of an eighteenth
century house with the tangle of water-pipes, gas-pipes, electric wires,
sewers, aerials, ventilators, heating and cooling systems that compose
a modern house, or when one compares the cobblestones of the old-
fashioned street, set directly on the earth, with the cave of cables,
pipes, and subway systems that run under the asphalt, one has no
doubt about the mechanical intricacy of modern existence.

But precisely because there are so many physical organs, and
because so many parts of our environment compete constantly for
our attention, we need to guard ourselves against the fatigue of dealing with too many objects or being stimulated unnecessarily by their
presence, as we perform the numerous offices they impose. Hence a
simplification of the externals of the mechanical world is almost a
prerequisite for dealing with its internal complications. To reduce
the constant succession of stimuli, the environment itself must be
made as neutral as possible. This, again, is partly in opposition to
the principle of many handicraft arts, where the effort is to hold the
eye, to give the mind something to play with, to claim a special
attention for itself. So that if the canon of economy and the respect
for function were not rooted in modern technics, it would have to be
derived from our psychological reaction to the machine: only by
esthetically observing these principles can the chaos of stimuli be
reduced to the point of effective assimilation.

Without standardization, without repetition, without the neutral-
izing effect of habit, our mechanical environment might well, by
reason of its tempo and its continuous impact, be too formidable: in
departments which have not been sufficiently simplified it exceeds
the limit of toleration. The machine has thus, in its esthetic manifestations, something of the same effect that a conventional code of
manners has in social intercourse: it removes the strain of contact
and adjustment. The standardization of manners is a psychological
shock-absorber: it permits intercourse between persons and groups
to take place without the preliminary exploration and understanding

that are requisite for an ultimate adjustment. In the province of esthetics, this simplification has still a further use: it gives small deviations and variations from the prevalent norm the psychological refreshment that would go only with much larger changes under a condition where variation was the expected mode and standardization was the exception. Mr. A. N. Whitehead has pointed out that one of our chief literary sins is in thinking of past and future in terms of a thousand years forward and backward, when really to experience the organic nature of past and future one should think of time in the order of a second, or a fraction of a second. One can make a similar remark about our esthetic perceptions: those who complain about the standardization of the machine are used to thinking of variations in terms of gross changes in pattern and structure, such as those that take place between totally different cultures or generations; whereas one of the signs of a rational enjoyment of the machine and the machine-made environment is to be concerned with much smaller differences and to react sensitively to them.

To feel the difference between two elemental types of window, with a slightly different ratio in the division of lights, rather than to feel it only when one of them is in a steel frame and the other is surmounted by a broken pediment, is the mark of a fine esthetic consciousness in our emerging culture. Good craftsmen have always had some of this finer sense of form: but it was confused by the snobbish taste and arbitrary literary standards of form that came into court life during the Renascence. As the various parts of our environment become more standardized, the senses must in turn become more acute, more refined: a hair's breadth, a speck of dirt, a faint wave in the surface will distress us as much as the pea hurt Hans Andersen's princess, and similarly pleasure will derive from delicacies of adaptation to which most of us are now indifferent. Standardization, which economizes our attention when our minds have other work to do, serves as the substratum in those departments where we deliberately seek esthetic satisfaction.

In creating the machine, we have set before ourselves a positively inhuman standard of perfection. No matter what the occasion, the criterion of successful mechanical form is that it should look *as if* no human hand had touched it. In that effort, in that boast, in that achievement the human hand shows itself, perhaps, in its most cunning manifestation. And yet ultimately it is to the human organism that we must return to achieve the final touch of perfection: the finest reproduction still lacks something that the original picture possessed: the finest porcelain produced with the aid of every mechanical accessory lacks the perfection of the great Chinese potters: the finest mechanical printing lacks that complete union of black and white that hand-printing produces with its slower method and its dampened paper. Very frequently, in machine work, the best structure is forfeited to the mere conveniences of production: given equally high standards of performance, the machine can often no more than hold its own in competition with the hand product. The pinnacles of handicraft art set a standard that the machine must constantly hold before it; but against this one must recognize that in a hundred departments examples of supreme skill and refinement have, thanks to the machine, become a commonplace. And at all levels, this esthetic refinement spreads out into life: it appears in surgery and dentistry as well as in the design of houses and bridges and high-tension power lines. The direct effect of these techniques upon the designers, workers, and manipulators cannot be over-estimated. Whatever the tags, archaicisms, verbalisms, emotional and intellectual mischiefs of our regnant system of education, the machine itself as a constant educator cannot be neglected. If during the paleotechnic period the machine accentuated the brutality of the mine, in the neotechnic phase it promises, if we use it intelligently, to restore the delicacy and sensitivity of the organism.

7: The Objective Personality

Granting these new instruments, this new environment, these new perceptions and sensations and standards, this new daily routine, these new esthetic responses—what sort of man comes out of modern technics? Le Play once asked his auditors what was the most important thing that came out of the mine; and after one had guessed coal and another iron and another gold, he answered: No, the most important thing that comes out of the mine is the miner. That is true for every occupation. And today every type of work has been affected by the machine.

I have already discussed, in terms of their limitations and renunciations, the type of man that influenced modern mechanization: the monk, the soldier, the miner, the financier. But the fuller experience of the machine does not necessarily tend to produce a repetition of these original patterns—although there is plenty of evidence to show that the soldier and the financier occupy a larger position in our world today than at perhaps any other time in the past. In the act of expressing themselves with the aid of the machine, the capacities of these original types have been modified and their character altered; moreover, what was once the innovation of a daring race of pioneers has now become the settled routine of a vast mass of people who have taken over the habits without having shared any of the original enthusiasm, and many of the latter still perhaps have no special bent toward the machine. It is difficult to analyze out such a pervasive influence as this: no single cause is at work, no single reaction can be attributed solely to the machine. And we who live in this medium, and who have been formed by it, who constantly breathe it and adapt ourselves to it, cannot possibly measure the deflection caused by the medium, still less estimate the drift of the machine, and all it carries with it, from other norms. The only partial corrective is to examine a more primitive environment, as Mr. Stuart Chase attempted to do; but even here one cannot correct for the way in which our very questions and our scale of values have been altered by our traffic with the machine.

But between the personality that was most effective in the technically immature environment of the tenth century and the type that is effective today, one may say that the first was subjectively conditioned, and that the second is more directly influenced by objective situations. These, at all events, seem to be the tendencies. In both types of personality there was an external standard of reference: but whereas the medieval man determined reality by the extent to which it agreed with a complicated tissue of beliefs, in the case of modern man the final arbiter of judgment is always a set of facts, recourse to which is equally open and equally satisfactory to all normally constituted organisms. With those that do not accept such a common substratum neither rational argument nor rational cooperation is possible. Moreover, matters that lie outside this verification in terms of fact have for the modern mind a lower order of reality, no matter how great the presumption, how strong the inner certainty, how passionate the interest. An angel and a high-frequency wave are equally invisible to the mass of mankind: but the reports of angels have come from only a limited number of human receptors, whereas by means of suitable apparatus communication between a sending and a receiving station can be inspected and checked up by any competent human being.

The technique of creating a neutral world of fact as distinguished from the raw data of immediate experience was the great general contribution of modern analytic science. This contribution was possibly second only to the development of our original language concepts, which built up and identified, with the aid of a common symbol, such as tree or man, the thousand confused and partial aspects of trees and men that occur in direct experience. Behind this technique, however, stands a special collective morality: a rational confidence in the work of other men, a loyalty to the reports of the senses, whether one likes them or not, a willingness to accept a competent and unbiased interpretation of the results. This recourse to a neutral judge and to a constructed body of law was a belated development in thought comparable to that which took place in morality when the blind conflicts between biassed men were replaced by the civil processes of justice. The collective process, even allowing for the accumulation of error and for the unconscious bias of the neutral instrument itself, gave a higher degree of certainty than the most forthright and subjectively satisfactory individual judgment.

The concept of a neutral world, untouched by man's efforts, indifferent to his activities, obdurate to his wish and supplication, is one of the great triumphs of man's imagination, and in itself it represents a fresh human value. Minds of the scientific order, even before Pythagoras, must have had intuitions of this world; but the habit of thought did not spread over any wide area until the scientific method and the machine technique had become common: indeed it does not begin to emerge with any clearness until the nineteenth century. The recognition of this new order is one of the main elements in the new objectivity. It is embodied in a common phrase which now rises to the lips of everyone when some accident or breakdown occurs in a process which lies outside everyone's immediate control: a leak in a gas tank in an airplane, a delay on a railroad: "That's that." "C'est ça." "So geht's." From machines that have broken down the same impersonal attitude begins to extend itself to the result of human negligence or human perversity: a badly cooked meal or the elopement of one's sweetheart. These events naturally often provoke stormy and uncontrollable emotional responses, but instead of magnifying the explosion and giving it more fuel, we tend to subject the response as well as the event to a common causal interpretation. The relative passiveness of machine-trained populations during periods when the industrial system itself has been disrupted, a passiveness that contrasts at times with the behavior of rural populations, is perhaps the less favorable side of the same objectivity.

Now in any complete analysis of character the "objective" personality is as much of an abstraction as the "romantic" personality. What we tend to call objective are those dispositions and attitudes which accord with the science and technics: but while one must take care not to confuse the objective or rational personality with the whole personality, it should be plain that the area of the first has increased—if only because it represents an adaptation indispensable to the running of the machine itself. And the adaptation in turn has further effects: a modulation of emphasis, a matter-of-factness, a reasonableness, a quiet assurance of a neutral realm in which the most obdurate differences can be understood, if not composed, is a mark of the emerging personality. The shrill, the violent, the vociferous, the purely animal tooth-baring and foot-stamping, paroxysms of uncritical self-love and uncontrolled hate—all these archaic qualities, which once characterized the leaders of men and their imitators, are now outside the style of our epoch: their recent revival and attempted sanctification is merely a symptom of that relapse into the raw primitive on which I dwelt a little while back. When one beholds these savage qualities today one has the sense of beholding a backward form of life, like the mastodon, or of witnessing the outburst of a demented personality. Between the fire of such low types and the ice of the machine one would have to choose the ice. Fortunately, our choice is not such a narrow one. In the development of the human character we have reached a point similar to that which we have attained in technics itself: the point at which we utilize the completest developments in science and technics to approach once more the organic. But here again: *our capacity to go beyond the machine rests upon our power to assimilate the machine. Until we have absorbed the lessons of objectivity, impersonality, neutrality, the lessons of the mechanical realm, we cannot go further in our development toward the more richly organic, the more profoundly human.*

CHAPTER VIII.   ORIENTATION

1: The Dissolution of "The Machine"

What we call, in its final results, "the machine" was not, we have seen, the passive by-product of technics itself, developing through small ingenuities and improvements and finally spreading over the entire field of social effort. On the contrary, the mechanical discipline and many of the primary inventions themselves were the result of deliberate effort to achieve a mechanical way of life: the motive in back of this was not technical efficiency but holiness, or power over other men. In the course of development machines have extended these aims and provided a physical vehicle for their fulfillment.

Now, the mechanical ideology, which directed men's minds toward the production of machines, was itself the result of special circumstances, special choices and interests and desires. So long as other values were uppermost, European technology had remained relatively stable and balanced over a period of three or four thousand years. Men produced machines partly because they were seeking an issue from a baffling complexity and confusion, which characterized both action and thought: partly, too, because their desire for power, frustrated by the loud violence of other men, turned finally toward the neutral world of brute matter. Order had been sought before, again and again in other civilizations, in drill, regimentation, inflexible social regulations, the discipline of caste and custom: after the seventeenth century it was sought in a series of external instruments and engines. The Western European conceived of the machine because he wanted regularity, order, certainty, because he wished to reduce the movement of his fellows as well as the behavior of the

364

environment to a more definite, calculable basis. But, more than an instrument of practical adjustment, the machine was, from 1750 on, a goal of desire. Though nominally designed to further the means of existence, the machine served the industrialist and the inventor and all the cooperating classes as an end. In a world of flux and disorder and precarious adjustment, the machine at least was seized upon as a finality.

If anything was unconditionally believed in and worshipped during the last two centuries, at least by the leaders and masters of society, it was the machine; for the machine and the universe were identified, linked together as they were by the formulae of the mathematical and physical sciences; and the service of the machine was the principal manifestation of faith and religion: the main motive of human action, and the source of most human goods. Only as a religion can one explain the compulsive nature of the urge toward mechanical development without regard for the actual outcome of the development in human relations themselves: even in departments where the results of mechanization were plainly disastrous, the most reasonable apologists nevertheless held that "the machine was here to stay"—by which they meant, not that history was irreversible, but that the machine itself was unmodifiable.

Today this unquestioned faith in the machine has been severely shaken. The absolute validity of the machine has become a conditioned validity: even Spengler, who has urged the men of his generation to become engineers and men of fact, regards that career as a sort of honorable suicide and looks forward to the period when the monuments of the machine civilization will be tangled masses of rusting iron and empty concrete shells. While for those of us who are more hopeful both of man's destiny and that of the machine, the machine is no longer the paragon of progress and the final expression of our desires: it is merely a series of instruments, which we will use in so far as they are serviceable to life at large, and which we will curtail where they infringe upon it or exist purely to support the adventitious structure of capitalism.

The decay of this absolute faith has resulted from a variety of causes. One of them is the fact that the instruments of destruction ingeniously contrived in the machine shop and the chemist's laboratory, have become in the hands of raw and dehumanized personalities a standing threat to the existence of organized society itself. Mechanical instruments of armament and offense, springing out of fear, have widened the grounds for fear among all the peoples of the world; and our insecurity against bestial, power-lusting men is too great a price to pay for relief from the insecurities of the natural environment. What is the use of conquering nature if we fall a prey to nature in the form of unbridled men? What is the use of equipping mankind with mighty powers to move and build and communicate, if the final result of this secure food supply and this excellent organization is to enthrone the morbid impulses of a thwarted humanity?

In the development of the neutral valueless world of science, and in the advance of the adaptive, instrumental functions of the machine, we have left to the untutored egoisms of mankind the control of the gigantic powers and engines technics has conjured into existence. In advancing too swiftly and heedlessly along the line of mechanical improvement we have failed to assimilate the machine and to co-ordinate it with human capacities and human needs; and by our social backwardness and our blind confidence that problems occasioned by the machine could be solved purely by mechanical means, we have outreached ourselves. When one subtracts from the manifest blessings of the machine the entire amount of energy and mind and time and resources devoted to the preparation for war—to say nothing of the residual burden of past wars—one realizes the net gain is dismayingly small, and with the advance of still more efficient means of inflicting death is becoming steadily smaller. Our failure here is the critical instance of a common failure all along the line.

The decay of the mechanical faith has, however, still another source: namely, the realization that the serviceability of machines has meant in the past serviceability to capitalist enterprise. We are now entering a phase of dissociation between capitalism and technics; and we begin to see with Thorstein Veblen that their respective interests, so far from being identical, are often at war, and that the human gains of technics have been forfeited by perversion in the interests of a pecuniary economy. We see in addition that many of the special gains in productivity which capitalism took credit for were in reality due to quite different agents—collective thought, cooperative action, and the general habits of order—virtues that have no necessary connection with capitalist enterprise. To perfect and extend the range of machines without perfecting and giving humane direction to the organs of social action and social control is to create dangerous tensions in the structure of society. Thanks to capitalism, the machine has been over-worked, over-enlarged, over-exploited because of the possibility of making money out of it. And the problem of integrating the machine in society is not merely a matter, as I have already pointed out, of making social institutions keep in step with the machine: the problem is equally one of altering the nature and the rhythm of the machine to fit the actual needs of the community. Whereas the physical sciences had first claim on the good minds of the past epoch, it is the biological and social sciences, and the political arts of industrial planning and regional planning and community planning that now most urgently need cultivation: once they begin to flourish they will awaken new interests and set new problems for the technologist. But the belief that the social dilemmas created by the machine can be solved merely by inventing more machines is today a sign of half-baked thinking which verges close to quackery.

These symptoms of social danger and decay, arising out of the very nature of the machine—its peculiar debts to warfare, mining, and finance—have weakened the absolute faith in the machine that characterized its earlier development.

At the same time, we have now reached a point in the development of technology itself where the organic has begun to dominate the machine. Instead of simplifying the organic, to make it intelligibly mechanical, as was necessary for the great eotechnic and paleotechnic inventions, we have begun to complicate the mechanical, in order to make it more organic: therefore more effective, more harmonious with our living environment. For our skill, perfected on the finger exercises of the machine, would be bored by the mere repetition of the scales and such childlike imbecilities: supported by the analytic methods and the skills developed in creating the machine, we can now approach the larger tasks of synthesis. In short, the machine is serving independently, in its neotechnic phase, as a point for a fresh integration in thought and social life.

While in the past the machine was retarded by its limited historic heritage, by its inadequate ideology, by its tendency to deny the vital and the organic, it is now transcending these limitations. And indeed, as our machines and our apparatus become more subtle, and the knowledge derived with their aid becomes more delicate and penetrating, the simple mechanical analysis of the universe made by the earlier physicists ceases to represent anything in which the scientist himself is now interested. The mechanical world-picture is dissolving. The intellectual medium in which the machine once spawned so rapidly is being altered at the same time that the social medium— the point of application—is undergoing a parallel change. Neither of these changes is yet dominant; neither is automatic or inevitable. But one can now say definitely, as one could not fifty years ago, that there is a fresh gathering of forces on the side of life. The claims of life, once expressed solely by the Romantics and by the more archaic social groups and institutions of society, are now beginning to be represented at the very heart of technics itself. Let us trace out some of the implications of this fact.

2: Toward an Organic Ideology

During the first period of mechanical advance, the application of simple mechanical analogies to complex organic phenomena helped the scientist to create a simple framework for experience in general, including manifestations of life. The "real" from this standpoint was that which could be measured and accurately defined; and the notion that reality might in fact be vague, complex, undefinable, perpetually a little obscure and shifty, did not go with the sure click and movement of machines.

Today this whole abstract framework is in process of reconstruction. Provisionally, it is as useful to say in science that a simple element is a limited kind of organism as it once was to say that an organism was a complicated kind of machine. "Newtonian physics," as Professor A. N. Whitehead says in Adventures of Ideas, "is based upon the independent individuality of every bit of matter. Each stone is conceived as fully describable apart from any reference to any other portion of matter. It might be alone in the universe, the sole occupant of uniform space. Also the stone could be adequately described without reference to past or future. It is to be conceived fully and adequately as wholly constituted within the present moment." These independent solid objects of Newtonian physics might move, touch each other, collide, or even, by a certain stretch of the imagination, act at a distance: but nothing could penetrate them except in the limited way that light penetrated translucent substances.

This world of separate bodies, unaffected by the accidents of history or of geographic location, underwent a profound change with the elaboration of the new concepts of matter and energy that went forward from Faraday and von Mayer through Clerk-Maxwell and Willard Gibbs and Ernest Mach to Planck and Einstein. The discovery that solids, liquids, and gases were phases of all forms of matter modified the very conception of substance, while the identification of electricity, light, and heat as aspects of a protean energy, and the final break-up of "solid" matter into particles of this same ultimate energy lessened the gap, not merely between various aspects of the physical world, but between the mechanical and the organic. Both matter in the raw and the more organized and internally self-sustaining organisms could be described as systems of energy in more or less stable, more or less complex, states of equilibrium.

In the seventeenth century the world was conceived as a series of independent systems. First, the dead world of physics, the world of matter and motion, subject to accurate mathematical description. Second, and inferior from the standpoint of factual analysis, was the world of living organisms, an ill-defined realm, subject to the intrusion of a mysterious entity, the vital principle. Third, the world of man, a strange being who was a mechanical automaton with reference to the world of physics, but an independent being with a destiny in heaven from the standpoint of the theologian. Today, instead of such a series of parallel systems, the world has conceptually become a single system: if it still cannot be unified in a single formula, it is even less conceivable without positing an underlying order that threads through all its manifestations. Those parts of reality that can be reduced to patent order, law, quantitative statement are no more real or ultimate than those parts which remain obscure and illusive: indeed, when applied at the wrong moment or in the wrong place or in a false context the exactness of the description may increase the error of interpretation.

All our really primary data are social and vital. One begins with life; and one knows life, not as a fact in the raw, but only as one is conscious of human society and uses the tools and instruments society has developed through history—words, symbols, grammar, logic, in short, the whole technique of communication and funded experience. The most abstract knowledge, the most impersonal method, is a derivative of this world of socially ordered values. And instead of accepting the Victorian myth of a struggle for existence in a blind and meaningless universe, one must, with Professor Lawrence Henderson, replace this with the picture of a partnership in mutual aid, in which the physical structure of matter itself, and the very distribution of elements on the earth's crust, their quantity, their solubility, their specific gravity, their distribution and chemical combination, are life-furthering and life-sustaining. Even the most rigorous scientific description of the physical basis of life indicates it to be internally teleological.

Now changes in our conceptual apparatus are rarely important or influential unless they are accompanied, more or less independently, by parallel changes in personal habits and social institutions. Mechanical time became important because it was re-enforced by the financial accountancy of capitalism: progress became important as a doctrine because visible improvements were being rapidly made in machines. So the organic approach in thought is important today because we have begun, here and there, to act on these terms even when unaware of the conceptual implications. This development has gone on in architecture from Sullivan and Frank Lloyd Wright to the new architects in Europe, and from Owen and Ebenezer Howard and Patrick Geddes in city design to the community planners in Holland, Germany, and Switzerland who have begun to crystallize in a fresh pattern the whole neotechnic environment. The humane arts of the physician and the psychologist and the architect, the hygienist and the community planner, have begun during the last few decades to displace the mechanical arts from their hitherto central position in our economy and our life. Form, pattern, configuration, organism, historical filiation, ecological relationship are concepts that work up and down the ladder of the sciences: the esthetic structure and the social relations are as real as the primary physical qualities that the sciences were once content to isolate. This conceptual change, then, is a widespread movement that is going on in every part of society: in part it arises out of the general resurgence of life—the care of children, the culture of sex, the return to wild nature and the re-newed worship of the sun—and in turn it gives intellectual re-enforcement to these spontaneous movements and activities. The very structure of machines themselves, as I pointed out in describing the neo-technic phase, reflects these more vital interests. We now realize that the machines, at their best, are lame counterfeits of living organisms. Our finest airplanes are crude uncertain approximations compared with a flying duck: our best electric lamps cannot compare in efficiency with the light of the firefly: our most complicated automatic telephone exchange is a childish contraption compared with the nervous system of the human body.

This reawakening of the vital and the organic in every department undermines the authority of the purely mechanical. Life, which has always paid the fiddler, now begins to call the tune. Like The Walker in Robert Frost's poem, who found a nest of turtle eggs near a railroad track, we are armed for war:

> *The next machine that has the power to pass*
> *Will get this plasm on its polished brass.*

But instead of being confined to a resentment that destroys life in the act of hurling defiance, we can now act directly upon the nature of the machine itself, and create another race of these creatures, more effectively adapted to the environment and to the uses of life. At this point, one must go beyond Sombart's so far excellent analysis. Sombart pointed out, in a long list of contrasting productions and inventions, that the clue to modern technology was the displacement of the organic and the living by the artificial and the mechanical.

Within technology itself this process, in many departments, is being reversed: we are returning to the organic: at all events, we no longer regard the mechanical as all-embracing and all-sufficient.

Once the organic image takes the place of the mechanical one, one may confidently predict a slowing down of the tempo of research, the tempo of mechanical invention, and the tempo of social change, since a coherent and integrated advance must take place more slowly than a one-sided unrelated advance. Whereas the earlier mechanical world could be represented by the game of checkers, in which a similar series of moves is carried out by identical pieces, qualitatively similar, the new world must be represented by chess, a game in which each order of pieces has a different status, a different value, and a different function: a slower and more exacting game. By the same token, however, the results in technology and in society will be of a more solid nature than those upon which paleotechnic science congratulated itself: for the truth is that every aspect of the earlier order, from the slums in which it housed its workers to the towers of abstraction in which it housed its intellectuals, was jerrybuilt— hastily clapped together for the sake of immediate profits, immediate practical success, with no regard for the wider consequences and implications. The emphasis in future must be, not upon speed and immediate practical conquest, but upon exhaustiveness, inter-relation- ship, and integration. The co-ordination of our technical effort—such co-ordination and adjustment as is pictured for us in the physiology of the living organism—is more important than extravagant advances along special lines, and equally extravagant retardations along other lines, with a disastrous lack of balance and harmony between the various parts.

The fact is then that, partly thanks to the machine, we have now an insight into a larger world and a more comprehensive intellectual synthesis than that which was originally outlined in our mechanical ideology. We can now see plainly that power, work, regularity, are adequate principles of action only when they cooperate with a humane scheme of living: that any mechanical order we can project must fit into the larger order of life itself. Beyond the necessary intellectual reconstruction, which is already going on in both science and technics, we must build up more organic centers of faith and action in the arts of society and in the discipline of the personality: this implies a re-orientation that will take us far beyond the immediate province of technics itself. These are matters—matters touching the building of communities, the conduct of groups, the development of the arts of communication and expression, the education and the hygiene of the personality—that I purpose to take up in another book. Here I will confine attention to co-ordinate readjustments which are clearly indicated and already partly formulated and enacted in the realm of technics and industry.

3: The Elements of Social Energetics

Let us examine the implications of neotechnic developments, within the machine itself, upon our economic objectives, upon the organization of work, upon the direction of industry and the goals of consumption, upon the emerging social purposes of the neotechnic phase of civilization.

First: the economic objectives.

In the course of capitalistic enterprise, which accompanied the widespread introduction of machines and machine-methods in the fifteenth and sixteenth centuries, the focus of industry shifted from the craft guild to the merchant guild or the livery company or the company of merchant adventurers, or to the special organization for exploiting patent monopolies. The means of exchange usurped the function and meaning of the things that were exchanged: money itself became a commodity and money-getting became a specialized form of activity. Under capitalism profit reigned as the main economic objective; and profit became the decisive factor in all industrial enterprise. Inventions that promised profits, industries that produced profits, were fostered. The reward of capital, if not the first claim upon productive enterprise, was at all events the dominating one: the service of the consumer and the support of the worker were entirely secondary. Even in a period of crisis and breakdown, such as that capitalism is still in the midst of at the moment I write, dividends continue to be paid to rentiers out of past accumulation while the industry itself often operates at a loss, or the mass of workers are turned out to starve. Sometimes profits were obtained by lowering the costs and spreading the product: but if they could be had only by offering inferior or adulterated goods—as in the sale of medical nostrums or the slum housing of the underpaid worker—health and well-being were sacrificed to gain. The community, instead of receiving a full return for its goods and services, permitted a portion of the product to be diverted for the private gratification of the holders of land and capital. These holders of land and capital, backed up by the law and all the instruments of government, determined privately and solely in accordance with the canon of profit what should be produced and how much and where and how and by whom and on what terms.

In the economic analysis of the society that grew up on this basis, the three main terms in industrial activity were production, distribution, and consumption. Profits were to be increased by cheaper production, by wider and multifold distribution, and by a steadily rising standard of consumptive expenditure, with—sometimes in lieu of that, sometimes accompanying it—an enlarging market of consumers. Saving labor, or cheapening labor by a superiority of bargaining power—obtained by withholding land from the laborer and monopolizing the new instruments of production—were the two chief means, from the capitalist's standpoint, of increasing the margin of profits. Saving labor by rationalization was a real improvement which bettered everything but the position of the laborer. The stimulation of the demand for goods was the chief means of increasing the turnover: hence the problem of capitalism was essentially not to satisfy needs but to create demands. And the attempt to represent this process of private aggrandizement and class-advantage as a natural and socially beneficent one was perhaps the main labor of political economists during the nineteenth century.

When one examines economic activities from the standpoint of the employment of energy and the service of human life, this whole financial structure of production and consumption turns out to have mainly a superstitious basis. At the bottom of the structure are farmer and peasant, who during the entire course of the industrial revolution, which their increase of the food supply has made possible, have scarcely ever received an adequate return for their products—at least on the basis of pecuniary accountancy by which the rest of this society was run. Furthermore: what are called gains in capitalist economics often turn out, from the standpoint of social energetics, to be losses; while the real gains, the gains upon which all the activities of life, civilization, and culture ultimately depend were either counted as losses, or were ignored, because they remained outside the commercial scheme of accountancy.

What are, then, the essentials of the economic processes in relation to energy and to life? The essential processes are conversion, production, consumption, and creation. In the first two steps energy is seized and prepared for the sustenance of life. In the third stage, life is supported and renewed in order that it may wind itself up, so to speak, on the higher levels of thought and culture, instead of being short-circuited at once back into the preparatory functions. Normal human societies exhibit all four stages of the economic processes: but their absolute quantities and their proportions vary with the social milieu.

Conversion has to do with the utilization of the environment as a source of energy. The prime fact of all economic activity, from that of the lower organisms up to the most advanced human cultures, is the conversion of the sun's energies: this transformation depends upon the heat-conserving properties of the atmosphere, upon the geological processes of uplift and erosion and soil-building, upon the conditions of climate and local topography, and—most important of all—upon the green leaf reaction in growing plants. This seizure of energy is the original source of all our gains: on a purely energetic interpretation of the process, all that happens after this is a dissipation of energy—a dissipation that may be retarded, that may be dammed up, that may be temporarily diverted by human ingenuity, but in the long run cannot be averted. All the permanent monuments of human culture are attempts, by using more attentuated physical means of preserving and transmitting this energy, to avert the hour of ultimate extinction. The most important conquest of energy was man's original discovery and utilization of fire; after that, the most significant transformation of the environment came through the cultivation of the grain-bearing grasses, the vegetables, and the domestic animals. Indeed, the enormous increase in population which took place at the beginning of the nineteenth century, *before* the machine had made any appreciable change in agriculture, was due to the opening of immense areas of free land for grain cultivation and cattle raising and the better provision of winter fodder crops, combined with the addition of three new energy crops—sugar cane, sugar beet, and potato—to the diet of the industrial population.

The mechanical conversion of energy is second in importance to the organic conversion. But in the development of technics the invention of the water-wheel, the water-turbine, the steam engine, and the gas engine multiplied the energies that were available to man through the use of foods grown for himself and his domestic animals. Without the magnification of human energy made possible through this series of prime movers, our apparatus of production and transport could not have reached the gigantic scale it attained in the nineteenth century. All the further steps in the economic process depend upon the original act of conversion: the level of achievement can never rise higher than the level of the energy originally converted, and just as only an insignificant part of the sun's energy available is utilized in conversion, so only a small part of this, in turn, finally is utilized in consumption and creation.

Conversion lifts the energy available to a peak: from that point on energy runs down hill, in gathering and shaping the raw materials, in transporting supplies and products, and in the processes of consumption itself. Not until the economic process reaches the stage of creation—not until it supplies the human animal with more energy than he needs to maintain his physical existence, and not until still other energies are transformed into the more durable media of art and science and philosophy, of books, buildings and symbols—is there anything that can be called, even within a limited span of time, a gain. At one end of the process is the conversion of the free energy of nature and its transformation into forms useable by agriculture and technology: at the other end of the process is the conversion of the intermediate, preparatory products into human

subsistence, and into those cultural forms that are useable by succeeding generations of men.

The amount of energy available for the final process depends upon
two facts: how much energy is converted by agriculture and technics
at the beginning, and how much of that energy is effectively applied
and conserved in transmission. Even the crudest society has some
surplus. But under the capitalist system the main use of this surplus
is to serve as profits which are incentives to capital investments,
which in turn increase production. Hence two massive and recurrent
facts in modern capitalism: first, an enormous over-expansion of
plant and equipment. Thus the Hoover Committee on the Elimination
of Waste in Industry found, for example, that clothing factories in
the United States are about 45 per cent larger than necessary; printing establishments are from 50 to 150 per cent over-equipped; and
the shoe industry has a capacity twice that of actual production.
Second: an excessive diversion of energy and man-power into sales
promotion and distribution. Whereas only ten per cent of the working population in the United States was engaged in transporting and
distributing the commodities produced in 1870, the proportion had
risen to 25 per cent in 1920. Other means of utilizing the surplus, such as the cultural and educational bequests of various phi-
lanthropies, relieve some of the burden of inane waste from both
the individual and from industrial society: but there is no capitalist
theory of non-profit-making enterprises and non-consumable goods.
These functions exist accidentally, by the grace of the philanthropist:
they have no real place in the system. Yet it should be plain that
as society becomes technically mature and civilized, the area occupied by the surplus must become progressively wider: it will be
greater than it occupied under capitalism or under those more
primitive non-capitalist civilizations which—as was pregnantly
demonstrated by Radhakamal Mukerjee—capitalist economics so
inadequately describes.

The permanent gain that emerges from the whole economic process
is in the relatively non-material elements in culture—in the social
heritage itself, in the arts and sciences, in the traditions and processes
of technology, or directly in life itself, in those real enrichments

that come from the free exploitation of organic energy in thought and action and emotional experience, in play and adventure and drama and personal development—gains that last through memory and communication beyond the immediate moment in which they are enjoyed. In short, as John Ruskin put it, *There is no Wealth but Life;* and what we call wealth is in fact wealth only when it is a sign of potential or actual vitality.

An economic process that did not produce this margin for leisure, enjoyment, absorption, creative activity, communication and transmission would completely lack human meaning and reference. In the histories of human groups there are of course periods, periods of starvation, periods of flood and earthquake and war, when man fights a losing fight with his environment, and does not even secure bare physical survival; and there are moments when the complete social process is brutally cut short. But even in the most perverse and degraded forms of life, there is an aspect that corresponds, vitally and psychally speaking, to "creation," and even in the most inadequate forms of production, such as that which prevailed during the paleotechnic phase, there remains a surplus not arrogated by industry. Whether this surplus goes to increase the preparatory processes, or whether it is to be spent on creation, is a choice that cannot be automatically decided; and the tendency in capitalist society to put it back quickly into the preparatory processes, and to make possible increased production by applying pressure to consumption, is merely a further indication of its absence of social criteria.

The real significance of the machine, socially speaking, does not consist either in the multiplication of goods or the multiplication of wants, real or illusory. Its significance lies in the gains of energy through increased conversion, through efficient production, through balanced consumption, and through socialized creation. The test of economic success does not, therefore, lie in the industrial process alone, and it cannot be measured by the amount of horsepower converted or by the amount commanded by an individual user: for the important factors here are not quantities but ratios: ratios of mechanical effort to social and cultural results. A society in which **production**

and consumption completely cancelled out the gains of conversion—
in which people worked to live and lived to work—would remain
socially inefficient, even if the entire population were constantly employed, and adequately fed, clothed, and sheltered.

The ultimate test of an efficient industry is the ratio between
productive means and the achieved ends. Hence a society with a low
scale of conversion but with a high amount of creation is humanly
speaking superior to a society with an enormous panoply of con-
verters and a small and inadequate army of creators. By the ruthless
pillage of the food-producing territories of Asia and Africa, the
Roman Empire appropriated far more energy than Greece, with its
sparse abstemious dietary and its low standard of living. But Rome
produced no poem, no statue, no original architecture, no work of
science, no philosophy comparable to the Odyssey, the Parthenon,
the works of sixth and fifth century sculptors, and the science of
Pythagoras, Euclid, Archimedes, Hero: and so the quantitative
grandeur and luxury and power of the Romans, despite their extraordinary capacity as engineers, remained relatively meaningless: even
for the continued development of technics the work of the Greek
mathematicians and physicists was more important.

This is why no working ideal for machine production can be
based solely on the gospel of work: still less can it be based upon an
uncritical belief in constantly raising the quantitative standard of
consumption. If we are to achieve a purposive and cultivated use of
the enormous energies now happily at our disposal, we must examine
in detail the processes that lead up to the final state of leisure, free
activity, creation. It is because of the lapse and mismanagement of
these processes that we have not reached the desirable end; and it is
because of our failure to frame a comprehensive scheme of ends
that we have not succeeded in achieving even the beginnings of social
efficiency in the preparatory work.

How is this margin to be achieved and how is it to be applied?
Already we are faced with political and moral problems as well as
technological ones. There is nothing in the nature of the machine
as such, nothing in the training of the technician as such, that will
provide us with a sufficient answer. We shall of course need his

help: but in turn *he* will need help from other quarters of the compass, far beyond the province of technology.

4: Increase Conversion!

Modern technics began in Western Civilization with an increased capacity for conversion. While society faces a fairly imminent shortage of petroleum and perhaps natural gas, and while the known coal beds of the world give no longer promise of life, at the present rates of consumption, than three thousand years, we face no serious energy problem that we cannot solve even with our present equipment, provided that we utilize to the full our scientific resources. Apart from the doubtful possibility of harnessing inter-atomic energy, there is the much nearer one of utilizing the sun's energy directly in sun-converters or of utilizing the difference in temperature between the lower depths and the surface of the tropical seas: there is likewise the possibility of applying on a wide scale new types of wind turbine, like the rotor: indeed, once an efficient storage battery was available the wind alone would be sufficient, in all probability, to supply any reasonable needs for energy.

Along with the renewed use through electricity of wind and water one must put the destructive distillation of coal, near the pitheads, in the new types of coke-oven. This not merely saves enormous amounts in energy now spent in transporting the fuel from the place where it is mined to the place where it is used, but it also conserves the precious compounds that now escape into the air in the wasteful individual furnaces. Theoretically, however, such economies of energy only lead to wider consumption and so to more rapid utilization of the very thing we wish to conserve: hence the necessity for making a socialized monopoly of all such raw materials and resources. The private monopoly of coal beds and oil wells is an intolerable anachronism—as intolerable as would be the monopoly of sun, air, running water. Here the objectives of a price economy and a social economy cannot be reconciled; and the common ownership of the means of converting energy, from the wooded mountain regions where the streams have their sources down to the remotest petroleum wells is the sole safeguard to their effective use and conservation. Only by increasing the amount of energy available, or, when the amount is restricted, by economizing more cunningly in its application, shall we be in a position to eliminate freely the basest forms of drudgery.

What is true for mechanical power production is likewise true for organic forms of power production, such as the growing of foods and the extraction of raw materials from the soil. In this department capitalistic society has confused ownership with security of tenure and continuity of effort, and in the very effort to foster ownership while maintaining the speculative market it has destroyed security of tenure. It is the latter condition that is necessary for conservative farming; and not until the community itself holds the land will the position of the farmer be a desirable one. The negative side of this socialization of the land—namely, the purchase of marginal land, unfit for any other purpose than forest growth—has already been taken up, for example, by the State of New York. It remains to accomplish a similar end on the positive side by taking over and appropriately planning for maximum cultivation and enjoyment the good agricultural lands.

Such ownership and planning by the community do not necessarily mean large-scale farming: for the efficient economic units differ with the type of farming, and the large mechanized units suitable to the cultivation of the wheatlands of the prairies are in fact inappropriate to other types of farming. Neither does such a system of rationalization inevitably mean the extinction of the small family farming group, with the skill and initiative and general intelligence that distinguishes the farmer favorably from the over-specialized factory worker of the old style. But the permanent zoning of certain areas for certain types of agriculture, and the experimental determination of the types of crop appropriate to a particular region or a particular section are matters that cannot be left to guess, chance, or blind individual initiative: they are, on the contrary, complicated technical questions in which objective answers are possible. In long-settled areas, like the various wine-growing sections of France, soil utilization surveys will probably only confirm existing types of effort: but wherever there is a question of choice between types of use, the decision cannot be left to the chance interests of individuals. The first step toward rationalization in agriculture is the common ownership of the land. Such ownership prevailed in Europe under customary forms down to the nineteenth century in certain regions; and its restoration involves no breach whatever with the essential foundations of rural life.

The private appropriation and exploitation of the land, indeed, must be looked upon as a transitory state, peculiar to capitalism, between customary local agriculture based upon the common needs of the small local community and a rationed world agriculture, based upon the cooperative resources of the entire planet, considered as a federation of balanced regions. The fact that, except in times of extreme scarcity, the farmer is pauperized or ruined by the abundance of his crops only emphasizes the point that a more stable basis for agricultural production must be found: a basis that does not rely upon the individual guesses of the farmer, the caprices of nature, and the speculative fluctuations of the world market. Within any given period price tends to vary inversely with the quantity available: here as elsewhere monetary values disappear toward zero as vital values and energies rise. Hence the need for rationing, for stable crops, and for an altogether new system of determining price and marketability. I shall go into this last point presently. It is enough to point out here that with the development of balanced eco-nomic regions, agricultural production will be related to a stable local market, the sudden gluts and shortages that arise with transportation to distant centers will disappear, and further to regularize production, a good part of the more delicate crops will be grown in small units, possibly, as in Holland, under glass, near the place of consumption.

To increase conversion, then, is no simple matter of merely mining coal or building more dynamos. It involves the social appropriation of natural resources, the replanning of agriculture and the maximum utilization of those regions in which kinetic energy in the form of sun, wind, and running water is abundantly available. The socialization of these sources of energy is a condition of their effective and purposive use.

5: Economize Production!

The application of power to production and the employment of quick and relatively tireless machines to perform manual movement and the organization of rapid transport and the concentration of work into factories were the chief means adopted during the nineteenth century to increase the quantity of commodities available. And the goal of this development within the factory was the complete substitution of non-human power for man power, of mechanical skill for human skill, of automatons for workers, in every department where this was possible. Where the absence of human feelings or intelligence did not manifest itself in an inferiority of the product itself, that goal was a legitimate one.

The mechanical elements in production were rationalized much more rapidly than the human elements. In fact, one might almost say that the human elements were irrationalized at the same time; for the stimuli to production, human fellowship, an *esprit de corps*, the hope of advancement and mastery, the appreciation of the entire process of work itself, were all reduced or wiped out at the very moment that the work itself, through its subdivision, ceased to give any independent gratification. Only the pecuniary interest in production remained; and the majority of mankind, unlike the avaricious and ambitious spirits who marched to the head of industry, are apparently so irresponsive to this pecuniary stimulus that the directing classes relied upon the lash of starvation, rather than upon the pleasures of surfeit, to drive them back to the machine.

Collective instruments of production were created and used, without the benefit of a collective will and a collective interest. That, to begin with, was a serious handicap upon productive efficiency. The workers grudged the efforts they gave to the machine, applied themselves with half a mind, loitered and loafed when there was an opportunity to escape the eye of the foreman or the taskmaster, sought to give as little as they could in return for as much wages as they could get. So far from attempting to combat these sources of inefficiency, the enterprisers sanctioned it by relieving the worker of such autonomy and responsibility as might naturally adhere to the job, by insisting upon speed for the sake of cheapness without regard for the excellence of workmanship, and by managing industry with an eye solely upon the maximum cash return. There were exceptions in every industry; but they did not establish the main line.

Not appreciating the gain to efficiency from collective loyalty and collective interest and a strong common drive, the great industrialists did their best to browbeat any of these incipient responses out of the worker: by lockouts, by ruthless warfare in strikes, by hard bargains in wages and by callous layoffs during periods of slack work the typical employers of labor did their ignorant best to decrease the efficiency of the workers and throw sand in the works. These tactics greatly increased the labor turnover and therefore lowered the internal efficiency of operation: even such a moderate improvement in the wage scale as Ford introduced in Detroit had a powerful effect in lessening such losses. But what shall one say to the efficiency of a productive system in which strikes and lockouts in the United States, according to Polakov, at the beginning of the last decade, averaged 54 million man-days of idleness per year? The loss and inefficiency due to the failure to create a cooperative pattern of human relations which would supplement that of the machine industry itself cannot be estimated: but the success of such occasional mutations within the capitalist system as the Cadbury Cocoa works at Bourneville, the Godin steel works at Guise—an adaptation of Fourier's scheme for a cooperative phalanstery—and the Dennison paper manufacturing works at Framingham, Massachusetts, gives a slight indication of what our total efficiency would have been had social relations themselves been rationalized at the time the machine was introduced. It is evident, at all events, that a good part of our mechanical adroitness has been annulled by social friction, waste, and unnecessary human wear and tear. Testimony to that effect comes from the production engineers themselves.

At the end of the nineteenth century a new attack upon the problem of efficiency in production was made within the factory: it was no accident perhaps that the distinguished engineer who initiated it was also the co-inventor of a new high-speed tool steel, a characteristic neotechnic advance. Instead of studying the machine as an isolated unit, Taylor studied the worker himself as an element in production. By a close factual study of his movements, Taylor was able to add to the labor output per man without adding to his physical burden. The time and motion studies that Taylor and his followers introduced have now, with the development of serial processes and greater automatism, become somewhat outmoded: their importance lay in the fact that they directed attention to the industrial process as a whole and treated the worker as an integral element in it. Their weakness lay in the fact that they accepted the aims of capitalist production as fixed, and they were compelled to rely upon a narrow pecuniary incentive—with piecework production and bonuses—to achieve the mechanical gains that were possible.

The next step toward the genuine rationalization of industry lies in widening the interests and increasing the social incentives to production. On one hand, this means the reduction of trivial and degrading forms of work: it likewise means the elimination of products that have no real social use, since there is no form of cruelty for a rational human being worse than making him produce goods that have no human value: picking oakum is by comparison an edifying task. In addition, the stimulation of invention and initiative within the industrial process, the reliance upon group activity and upon intimate forms of social approval, and the transformation of work into education, and of the social opportunities of factory production into effective forms of political action—all these incentives toward a humanly controlled and effectively directed industrial production await the formulation of non-capitalist modes of enterprise. Taylorism, though it had within its technique the germ of a revolutionary change in industry, was reduced to a minor instrument in almost every country except Russia. But it is precisely in the political and psychological relations of the worker to the industry that the most effective economies have still to be made. This has been excellently illustrated in an experiment in a Westinghouse plant described by Professor Elton Mayo. By paying attention to the conditions of work and by providing rest periods, the efficiency of a group of workers was steadily raised. After a certain period of experiment, the group was put back in the original condition of work without rest periods: still the output was greater than it had been originally. What had happened? There was a feeling among the operatives, according to the observer, that "better output is in some way related to the distinctively pleasanter, freer, and happier working conditions." This is a long stage beyond Taylor's original mechanical motion study. And it points to a factor of efficiency in socialized industry, in which the worker himself is fully respected, which capitalism at its most enlightened best can scarcely more than touch. (Is not this human factor perhaps one of the reasons why small scale industry—in addition to its lower overhead—can still often compete with large scale industry, where monopoly does not favor the latter?)

Meanwhile, modern production has added enormously to the productive output without adding a single horsepower or a single machine or a single workman. What have been the means? On one hand there have been great gains through mechanical articulation within the factory, and through the closer organization of raw materials, transport, storage, and utilization in the factory itself. By timing, working out economic sequences, creating an orderly pattern of activity, the engineer has added enormously to the collective product. By transferring power from human organisms to machines, he has decreased the number of variable factors and integrated the process as a whole. These are the gains of organization and administration. The other set of gains has come through standardization and serial production. This involves the reduction of a whole group of different articles, in which differences did not correspond to essential qualities, to a limited number of types: once these types can be established and suitable machines devised to processing and manufacturing them the process can approach more and more closely to automatism. The dangers here lie in premature standardization; and in making assembled objects—like automobiles—so completely standardized that they cannot be improved without a wholesale scrapping of the plant. This was the costly mistake that was made in the Ford Model T. But in all the ranges of production where typification is possible large productive economies can be achieved by that method alone.

One returns to the illustration originally used by Babbage. The stone could be moved without skill or organized effort by exerting 753 pounds of effort: or it might be moved, by adapting appropriately every part of the environment, by using only twenty-two pounds. In its crude state, industry prides itself upon its gross use of power and machinery. In its advanced state it rests upon rational organization, social control, physiological and psychological understanding. In the first case, it relies upon the external exercise of power in its political relations: indeed, it prides itself upon surmounting the friction which with such superb ineptitude it creates. In the second state, no part of the works can remain immune to criticism and rational criteria: the goal is no longer as much production as is compatible with the canons of private enterprise and private profit and individual money-incentives: it is rather efficient production for social uses no matter how drastically these sacred canons must be revised or extirpated.

In a word, to economize production, we cannot begin or end with the physical machines and utilities themselves, nor can efficient production begin and end in the individual factory or industry. The process involves an integration of the worker, the industrial function, and the product, just as it involves a further co-ordination between the sources of supply and the final consumptive outlets. At hardly any point in our present system of production have we begun to utilize the latent energies that are available through organization and social control: at best, here and there, we have just begun to sample such efficiencies.

If we have only begun to utilize the latent energies of the personnel, it is equally true that the geographic distribution of industries, hitherto governed by accidental choices and opportunities, has still to be worked out rationally in terms of the world's resources and the re-settlement of the world's population into the areas marked as favorable for human living. Here, through economic regionalism, a new series of economies offers itself.

The accidents of original manufacture or of the original location of resources cannot continue as guiding factors in growth when new sources of supply and new distribution of markets are recognized.

Moreover, the neotechnic distribution of power makes for economic regionalism: the concentration of population in the coal towns and the port towns was a mark of a haphazardly organized labor supply and of the high cost of coal transportation. One of the large possibilities for economy here lies in the abolition of cross-hauls: the familiar process of carrying coals to Newcastle. Traders and middlemen gain by lengthening the distance in space and time between the producer and the ultimate consumer. Under a rationally planned distribution of industry, this parasitism in transit would be reduced to a minimum. And as the knowledge of modern technics spreads, the special advantages in skill and organization and science, once enjoyed by a few countries alone, by England during the nineteenth century above all, tend to become the common property of mankind at large: for ideas are not stopped by customs barriers or freight rates. Our modern world, transporting knowledge and skill, has diminished the need for transporting goods: St. Louis's shoes are as good as New England's, and French textiles are as good as English. In a balanced economy, regional production of commonplace commodities becomes rational production; and inter-regional exchange becomes the export of the surplus from regions of increment to regions of scarcity, or the exchange of special materials and skills—like Tungsten, manganese, fine china, lenses—not universally found or developed throughout the world. But even here the advantages of a particular place may remain temporary. While American and German camembert cheese is still vastly inferior to the French variety, the gruyère cheese produced in Wisconsin compares favorably with that produced in Switzerland. With the growth of economic regionalism, the advantages of modern industry will be spread, not chiefly by transport—as in the nineteenth century—but by local development.

The prime examples of conscious economic regionalism up to the present have come from countries like Ireland and Denmark, or states like Wisconsin, where the occupations were predominantly agricultural, and where a flourishing economic life depended upon an intelligent exploitation of all the regional resources. But economic regionalism does not aim at complete self-sufficiency: even under the most primitive conditions no region has ever been economically self-sufficient in all respects. On the other hand, economic regionalism does aim at combating the evil of over-specialization: since whatever the temporary commercial advantages of such specialization it tends to impoverish the cultural life of a region and, by placing all its eggs in one basket, to make precarious ultimately its economic existence. Just as every region has a potential balance of animal life and vegetation, so it has a potential social balance between industry and agriculture, between cities and farms, between built-up spaces and open spaces. A region entirely specialized for a single resource, or covered from boundary line to boundary line by a solid area of houses and streets, is a defective environment, no matter how well its trade may temporarily flourish. Economic regionalism is necessary to provide for a varied social life, as well as to provide for a balanced economy.

Plainly, a good part of the activity and business and power of the modern world, in which the nineteenth century took so much pride, was the result of disorganization, ignorance, inefficiency and social ineptitude. But the spread of technical knowledge, standardized methods, and scientifically controlled performances diminishes the need for transportation: in the new economy the old system of regional over-specialization will become the exception rather than the rule. Even today England is no longer the workshop of the world, and New England is no longer the workshop of America. And as mechanical industry becomes more highly rationalized and more finely adapted to the environment, a varied and many-sided indus- trial life tends to develop within each natural human region.

To achieve all these possible gains in production takes us far beyond the individual factory or industry, far beyond the current tasks of the administrator or engineer: it requires the services of the geographer and the regional planner, the psychologist, the educator, the sociologist, the skilled political administrator. Perhaps Russia alone at present has the necessary framework for this planning in its fundamental institutions; but to one degree or another, pushed by the necessity for creating order out of the existing chaos and dis-organization, other countries are moving in the same direction: the

Zuyder Zee reclamation in Holland, for example, is an example of the multifold rationalization of industry and agriculture and the building up of economic regional units here indicated.

The older modes of production have exploited only the superficial processes that were capable of being mechanized and externally ordered: whereas a bolder social economy will touch every aspect of the industrial complex. Complete organization of the mechanical elements, with ignorance, accident, and uncriticized custom dominant in society as a whole, was the formula of capitalistic enterprise during its earlier phases. That formula belongs to the past. It achieved only a small part of the potential production that even the crude machine age of the past was capable of, provided that it could have removed the frictions and contradictions and cross-purposes that perpetually impeded the flow of goods from source to mouth. To achieve efficiency in the past was as self-defeating a task as Carlyle's famous dilemma—given a band of thieves to produce an honesty out of their united action. In detail, we will doubtless carry over many admirable practices and rational arrangements derived from capitalism: but it is entirely doubtful, so deep are the dissonances, so inevitable are the frictions, that we shall carry over capitalist society itself. Humanly speaking, it has worn out its welcome. We need a system more safe, more flexible, more adaptable, and finally more life-sustaining than that constructed by our narrow and one-sided financial economy. Its efficiency was a mere shadow of real efficiency, its wasteful power was a poor substitute for order; its feverish productivity and its screaming breakdowns, wastes, and jams were low counterfeits of a functional economy that could really profit by modern technics.

6: Normalize Consumption!

Whereas we must maximize conversion, in order to have surplus energies ready to fulfill existing wants, and to be prepared for unexpected needs, it does not follow that we must also maximize production along the existing lines of effort. The aimless expansion of production is in fact the typical disease of capitalism in its application of modern technics: for since it failed to establish norms it had no definite measure for its productive achievement and no possible goals, except those erected by custom and accidental desire.

The expansion of the machine during the past two centuries was accompanied by the dogma of increasing wants. Industry was directed not merely to the multiplication of goods and to an increase in their variety: it was directed toward the multiplication of the desire for goods. We passed from an economy of need to an economy of acquisition. The desire for more material satisfactions of the nature furnished by mechanized production kept up with and partly cancelled out the gains in productivity. Needs became nebulous and indirect: to satisfy them appropriately under the capitalist criterion one must gratify them with profitable indirectness through the channels of sale. The symbol of price made direct seizure and gratification vulgar: so that finally the farmer who produced enough fruit and meat and vegetables to satisfy his hunger felt a little inferior to the man who, producing these goods for a market, could buy back the inferior products of the packing house and the cannery. Does that exaggerate the reality? On the contrary, it hardly does justice to it. Money became the symbol of reputable consumption in every aspect of living, from art and education to marriage and religion.

Max Weber pointed out the extraordinary departure of the new doctrines of industrialism from the habits and customs of the greater part of mankind under the more parsimonious system of production that prevailed in the past. The aim of traditional industry was not to increase the number of wants, but to satisfy the standards of a particular class. Even today, among the poor, the habits of this past linger on along with relics of magic and primitive medicine: for an increase in wages, instead of being used to raise the worker's standard of expenditure, is sometimes used to secure respite from work, or to provide the wherewithal for a spree which leaves the worker in exactly the same physical and social state he was in before beginning it. The notion of employing money to escape one's class, and of spending money conspicuously in order to register the fact that one has escaped, did not come into existence in society at large until a fairly late stage in the development of capitalism, although it manifested itself in the upper ranks at the very beginning of the modern régime.

The dogma of increasing wants, like so many other dogmas of industrialism and democracy, first appeared in the counting house and the court, and then filtered down into the rest of society. When abstract counters in gold or paper became the symbols of power and wealth, men began to value a form of commodity that had in fact no natural limits. The absence of normal standards of acquisition first manifested itself among the successful bankers and merchants; yet even here these standards lingered on far into the nineteenth century in the conception of retiring from business after achieving a competence—that is, the standards of one's class. The absence of a customary norm of consumption was most conspicuous in the extravagant life of the courts. To externalize the desire for power, wealth, and privilege, the princes of the Renascence lavished upon private luxury and display enormous amounts of money. They themselves, unless they happened to rise from the merchant class, did not earn this money: they were forced therefore to beg, borrow, extort, steal, or pillage it; and truth to tell, they left none of these possibilities unexplored. Once the machine began to increase the money-making capacities of industry, these limits were extended and the level of expenditure was raised for the entire society. This phase of capitalism was accompanied, as I have already pointed out, by a widespread breakdown of social institutions: hence the private individual often sought to compensate by egocentric getting and spending for the absence of collective institutions and a collective aim. The wealth of nations was devoted to the private gratification of individuals: the marvels of collective enterprise and cooperation that the machine brought into play left the community itself impoverished.

Despite the natural egalitarian tendency of mass production, a great gap continued to exist between the various economic classes: this gap was glibly accounted for, in terms of Victorian economics, by a differentiation between necessities, comforts, and luxuries. The bare necessities were the lot of the mass of workers. The middle classes, in addition to having their necessities satisfied on an ampler scale than the workers, were supported by comforts: while the rich possessed in addition—and this made them more fortunate— luxuries. Yet there was a contradiction. Under the doctrine of increasing wants the mass of mankind was supposed to adopt for itself the ultimate goal of a princely standard of expenditure. There existed nothing less than a moral obligation to demand larger quantities and more various kinds of goods—the only limit to this obligation being the persistent unwillingness of the capitalist manufacturer to give the worker a sufficient share of the industrial income to make an effective demand. (At the height of the last wave of financial expansion in the United States the capitalist sought to solve this paradox by loaning money for increased consumption—installment purchase —without raising wages, lowering prices, or decreasing his own excessive share in the national income: a device which would never have occurred to the more sober Harpagons of the seventeenth century.)

The historic mistakes of men are never so plausible and so dangerous as when they are embodied in a formal doctrine, capable of being expressed in a few catchwords. The dogma of increasing wants, and the division of consumption into necessities, comforts, and luxuries, and the description of the economic process as leading to the universalizing of more expensive standards of consumption *in terms of machine-made goods*—all these beliefs have been largely taken for granted, even by many of those who have opposed the outright injustices and the more flagrant inequalities of the capitalist economic system. The doctrine was put, with a classic fatuousness and finality, by the Hoover Committee's report on Recent Economic Changes in the United States. "The survey has proved conclusively," says the report, "what has long been held theoretically to be true, that wants are almost insatiable; that one want makes way for another. The conclusion is that economically we have a boundless field before us; that there are new wants which will make way endlessly for newer wants, as fast as they are satisfied."

When one abandons class standards of consumption and examines the facts themselves from the standpoint of the vital processes that are to be served, one finds that there is not a single element in these doctrines that can be retained.

First of all: vital wants are all necessarily limited. Just as the organism itself does not continue to grow beyond the norm of its species, a norm established within relatively narrow limits, so neither can any particular function of life be satisfied by limitless indulgence. The body does not require more than a limited number of calories of food per day. If it functions adequately on three meals a day, it does not become three times as strong or effective on nine meals: on the contrary, it is likely to suffer from indigestion and constipation. If the intensity of amusement is tripled in a circus by the use of three rings instead of one there are few other circumstances in which this rule holds: the value of various stimuli and interests is not increased by quantitative multiplication, nor yet, beyond a certain point, by endless variety. A variety of products which perform similar functions is like omnivorousness in diet: a useful factor of safety. But this does not alter the essential fact of stability of desire and demand. A harem of a thousand wives may satisfy the vanity of an oriental monarch; but what monarch is sufficiently well endowed by nature to satisfy the harem?

Healthy activity requires restriction, monotony, repetition, as well as change, variety, and expansion. The querulous boredom of a child that possesses too many toys is endlessly repeated in the lives of the rich who, having no pecuniary limit to the expression of their desires, are unable without tremendous force of character to restrict themselves to a single channel long enough to profit by its trenching and deepening and wearing through. While the man of the twentieth century has use for instruments, like the radio and the phonograph and the telephone, which have no counterpart in other civilizations, the number of such commodities is in itself limited. No one is better off for having furniture that goes to pieces in a few years or, failing that happy means of creating a fresh demand, "goes out of style." No one is better dressed for having clothes so shabbily woven that they are worn out at the end of the season. On the contrary, such rapid consumption is a tax on production; and it tends to wipe out the gains the machine makes in that department. To the extent that people develop personal and esthetic interests, they are immune to trivial changes in style and they disdain to foster such low demands. Moreover, as Mr. J. A. Hobson has wisely pointed out, "if an undue amount of individuality be devoted to the production and consumption of food, clothing, etc., and the conscious, refined cultivation of these tastes, higher forms of individual expression in work and life will be neglected."

The second characteristic of vital wants is that they cannot be restricted to the bare elements of food enough to forestall starvation and clothing and shelter enough to satisfy convention and to ward off death by exposure. Life, from the very moment of birth on, requires for its fulfillment goods and services that are usually placed in the department of "luxuries." Song, story, music, painting, carving, idle play, drama—all these things lie outside the province of animal necessities: but they are not things which are to be included after the belly is satisfied: they are functions which must be included in human existence even to satisfy the belly, to say nothing of the emotional and intellectual and imaginative needs of man. To put these functions at a distance, to make them the goal of an acquisitive life, or to accept only so much of them as can be canalized into machine goods and sold at a profit—to do this is to misinterpret the nature of life as well as the possibilities of the machine.

The fact is that every vital standard has its own necessary luxuries; and the wage that does not include them is not a living wage, nor is the life made possible by bare subsistence a humane life. On the other hand, to set as a goal for universal economic effort, or at least to bait as a temptation, the imbecile standard of expenditure adopted by the rich and the powerful is merely to dangle a wooden carrot before the donkey; he cannot reach the carrot, and if he could, it still would not nourish him. A high scale of expense has no essential relation whatever to a high standard of living; and a plethora of machine-made goods has no essential relation, either, since one of the most essential elements of a good life—a pleasant and stimulating natural environment, both cultivated and primitive—is not a machine-made product. The notion that one implies the other is a figment of the business man's will-to-believe. As for what is called comfort, a good part of it, freedom from exertion, the extensive use of mechanical and personal service, leads in fact to an atrophy of function: the ideal is at best a valetudinarian one. The reliance for sensual pleasure upon inanimate objects—sofa pillows, upholstered furniture, sweetmeats, and soft textiles—was one of those devices whereby a bourgeois Puritanism, affecting to renounce the flesh and to castigate the body, merely acknowledged them in their most decadent forms, transferring attention from the animate bodies of men and women to objects that simulated them. The Renascence, which celebrated a vigorous sensual life, scarcely produced a comfortable chair in two hundred years: but one has only to look at the women painted by Veronese and Rubens to see how little such inorganic upholstery was needed.

As mechanical methods have become more productive, the notion has grown up that consumption should become more voracious. In back of this lies an anxiety lest the productivity of the machine create a glut in the market. The justification of labor-saving devices was not that they actually saved labor but that they increased consumption: whereas, plainly, labor-saving can take place only when the standard of consumption remains relatively stable, so that increases in conversion and in productive facility will be realized in the form of actual increments of leisure. Unfortunately, the capitalistic industrial system thrives by a denial of this condition. It thrives by stimulating wants rather than by limiting them and satisfying them. To acknowledge a goal of consummation would be to place a brake upon production and to lessen the opportunities for profit.

Technically speaking, changes in form and style are symptoms of immaturity; they mark a period of transition. The error of capitalism as a creed lies in the attempt to make this period of transition a permanent one. As soon as a contrivance reaches technical perfection, there is no excuse for replacement on the ground of increased efficiency: hence the devices of competitive waste, of shoddy workmanship, and of fashion must be resorted to. Wasteful consumption and shoddy craftmanship go hand in hand: so that if we value soundness and integrity and efficiency within the machine system, we must create a corresponding stability in consumption.

Speaking in the broadest terms this means that once the major wants of mankind are satisfied by the machine process, our factory system must be organized on a basis of regular annual replacement instead of progressive expansion—not on a basis of premature replacement through debauched workmanship, adulterated materials, and grossly stimulated caprice. "The case," as Mr. J. A. Hobson again puts it, "is a simple one. A mere increase in the variety of our material consumption relieves the strain imposed upon man by the limits of the material universe, for such variety enables him to utilize a larger proportion of the aggregate of matter. But in proportion as we add to mere variety a higher appreciation of those adaptations of matter which are due to human skill, which we call Art, we pass outside the limit of matter and are no longer the slaves of roods and acres and a law of diminishing returns." In other words: a genuine standard, once the vital physical wants are satisfied, tends to change the *plane* of consumption and therefore to limit, in a considerable degree, the extent of further mechanical enterprise.

But mark the vicious paradox of capitalist production. Although the factory system has been based on the doctrine of expanding wants and upon an expanding body of consumers, it has universally fallen short of supplying the normal wants of mankind. Horrified at the "utopian" notion of limited and normalized wants, and proudly proclaiming on the contrary that wants are insatiable, *capitalism has not come within miles of satisfying the most modest standard of normalized consumption.* Capitalism, with respect to the working mass of humanity, has been like a beggar that flaunts a hand covered with jewels, one or two of them genuine, whilst it shivers in rags and grabs at a crust of bread: the beggar may have money in the bank, too, but that does not improve his condition. This has been brought out clearly in every factual study that has been made of "advanced" industrial communities, from Charles Booth's classic survey of London to the thoroughly documented Pittsburgh survey: it has been re-enforced once more by Robert Lynd's study of the fairly representative community of "Middletown." What does one find? While the poorer inhabitants of Middletown often boast a motor car or a radio set, the houses they lived in during their period of putative prosperity often did not have even ordinary sanitary toilet facilities, while the state of the house and the general environment was, factually speaking, that of a slum.

When one says that the doctrine of increasing wants must be thrown overboard and the standard of consumption normalized, one does not in fact call for a contraction of our present industrial facilities. In many departments, on the contrary, we are urgently in need of an expansion of them. For the truth is that, despite all boasts of progress and mechanical achievement, despite all fears of surpluses and gluts, the mass of mankind, even in the countries that are technically the most advanced and financially the most prosperous, do not have—and apart from the agricultural population never have had—an adequate diet, proper facilities for hygiene, decent dwellings, sufficient means and opportunities for education and recreation. Indeed, in terms of vital norm a good part of these things have been equally lacking in the spurious standard of expenditure secured by the rich. In most great cities the urban dwellings of the upper classes, for example, are lacking in sunlight and open spaces, and are almost as inadequate as those of the very poor: so that, under a normalized standard of life, they would in many cases be healthier and happier than they are at present even though they would lack the illusion of success and power and distinction.

To normalize consumption is to erect a standard that no single class, whatever its expenditures, possesses today. But that standard cannot be expressed in terms of any arbitrary sum of money—the five thousand dollars per individual yearly suggested by Bellamy in the eighties, or the twenty thousand dollars suggested by a recent group of technocrats: for the point is that what five or twenty thousand dollars could purchase today for any single individual would not necessarily fulfill the more exacting vital requirements of this standard. And indeed, *the higher the vital standard, the less can it be expressed adequately in terms of money:* the more must it be expressed in terms of leisure, and health, and biological activity, and esthetic pleasure, and the more, therefore, will it tend to be expressed in terms of goods and environmental improvements that lie outside of machine production.

At the same time, the conception of a normalized consumption acknowledges the end of those princely capitalistic dreams of limitless incomes and privileges and sensuous vulgarities whose possession by the masters of society furnished endless vicarious gratification to their lackeys and imitators. Our goal is not *increased* consumption but a vital standard: less in the preparatory means, more in the ends, less in the mechanical apparatus, more in the organic fulfillment. When we have such a norm, our success in life will not be judged by the size of the rubbish heaps we have produced: it will be judged by the immaterial and non-consumable goods we have learned to enjoy, and by our biological fulfillment as lovers, mates, parents and by our personal fulfillment as thinking, feeling men and women. Distinction and individuality will reside in the personality, where it belongs, not in the size of the house we live in, in the expense of our trappings, or in the amount of labor we can arbitrarily command. Handsome bodies, fine minds, plain living, high thinking, keen perceptions, sensitive emotional responses, and a group life keyed to make these things possible and to enhance them—these are some of the objectives of a normalized standard.

While the animus that led to the expansion of the machine was narrowly utilitarian, the net result of such an economy is to create an antithetical stage, paralleled by the slave civilizations of old, endowed with an abundance of leisure. This leisure, if not vilely misused in the thoughtless production of more mechanical work, either through misplaced ingenuity or a vain consumptive ritual, may eventuate in a non-utilitarian form of society, dedicated more fully to play and thought and social intercourse and all those adventures and pursuits that make life more significant. The maximum of machinery and organization, the maximum of comforts and luxuries, the maximum of consumption, do not necessarily mean a maximum of life-efficiency or life-expression. The mistake consists in thinking that comfort, safety, absence of physical disease, a plethora of goods are the greatest blessings of civilization, and in believing that as they increase the evils of life will dissolve and disappear. But comfort and safety are not unconditioned goods; they are capable of defeating life just as thoroughly as hardship and uncertainty; and the notion that every other interest, art, friendship, love, parenthood, must be subordinated to the production of increasing amounts of comforts and luxuries is merely one of the superstitions of a money-bent utilitarian society.

By accepting this superstition the utilitarian has turned an elementary condition of existence, the necessity for providing a physical basis for life, into an end. As a result, our machine-dominated society is oriented solely to "things," and its members have every kind of possession except self-possession. No wonder that Thoreau observed that its members, even in an early and relatively innocent state of commerce and industry, led lives of quiet desperation. By putting business before every other manifestation of life, our mechanical and financial leaders have neglected the chief business of life: namely, growth, reproduction, development, expression. Paying infinite attention to the invention and perfection of incubators, they have forgotten the egg, and its reason for existence.

7: Basic Communism

A normalized mode of consumption is the basis of a rationalized mode of production. If one begins with production as an end in itself there is nothing within the machine system or the price system to guarantee a sufficient supply of vital goods. The capitalist economy attempted to avoid the necessity for erecting a real standard of life by relying upon the automatic operation of men's private interests, under the spell of the profit motive. All the necessary gains in production, along with a cheapening of the objects sold, were supposed to be an inevitable by-product of the business of buying cheap and selling where the demand was strongest and the supply scantest. The enlightened self-interest of individual buyers was the guarantee that the right things would be produced, in the right order, at the right time.

Lacking any standard for distributing income except on the basis of the gross labor performed and on the bare subsistence necessary to enable the worker to return each day to his job, this system never succeeded in its best days even on its own terms. The history of capitalism is the history of quantity production, over-expansion, greedy private over-capitalization on the basis of an increasing prospective income, the private appropriation of profits and dividends at the expense of the workers and the vast body of non-capitalist ultimate consumers—all followed, again and again, by a glut of unbought goods, a breakdown, bankruptcy, deflation, and the bitter starvation and depression of the working classes whose original inability to buy back the goods they had produced was always the major factor in this debacle.

This system is necessarily unworkable upon its own premises except perhaps under a pre-machine mode of production. For upon capitalist terms, the price of any commodity, roughly speaking, varies inversely as the quantity available at a given moment. This means that as production approaches infinity, the price of a single article must fall correspondingly toward zero. Up to a certain point, the fall in prices expands the market: beyond that point, the increase in real wealth for the community means a steady decrease in profits per unit for the manufacturer. If the prices are kept up without an expansion of real wages, an overplus occurs. If the price is lowered far enough, the manufacturer cannot, no matter how great his turnover, produce a sufficient margin of profit. Whereas mankind as a whole gains in wealth to the extent that the necessaries of life can, like the air, be had for the asking, the price system crashes into disaster long before this ideal point has been reached. Thus the gains in production under the price system must be diminished or cancelled out, as Veblen mordantly pointed out, by deliberate sabotage on the part of the financier and the business man. But this strategy has only a temporary effect: for the burden of debt, especially when recapitalized on the basis of a prospective expansion of the population and the market, ultimately outruns the curtailed productive capacities and subjects them to a load they cannot meet.

Now, the chief meaning of power conversion and mechanized production lies in the fact that they have created an economy of surplus —which is to say, an economy not adapted to the price system. As more and more work is transferred to automatic machines, the process of displacing workers from industry under this system is the equivalent of disfranchising them as consumers, since, unlike the holders of stock, bonds, and mortgages, they have no claim upon industry under capitalist conventions other than that resulting from their labor. It is useless to talk about temporary absorptions of labor by this or that industry: part of this absorption by the industries concerned with distribution only increases the overhead and the waste. And apart from this, under the system itself labor has lost both its bargaining power and its capacity to obtain subsistence: the existence of substitute industries sometimes postpones the individual but does not avert the collective day of reckoning. Lacking the power to buy the necessaries of life for themselves, the plight of the displaced workers reacts upon those who remain at work: presently the whole structure collapses, and even financiers and enterprisers and managers are sucked into the whirlpool their own cupidity, short-sightedness and folly have created. All this is a commonplace: but it rises, not as a result of some obscure uncontrollable law, like the existence of spots on the sun, but as the outcome of our failure to take advantage by adequate social provision of the new processes of mechanized production.

The problem presses for solution: but in one sense it has already been solved. For the better part of a thousand years, widows, orphans, and prudent sedentary people have been living at ease, buying food, drink, and shelter, without performing any work for the community. Their shares and their insurance payments constitute a first claim upon industry; and as long as there is any production of goods at all, and as long as the present legal conventions are maintained, they are sure of their means of existence. No capitalist talks about this system as one that demoralizes or undermines the self-respect of those who are so supported: indeed, the small incomes of the rentier classes have been an obvious help in the arts and sciences to their recipients: a Milton, a Shelley, a Darwin, a Ruskin existed by such grace; and one might even show, perhaps, that they had been more beneficial to society at large than the swollen fortunes of the more active capitalists. On the other hand, the small fixed income, though it sets at a distance the worst torments of economic distress, does not completely meet every economic requirement: so, in the case of the young and the ambitious, there is an incentive to productive and professional enterprise, even though the sting of starvation be absent.

The extension of this system to the community as a whole is what I mean by basic communism. In recent times, it was first seriously proposed by Edward Bellamy, in a somewhat arbitrary form, in his utopia, Looking Backward; and it has become plain during the last fifty years that an efficient mechanized system of production can be made serviceable to humanity at large in no other fashion. To make the worker's share in production the sole basis for his claim to a livelihood—as was done even by Marx in the labor theory of value he took over from Adam Smith—is, as power-production approaches perfection, to cut the ground from under his feet. In actuality, the claim to a livelihood rests upon the fact that, like the child in a family, one is a member of a community: the energy, the technical knowledge, the social heritage of a community belongs equally to every member of it, since in the large the individual contributions and differences are completely insignificant.

[*The classic name for such a universal system of distributing the essential means of life—as described by Plato and More long before Owen and Marx—is communism, and I have retained it here. But let me emphasize that this communism is necessarily post-Marxian, for the facts and values upon which it is based are no longer the paleotechnic ones upon which Marx founded his policies and programs. Hence communism, as used here, does not imply the particular nineteenth century ideology, the messianic absolutism, and the narrowly militarist tactics to which the official communist parties usually cling, nor does it imply a slavish imitation of the political methods and social institutions of Soviet Russia, however admirable soviet courage and discipline may be.*]

Differentiation and preference and special incentive should be taken into account in production and consumption only after the security and continuity of life itself is assured. Here and there we have established the beginnings of a basic communism in the provision of water and education and books. There is no rational

reason for stopping short any point this side of a normal standard of consumption. Such a basis has no relation to individual
capacities and virtues: a family of six requires roughly three times
as much goods as a family of two, although there may be but
one wage-earner in the first group and two in the second. We give at
least a minimum of food and shelter and medical attention to crim-
inals who have presumably behaved against the interests of society:
why then should we deny it to the lazy and the stubborn? To assume
that the great mass of mankind would belong to the latter category
is to forget the positive pleasures of a fuller and richer life.

Moreover, under a scientific economy, the amount of grain, fruit,
meat, milk, textiles, metals and raw materials, like the number of
houses needed annually for replacement and for the increase of population, can be calculated in the gross in advance of production. It
needs only the insurance of consumption to make the tables of production progressively more accurate. Once the standard was established,
gains beyond those calculated would be bonuses for the whole community: such gains, instead of stopping the works, as they do now,
would lubricate them, and so far from throwing the mechanism out
of gear they would lighten the load for the whole community and
increase the margin of time or energy available for the modes of life,
rather than for the means.

To speak of a "planned economy," without such a basic standard
of consumption and without the political means of making it prevail,
is to mistake the monopolistic sabotage of large-scale capitalist industry for intelligent social control.

The foundations of this system of distribution already, I repeat,
exist. Schools, libraries, hospitals, universities, museums, baths, lodg-
ing houses, gymnasia, are supported in every large center at the expense of the community as a whole. The police and the fire services,
similarly, are provided on the basis of need instead of on the ability
to pay: roads, canals, bridges, parks, playgrounds, and even—in
Amsterdam—ferry services are similarly communized. Furthermore,
in the most jejune and grudging form, a basic communism is in existence in countries that have unemployment and old-age insurance. But
the latter measures are treated as means of salvage, rather than as a

salutary positive mechanism for rationalizing the production and normalizing the consumptive standards of the whole community.

A basic communism, which implies the obligation to share in the work of the community up to the amount required to furnish the basis, does not mean the complete enclosure of every process and the complete satisfaction of every want in the system of planned production. Careful engineers have figured that the entire amount of work of the existing community could be carried on with less than twenty hours work per week for every existing worker: with complete rationalization all along the line, and with the elimination of duplications and parasitisms, probably less than twenty hours would suffice to produce a far greater quantity of goods than is produced at present. As it is, some 15 million industrial workers supply the needs of 120 million inhabitants of the United States. Limiting rationed production and communized consumption to basic requirements, the amount of compulsory labor would be even less. Under such provisions, technological unemployment would be a boon.

Basic communism would apply to the calculable economic needs of the community. It would touch those goods and services which can be standardized, weighed, measured, or about which a statistical computation can be made. Above such a standard the desire for leisure would compete with the desire for more goods: and here fashion, caprice, irrational choice, invention, special aims, would still perhaps have a part to play: for although all these elements have been grossly over-stimulated by capitalism, a residue of them would remain and would have to be provided for in any conceivable economic system. But under a basic communism, these special wants would not operate so as to disorganize production and paralyze distribution. With regard to the basic commodities there would be complete equality of income: and as consumption became normalized, the basic processes would care, in all probability, for a larger and larger part of the community's needs. On this basis—and so far as I can see on no other basis—can our gains in production and our growing displacement of human labor be realized in benefits for society at large. The alternative to basic communism is the toleration of chaos: either the closing down periodically of the productive plant and the destruction—quaintly called valorization—of essential goods, with shifty efforts at imperialist conquest to force open foreign markets; either that or a complete retreat from the machine into a sub-agriculture (subsistence farming) and a sub-industry (subsistence manufacture) which would be far lower in every way than what handicraft industry had provided in the eighteenth century. If we wish to retain the benefits of the machine, we can no longer afford to deny its chief social implication: namely, basic communism.

Not the least advantage of basic communism would be the fact that it would tend to put a brake upon industrial enterprise. But such a brake, instead of being in the form of capitalist sabotage, or in the shocking dislocation of a commercial crisis, would be a gradual lessening of the speed of individual parts and a gearing of the whole organization into a steady routine of productivity. Mr. J. A. Hobson has again put this matter with his usual insight and wisdom: "Industrial progress," he says, "would undoubtedly be slower under State-control, because the very object of such control is to divert a larger proportion of human genius and effort from these occupations [preparatory production] to apply them in producing higher forms of wealth. It is not, however, right to assume that progress in the industrial arts would cease under state-industry: such progress would be slower, and would itself partake of a routine character—a slow, continuous adjustment of the mechanism of production and distribution to the slowly changing needs of the community." However forbidding such a prospect looks to the enterpriser of the old order, humanly speaking it would represent a tremendous gain.

8: Socialize Creation!

During a great part of the history of mankind, from neolithic times onward, the highest achievements of the race in art and philosophy and literature and technics and science and religion were in the possession of a small caste of people. The technical means of multiplying these achievements were so cumbrous—the hieroglyphics of the Egyptians, the baked slabs of the Babylonian texts, even the handwritten letters on the papyrus or parchment of a later period—that the mastery of the implements of thought and expression was the work of the better part of a lifetime. Those who had manual tasks to perform were automatically excluded from most of the avenues of creation outside their tasks, though they might eventually share in the product created, at second or third hand. The life of the potter or the smith, as Jesus ben Sirach pointed out with priggish but realistic self-justification, unfitted him for the offices of the creative life.

This caste-monopoly was seriously disrupted during the Middle Ages, partly because Christianity itself was in origin the religion of the lowly and the downtrodden. Not merely was every human creature a worthy subject of salvation, but within the monastery and the church and the university there was a steady recruitment of novices and students from every rank in society; and the powerful Benedictine order, by making manual work itself one of the obligations of a disciplined life, broke down an ancient and crippling prejudice against participation and experiment, as complementary to observation and contemplation, in creative activity. Within the craft guilds the same process took place in reverse direction: not merely did the journeyman, in qualifying for his craft, get an opportunity to view critically the arts and achievements of other cities, not merely was he encouraged to rise from the menial and mechanical operations of his craft to such esthetic mastery as it offered, but in the performance of the mysteries and the moralities the worker participated in the esthetic and religious life of the whole community. Indeed the writer, like Dante, could have a political status in this society only as the member of a working guild.

The humanist movement, by placing an emphasis upon textual scholarship and the dead languages to which this scholarship applied, re-enforced the widened separation of classes under capitalism. Unable to obtain the necessary preparatory training, the worker was excluded from the higher culture of Europe: even the highest type of eotechnic worker, the artist, and even one of the proudest figures among these artists, Leonardo, felt obliged in his private notes to defend himself against the assumption of the merely literate that his interests in painting and science were somehow inferior.

Indifferent to the essential life of men as workers, this culture developed primarily as an instrument of caste-power, and only in a feeble and secondary way for the benefit of mankind as a whole. From one end to the other some of the very best minds of the last three centuries, in the midst of their most vigorous creative efforts, have been apologizing for the injustices and perversions of their masters. Thorndike in his History of Science and Medicine in the Fifteenth Century notes the degradation that overcame thought when the free cities that Petrarch had known in his youth were enslaved by conquering armies: but the same fact is equally plain in Macchiavelli, Hobbes, Leibniz, Hegel; and this tendency of thought reached a certain climax in the misapplication of the Malthus-Darwin theory of the struggle for existence, to justify warfare, the nordic race, and the dominant position of the bourgeoisie.

But while the humanist side of this new culture was fostered on individualistic and caste lines, with a marked bias in favor of the possessing classes, science worked in an opposite direction. The very growth of scientific knowledge made it impossible to confine it, as a secret, to a small group, as astronomy was maintained in earlier civilizations. Not merely this, but science, by systematically utilizing the practical knowledge of artists and physicians in anatomy, of miners and metallurgists in chemistry, kept in touch with the working life of the community: was it not the predicament of vintners, brewers, and silkworm growers that roused Pasteur to his productive researches in bacteriology? Even when science was remote and by nature esoteric, it was not snobbish. Socialized in method, international in scope, impersonal in animus, performing some of its most hazardous and fruitful feats of thought by reason of its very divorce from immediate responsibility, the sciences have been slowly building up a grand cosmogony in which only one element is still lacking—the inclusion of the spectator and experimenter in the final picture.

Unfortunately, the dulling and depressing of the mind that inevitably followed from the division of labor and the bare routine of factory life, have opened an unnatural breach between science and technics and common practice and all the arts that lie outside the machine system. The workers themselves were thrown back upon the rubbish of earlier cultures, lingering in tradition and memory, and they clung to superstitious forms of religion which kept them in a state of emotional tutelage to the very forces that were exploiting them, or else they forfeited altogether the powerful emotional and moral stimulus that a genuine religion contributes to life. This applies likewise to the arts. The peasant and handworker of the Middle Age was the equal of the artists who carved and painted in his churches and his public halls: the highest art of that time was not too high for the common people, nor was there, apart from the affectations of court poesy, one kind of art for the few and another kind for the many. There were high and low levels in all this art: but the division was not marked by status or pecuniary condition.

During the last few centuries, however, popular means "vulgar" and "vulgar" means not simply the broadly human, but something inferior and crass and a little dehumanized. In short, instead of socializing the creative activities of society, we have socialized on a great scale only the low counterfeits of those activities: counterfeits that limit and stultify the mind. A Millet, a van Gogh, a Daumier, a Whitman, a Tolstoy naturally seek the working class for companionship: but they were actually kept alive and rewarded and appreciated chiefly by the very bourgeoisie whose manners they loathed and whose patronage they wished to escape. On the other hand, the experience of New England and New York between 1830 and 1860, when there was still to the westward a great sweep of unappropriated land, shows how fruitful an essentially classless society can be when it is nourished by the very occupations that a caste-culture disdains. It is no accident that the epic of Moby Dick was written by a common sailor, that Walden was written by a pencil-maker and surveyor, and that Leaves of Grass was written by a printer and carpenter. Only when it is possible to move freely from one aspect of experience and thought and action to another can the mind follow its complete trajectory. Division of labor and specialization, specialization between occupations, specialization in thought, can be justified only as temporary expedients: beyond that, as Kropotkin pointed out, lies the necessity of integrating labor and restoring its unity with life.

What we need, then, is the realization that the creative life, in all its manifestations, is necessarily a social product. It grows with the aid of traditions and techniques maintained and transmitted by society

at large, and neither tradition nor product can remain the sole possession of the scientist or the artist or the philosopher, still less of
the privileged groups that, under capitalist conventions, so largely
support them. The addition to this heritage made by any individual,
or even by any generation, is so small in comparison with the accumulated resources of the past that the great creative artists, like
Goethe, are duly humble about their personal importance. To treat
such activity as egoistic enjoyment or as property is merely to brand
it as trivial: for the fact is that creative activity is finally the only
important business of mankind, the chief justification and the most
durable fruit of its sojourn on the planet. The essential task of all
sound economic activity is to produce a state in which creation will
be a common fact in all experience: in which no group will be denied,
by reason of toil or deficient education, their share in the cultural
life of the community, up to the limits of their personal capacity.
Unless we socialize creation, unless we make production subservient
to education, a mechanized system of production, however efficient,
will only harden into a servile byzantine formality, enriched by
bread and circuses.

9: Work for Automaton and Amateur

Not work, not production for its own sake or for the sake of ul-
terior profit, but production for the sake of life and work as the
normal expression of a disciplined life, are the marks of a rational
economic society. Such a society brings into existence choices and possibilities that scarcely existed so long as work was considered extraneous, and profit—or terror of starvation—was the chief impetus
to labor.

The tendency of mechanization, from the seventeenth century on,
has been to standardize the processes of work and to make them capable of machine operation. In power plants with automatic stokers, in
advanced textile mills, in stamping factories, in various chemical
works, the worker has scarcely any direct part in the process of production: he is, so to say, a machine-herd, attending to the welfare of
a flock of machines which do the actual work: at best, he feeds them,
oils them, mends them when they break down, while the work itself

is as remote from his province as is the digestion which fattens the sheep looked after by the shepherd.

Such machine-tending often calls for alertness, non-repetitious movement, and general intelligence: in discussing neotechnics I pointed out that in industries that have advanced to this level the worker has recovered some of the freedom and self-direction that were frustrated in the more incomplete mechanical processes where the worker, instead of being general mechanic and overseer, is merely a substitute for the hand or eye that the machine has not yet developed. But in other processes, such as the straight line assemblage of the motor factory, for example, the individual worker is part of the process itself, and only a small fraction of him is engaged. Such labor is necessarily servile in character, and no amount of apology or psychological rationalization can make it otherwise: nor can the social necessity for the product mollify the process itself.

Our disregard for the quality of work itself, for work as a vital and educational process, is so habitual that it scarcely ever enters into our social demands. Yet it is plain that in the decision as to whether to build a bridge or a tunnel there is a human question that should outweigh the question of cheapness or mechanical feasability: namely, the number of lives that will be lost in the actual building or the advisability of condemning a certain number of men to spend their entire working days underground supervising tunnel traffic. As soon as our thought ceases to be automatically conditioned by the mine, such questions become important. Similarly the social choice between silk and rayon is not one that can be made simply on the different costs of production, or the difference in quality between the fibres themselves: there also remains, to be integrated in the decision, the question as to difference in working-pleasure between tending silkworms and assisting in rayon production. What the product contributes to the laborer is just as important as what the worker contributes to the product. A well-managed society might alter the process of motor car assemblage, at some loss of speed and cheapness, in order to produce a more interesting routine for the worker: similarly, it would either go to the expense of equipping dry-process cement making plants with dust removers—or replace the product itself with a less noxious substitute. When none of these alternatives was available, it would drastically reduce the demand itself to the lowest possible level.

Now, taken as a whole, including the preparatory processes of scientific investigation and mechanical design, to say nothing of the underlying political organization, industry is potentially a valuable instrument of education. This point, originally stressed by Karl Marx, was well put by Helen Marot when she said: "Industry offers opportunities for creative experience which is social in its processes as well as in its destination. The imaginative end of production does not terminate with the possession of an article; it does not center in the product or in the skill of this or that man, but in the development of commerce and technological processes and the evolution of world acquaintanceship and understanding. Modern machinery, the division of labor, the banking system, methods of communication, *make possible* real association. But they are real and possible only as the processes are open for the common participation, understanding, and judgment of those engaged in industrial enterprise; they are real and possible as the animus of industry changes from exploitation to a common and associated desire to create; they are real and possible as the individual character of industry gives way before the evolution of social effort."

Once the objective of industry is diverted from profit-making, private aggrandizement, crude exploitation, the unavoidable monotonies and restrictions will take a subordinate place, for the reason that the process will be humanized as a whole. This means that compensations for the repressive elements in the industrial routine will take place by adjustments within industry itself, instead of being permitted to heap up there, and to explode disastrously and anti-socially in other parts of society. To fancy that such a non-profit system is an impossibility is to forget that for thousands of years the mass of mankind knew no other system. The new economy of needs, replacing the capitalist economy of acquisition, will put the limited corporations and communities of the old economy on a broader and more intelligently socialized basis: but at bottom it will draw upon and canalize similar impulses. Despite all its chequered features and internal contradictions, this is to date perhaps the chief promise held out by Soviet Russia.

To the extent that industry must still employ human beings as machines, the hours of work must be reduced. We must determine the number of hours of blank routine per week that is within the limits of human tolerance, beyond which obvious deterioration of mind and spirit sets in. The very fact that purely repetitious work, without choices or variations, seems to agree with morons is enough to warn us of its dangers in relation to human beings of higher grade. But there remain occupations, machine-crafts as well as hand-crafts, which are interesting and absorbing in their own right, provided that they are not regimented too strictly in the interests of superficial efficiency. In the act of rationalizing and standardizing the methods of production, human engineering will have to weigh the social benefits of increased production with automatic machinery, with a lessened participation and satisfaction upon the part of the worker, against a lower level of production, with a larger opportunity for the worker. It is a shallow technicism to enforce the cheaper product at any price. Where the product is socially valuable and where the worker himself can be completely eliminated the answer will often, perhaps, favor automatism: but short of this state the decision cannot be lightly made. For no gain in production will justify the elimination of a humane species of work, unless other compensations in the way of work itself are at the same time provided. Money, goods, vacant leisure, cannot possibly make up for the loss of a life-work; although it is plain that money and goods, under our present abstract standards of success, are called upon often to do precisely this.

When we begin to rationalize industry organically, that is to say, with reference to the entire social situation, and with reference to the worker himself in all his biological capacities—not merely with reference to the crude labor product and an extraneous ideal of mechanical efficiency—the worker and his education and his environment become quite as important as the commodity he produces. We already acknowledge this principle on the negative side when we prohibit cheap lead glazes in pottery manufacture because the worker's health is undermined by their use: but it has a positive application as well. Not merely should we prohibit work that is bad for the health: we should promote work that is good for the health. It is on these grounds that agriculture and our rural regions may presently get back part of the population that was originally sucked into the *villes tentaculaires* by the machine.

Labor itself, from spading a garden to mapping the stars, is one of the permanent joys of life. A machine economy that permitted mankind the inane and trivial leisure Mr. H. G. Wells once depicted in The Time Machine, and that most city dwellers are condemned to under capitalist society, particularly during periods of unemployment, would scarcely be worth the effort necessary to lubricate it: such vacuity, such boredom, such debilitating lack of function do not represent a gain of any kind. The chief benefit the rational use of the machine promises is certainly not the elimination of work: what it promises is something quite different—the elimination of *servile* work or *slavery:* those types of work that deform the body, cramp the mind, deaden the spirit. The exploitation of machines is the alternative to that exploitation of degraded men that was practiced throughout antiquity and that was challenged on a large scale, for the first time, in the power economy evolved in the eotechnic phase.

By the completion of our machine organization, we can recover for work the inherent values which it was robbed of by the pecuniary aims and class animosities of capitalist production. The worker, properly extruded from mechanical production as slave, comes back as director: if his instincts of workmanship are still unsatisfied by these managerial tasks, he has by reason of the power and leisure he now potentially commands a new status within production as an amateur. The gain in freedom here is a direct compensation for the pressure and duress, for the impersonality, the anonymity, the collective unity of machine production.

Beyond the basic needs of production, beyond a normalized—and therefore moralized—standard of life, beyond the essential communism in consumption I have posited, there lie wants which the individual or the group has no right to demand from society at large, and which, in turn, society has no need to curtail or arbitrarily re-

press in the individual, so long as the motive of exploitation is removed. These wants may be satisfied by direct effort. To weave or
knit clothes by hand, to produce a necessary piece of furniture, to
experimentally build an airplane on lines that have not won official
approval—these are samples of occupations open to the individual,
the household, the small working group, apart from the regular channels of production. Similarly, while the great staples in agriculture,
like wheat, corn, hogs, beef, will possibly tend to be the work of large
cooperatives, green vegetables and flowers may be raised by individuals on a scale impossible so long as land was privately appropriated and the mass of industrial mankind was packed together in
solid areas of house and pavement.

As our basic production becomes more impersonal and routinized,
our subsidiary production may well become more personal, more
experimental, and more individualized. This could not happen under
the older régime of handicraft: it was a development not possible
before the neotechnic improvements of the machine with electricity
as a source of power. For the acquisition of skill necessary for efficient
production on a handicraft basis was a tedious process, and the slow
tempo of handicraft in the essential occupations did not give a sufficient margin of time for achievement along other lines. Or rather, the
margin was achieved by the subordination of the working class and
the elevation of a small leisure class: the worker and the amateur
represented two different strata. With electric power a small machine shop may have all the essential devices and machine tools—
apart from specialized automatic machines—that only a large plant
could have afforded a century ago: so the worker can regain, even
within the machine occupations, most of the pleasure that the machine
itself, by its increasing automatism, has been taking away from him.
Such workshops connected with schools should be part of the public
equipment of every community.

The work of the amateur, then, is a necessary corrective to the impersonality, the standardization, the wholesale methods and products
of automatic production. But it is likewise an indispensable educational preparation for the machine process itself. All the great advances in machines have been on the basis of the handicraft opera-

tions or scientific thought—itself aided and corrected by small-scale manual operations called experiments. As "technological tenuousness" increases, the diffusion of handicraft knowledge and skill as a mode of education is necessary, both as a safety device and as a means to further insight, discovery, and invention. For the machine cannot know more or do more than the human eye or hand or mind that designs or operates it. Given knowledge of the essential operations, one could reconstruct every machine in the world. But let that knowledge be cut off for so much as a single generation, and all the complicated derivatives would be so much junk. If parts broke and rusted without being immediately replaced, the whole fabric would be in ruins. And there is still a further reason to give an important position to the hand-crafts and machine-crafts, as subsidiary forms of production, run on a domestic scale. For both safety and flexibility in all forms of industrial production it is important that we learn to travel light. Our specialized automatic machines, precisely because of their high degree of specialization, lack adaptability to new forms of production: a change in demand, a change in pattern, leads to the wholesale scrapping of very expensive equipment. Wherever demand for products is of an uncertain or variable nature, it is an economy in the long run to use non-specialized machines: this decreases the burden of wasted effort and idle machinery. What is true of the machine is equally true of the worker: instead of a high degree of specialized skill, an all-round competence is better preparation for breaking through stale routines and for facing emergencies.

It is the basic skills, the basic manual operations, the basic discoveries, the basic formulas which must be transmitted from generation to generation. To maintain the superstructure whilst we let the foundations moulder away is to endanger not alone the existence of our complicated civilization but its further development and refinement. For critical changes and adaptations in machines, as in organisms, come not from the differentiated and specialized stock, but from the relatively undifferentiated common ancestor: it was the foot-treadle that served Watt's need for transmitting power in a steam engine. Automatic machines may conquer an ever-larger province in basic production: but it must be balanced by the hand-crafts and the

machine-crafts for education, recreation, and experiment. Without
the second, automatism would ultimately be a blight on society, and
its further existence would be imperilled.

10: Political Control

Plan and order are latent in all modern industrial processes, in
the working drawing, in the preliminary calculations, in the organization chart, in the time-schedule, in the graphs that keep track of
production day by day, and even hour by hour, as in a power plant.
This graphic and ordered procedure, originating in the separate
techniques of the civil engineer, the architect, the mechanical engineer, the forester, and other types of technician, is particularly evident in the neotechnic industries. (See, for example, the elaborate
economic and social surveys of the Bell Telephone Company, in
preparation for establishing or extending services.) What is still
lacking is the transference of these techniques from industry to the
social order at large. The order so far established is too local to be
socially effective on a great scale, and apart from Soviet Russia the
social apparatus is either antiquated, as in the "democratic" countries, or renovated in archaic forms, as in the even more backward
Fascist countries. In short, our political organization is either paleotechnic or pre-technic. Hence the hiatus between the mechanical
achievements and the social results. We have now to work out the
details of a new political and social order, radically different by
reason of the knowledge that is already at our command from any
that now exists. And to the extent that this order is the product of
scientific thought and humanistic imagination, it will leave a place
for irrational and instinctive and traditional elements in society
which were flouted, to their own ultimate peril, by the narrow forms
of rationalism that prevailed during the past century.

The transformation of the worker's status in industry can come
about only through a three-fold system of control: the functional
political organization of industry from within, the organization of
the consumers as active and self-regulating groups, giving rational
expression to collective demands, and the organization of industries
as units within the political framework of cooperating states.

The internal organization implies the transformation of the trade union from a bargaining organization, seeking special privileges apart from the industry or the working class as a whole, into a producing organization, concerned with establishing a standard of production, a humane system of management, and a collective discipline which will include every member, from such unskilled workers who may enter as apprentices up to the administrators and engineers. In the nineteenth century the mass of workers, cowed, uneducated, unskilled in cooperation, were only too willing to permit the capitalists to retain the responsibilities for financial management and production: their unions sought for the most part merely to obtain for the worker a greater share of the income, and somewhat more favorable conditions of labor.

The enterpriser, in turn, looked upon the management of his industry as a god-given right of ownership: to hire and fire, to stop and start, to build and destroy were special rights which neither the worker nor the government could encroach upon. The development of laws restricting the hours of labor and establishing minimum sanitary conditions, the development of public control of important public utilities, the growth of cartels and semi-monopolistic trade organizations under government supervision, have broken down this self-sufficiency of the manufacturer. But these measures, though struggled for by the worker, have done little to increase his dynamic participation in the management of industry itself. While here and there moves have been made toward a more positive integration of labor, as in the Baltimore and Ohio Railroad machine shops and in certain sections of the Garment Industry in America, for the most part the worker has no responsibility beyond his detailed job.

Until the worker emerges from a state of spiritless dependence there can be no large gain either in collective efficiency or in social direction: by its nature autonomy is something that cannot be handed from above. For the functional organization of industry there must be collective discipline, collective efficiency, above all collective responsibility: along with this must go a deliberate effort to produce engineering and scientific and managerial talent from within the ranks of the workers themselves, in addition to enlisting the services of more socialized members of this group, who are already spiritually developed beyond the lures and opportunities of the financial system to which they are attached. Without growth within the factory of effective units for work, the position of the worker, no matter what the ostensible nature of the political system, must remain a precarious and servile one; for the increase of mechanization vitiates his bargaining power, the increasing ranks of the unemployed tend automatically to beat down his wages, and the periodical disorganization of industry cancels out any small gains he may momentarily make. Plainly, such control, such autonomy, will not be achieved without a struggle—internal struggle for training and knowledge, and an external struggle against the weapons and the instruments handed down from the past. In the long run this struggle involves a fight not only against a sessile administrative bureaucracy within the trade unions themselves; more importantly, it involves an outright battle with the guardians of capitalism. Fortunately, the moral bankruptcy of the capitalist system is an opportunity as well as an obstacle: a decayed institution, though more dangerous to live with than a sound one, is easier to remove. The victory over the possessing classes is not the goal of this struggle: that is but a necessary incident in the effort to achieve a solidly integrated and socialized basis for industry. The struggle for power is a futile one, no matter who is victorious, unless it is directed by the will-to-function. Fascism has effaced the workers' attempts to overwhelm the capitalist system in Italy and Germany because ultimately the workers had no plan for carrying the fight beyond the stage of fighting.

The point to remember, however, is that the power needed to operate and to transform our modern technics is something other than physical force. The whole organization of modern industry is a complicated one, dependent upon a host of professionalized skills that link into each other, dependent likewise upon the faith and good will of those interchanging services, data, and calculations. Unless there is an inner coherence here, no amount of supervision will ensure against knavery and non-cooperation. This society cannot be run by brute force or by servile truculent skill backed by brute force: in the long run such habits of action are self-defeating. The principle

of functional autonomy and functional responsibility must be observed at every stage of the process, and the contrary principle of
class domination, based upon a privileged status—whether that class
be aristocratic or proletarian—is technically and socially inefficient.
Moreover, technics and science demand autonomy and self-control,
that is, freedom, in the realm of thought. The attempt to limit this
functional autonomy by the erection of special dogmas, as the
Christians limited it in the early days of Christianity, will cause a
fall into cruder methods of thinking, inimical to the essential basis
of both technics and modern civilization.

As industry advances in mechanization, a greater weight of political power must develop outside it than was necessary in the past. To
counterbalance the remote control and the tendency to continue along
the established grooves of industrial effort there must arise a collective organization of consumers for the sake of controlling the
kind and quantity and distribution of the product itself. In addition
to the negative check to which all industry is subject, the struggle for
existence between competing commodities, there must be a positive
mode of regulation which will ensure the production of desirable
types of commodities. Without such organization even our semi-com-
petitive commercial régime is slow in adapting itself to demand: at
the very moment that it changes, from month to month and year to
year, the superficial styles of its products, it resists the introduction
of fresh ideas, as the American furniture industry for long and stubbornly resisted the introduction of non-period furniture. Under a
more stable noncompetitive organization of industry, consumers'
groups for formulating and imposing demands will be even more
important for rational production: without such groups any central
agency for determining lines of production and quotas must necessarily be arbitrary and inefficient. Meanwhile the erection of scientific scales of performance and material quality—so that goods will
be sold on the basis of actual value and service, rather than on the
basis of clever packaging and astute advertising—is a natural corol-
lary on the consumer's side to the rationalization of industry. The
failure to use the existing laboratories for determining such standards—like the National Bureau of Standards in the United States—

for the benefit of the entire body of consumers is one of the most impudent miscarriages of knowledge under the capitalist system.

The third necessary element of political control lies in the possession of land, capital, credit, and machines. In America, which has reached an advanced stage of both mechanical improvement and financial organization, almost fifty per cent of the capital invested in industry, and something over forty per cent of the income of the nation, is concentrated in two hundred corporations. These corporations are so huge and have their capital in so many shares, that in no one of them does any particular person control by ownership more than five per cent of the capital invested. In other words, administration and ownership, which had a natural affiliation in small-scale enterprise, are now almost completely divorced in the major industries. (This condition was astutely used during the last two decades, by the bankers and administrators of American industry, for example, to appropriate for their private advantage a lion's share of the income, by a process of systematic pillage through recapitalization and bonuses.) Since the present shareholders of industry have already been dispossessed by the machinations of capitalism itself, there would be no serious jar if the system were put on a rational basis, by placing the banking functions directly under the state, and collecting capital directly out of the earnings of industry instead of permitting it to be routed in a roundabout fashion through acquisitive individuals, whose knowledge of the community's needs is empirical and unscientific and whose public interest is vitiated by private concerns—if not by outright anti-social animus. Such a change in the financial structure of our major instruments of production is a necessary prelude to humanizing the machine. Naturally, this means a revolution: whether it shall be humane or bloody, whether it shall be intelligent or brutal, whether it shall be accomplished smoothly, or with a series of violent shocks and jerks and catastrophes, depends to a large extent upon the quality of mind and the state of morals that exists among the present directors of industry and their opponents.

Now, the necessary impulses toward such a change are already apparent within the bankrupt structure of capitalist society: during its seizures of paralysis, it openly begs for the state to come in and rescue it and put it once more on its feet. Once the wolf is driven away, capitalism becomes brave again: but at scarcely any point during the last century has it been able to live without the help of state subsidies, state privileges, state tariffs, to say nothing of the aid of the state in subduing and regimenting the workers when the two groups have broken out into open warfare. Laissez-faire is in fact advocated and preached by capitalism only during those rare moments when it is doing well without the help of the state: but in its imperialist phase, laissez-faire is the last thing that capitalism desires. What it means by that slogan is not Hands off Industry— but Hands off Profits! In concluding his monumental survey of Capitalism Sombart looks upon 1914 as a turning-point for capitalism itself. The signs of the change are the impregnation of capitalistic modes of existence with normative ideas: the displacement of the struggle for profit as the sole condition of orientation in industrial relations, the undermining of private competition through the principle of understandings, and the constitutional organization of industrial enterprise. These processes, which have actually begun under capitalism, have only to be pushed to their logical conclusions to carry us beyond the capitalist order. Rationalization, standardization, and above all, rationed production and consumption, on the scale necessary to bring up to a vital norm the consumptive level of the whole community—these things are impossible on a sufficient scale without a socialized political control of the entire process.

If such a control cannot be instituted with the cooperation and intelligent aid of the existing administrators of industry, it must be achieved by overthrowing them and displacing them. The application of new norms of consumption, as in the housing of workers, has during the last thirty years won the passive support, sometimes subsidies drawn from taxation, of the existing governments of Europe, from conservative London to communistically bent Moscow. But such communities, while they have challenged and supplemented capitalist enterprise, are merely indications of the way in which the wind is blowing. Before we can replan and reorder our entire environment, on a scale commensurate with our human needs, the moral and legal and political basis of our productive system will have to be sharply revised. Unless such a revision takes place, capitalism itself will be eliminated by internal rot: lethal struggles will take place between states seeking to save themselves by imperialist conquest, as they will take place between classes within the state, jockeying for a power which will take the form of brute force just to the extent that society's grip on the productive mechanism itself is weakened.

11: The Diminution of the Machine

Most of the current fantasies of the future, which have been suggested by the triumph of the machine, are based upon the notion that our mechanical environment will become more pervasive and oppressive. Within the past generation, this belief seemed justified: Mr. H. G. Wells's earlier tales of The War of the Worlds and When the Sleeper Wakes, predicted horrors, great and little, from gigantic aerial combats to the blatant advertisement of salvation by go-getting Protestant churches—horrors that were realized almost before the words had left his mouth.

The belief in the greater dominance of mechanism has been reenforced by a vulgar error in statistical interpretation: the belief that curves generated by a past historic complex will continue without modification into the future. Not merely do the people who hold these views imply that society is immune to qualitative changes: they imply that it exhibits uniform direction, uniform motion, and even uniform acceleration—a fact which holds only for simple events in society and for very minor spans of time. The fact is that social predictions that are based upon past experience are always retrospective: they do not touch the real future. That such predictions have a way of justifying themselves from time to time is due to another fact: namely that in what Professor John Dewey calls judgments of practice the hypothesis itself becomes one of the determining elements in the working out of events: to the extent that it is seized and acted upon it weights events in its favor. The doctrine of mechanical progress doubtless had such a rôle in the nineteenth century.

What reason is there to believe that the machine will continue to multiply indefinitely at the rate that characterized the past, and that it will take over even more territory than it has already conquered? While the inertia of society is great, the facts of the matter lend themselves to a different interpretation. The rate of growth in all the older branches of machine production has in fact been going down steadily: Mr. Bassett Jones even holds that this is generally true of all industry since 1910. In those departments of mechanical industry that were well-established by 1870, like the railroad and the textile mill, this slowing down applies likewise to the critical inventions. Have not the conditions that forced and speeded the earlier growth—namely, the territorial expansion of Western Civilization and the tremendous increase in population—been diminishing since that point?

Certain machines, moreover, have already reached the limit of their development: certain areas of scientific investigation are already completed. The printing press, for example, reached a high pitch of perfection within a century after its invention: a whole succession of later inventions, from the rotary press to the linotype and monotype machines, while they have increased the pace of production, have not improved the original product: the finest page that can be produced today is no finer than the work of the sixteenth century printers. The water turbine is now ninety per cent efficient; we cannot, on any count, add more than ten per cent to its efficiency. Telephone transmission is practically perfect, even over long distances; the best the engineers can now do is to multiply the capacity of the wires and to extend the inter-linkages. Distant speech and vision cannot be transmitted faster than they are transmitted today by electricity: what gains we can make are in cheapness and ubiquity. In short: there are bounds to mechanical progress within the nature of the physical world itself. It is only by ignoring these limiting conditions that a belief in the automatic and inevitable and limitless expansion of the machine can be retained.

And apart from any wavering of interest in the machine, a general increase in verified knowledge in other departments than the physical sciences already threatens a large curtailment of mechanical practices and instruments. It is not a mystic withdrawal from the practical concerns of the world that challenges the machine so much as a more comprehensive knowledge of phenomena to which our mechanic contrivances were only partial and ineffective responses. Just as, within the domain of engineering itself, there has been a growing tendency toward refinement and efficiency through a nicer inter-relation of parts, so in the environment at large the province of the machine has begun to shrink. When we think and act in terms of an organic whole, rather than in terms of abstractions, when we are concerned with life in its full manifestation, rather than with the fragment of it that seeks physical domination and that projects itself in purely mechanical systems, we will no longer require from the machine alone what we should demand through a many-sided adjustment of every other aspect of life. A finer knowledge of physiology reduces the number of drugs and nostrums in which the physician places confidence: it also decreases the number and scope of surgical operations—those exquisite triumphs of machine-technics!—so that although refinements in technique have increased the number of potential operations that can be resorted to, competent physicians are tempted to exhaust the resources of nature before utilizing a mechanical shortcut. In general, the classic methods of Hippocrates have begun to displace, with a new certitude of conviction, both the silly potions prescribed in Molière's Imaginary Invalid and the barbarous intervention of Mr. Surgeon Cuticle. Similarly, a sounder notion of the human body has relegated to the scrapheap most of the weight-lifting apparatus of late Victorian gymnastics. The habit of doing without hats and petticoats and corsets has, in the past decade, thrown whole industries into limbo: a similar fate, through the more decent attitude toward the naked human body, threatens the bathing suit industry. Finally, with a great part of the utilities, like railroads, power lines, docks, port facilities, automobiles, concrete roads which we constructed so busily during the last hundred years, we are now on a basis where repair and replacement are all that is required. As our production becomes more rationalized, and as population shifts and regroups in better relationship to industry and recreation, new communities designed to the human scale are being constructed. This movement which has been taking place in Europe during the last generation is a result of pioneering work done over a century from Robert Owen to Ebenezer Howard. As these new communities are built up the need for the extravagant mechanical devices like subways, which were built in response to the disorganization and speculative chaos of the megalopolis, will disappear.

In a word, *as social life becomes mature, the social unemployment of machines will become as marked as the present technological unemployment of men.* Just as the ingenious and complicated mechanisms for inflicting death used by armies and navies are marks of international anarchy and painful collective psychoses, so are many of our present machines the reflexes of poverty, ignorance, disorder. The machine, so far from being a sign in our present civilization of human power and order, is often an indication of ineptitude and social paralysis. Any appreciable improvement in education and culture will reduce the amount of machinery devoted to multiplying the spurious mechanical substitutes for knowledge and experience now provided through the channels of the motion picture, the tabloid newspaper, the radio, and the printed book. So, too, any appreciable improvement in the physical apparatus of life, through better nutrition, more healthful housing, sounder forms of recreation, greater opportunities for the natural enjoyments of life, will decrease the part played by mechanical apparatus in salvaging wrecked bodies and broken minds. Any appreciable gain in personal harmony and balance will be recorded in a decreased demand for compensatory goods and services. The passive dependence upon the machine that has characterized such large sections of the Western World in the past was in reality an abdication of life. Once we cultivate the arts of life directly, the proportion occupied by mechanical routine and by mechanical instruments will again diminish.

Our mechanical civilization, contrary to the assumption of those who worship its external power the better to conceal their own feeling of impotence, is not an absolute. All its mechanisms are dependent upon human aims and desires: many of them flourish in direct proportion to our failure to achieve rational social cooperation and integrated personalities. Hence we do not have to renounce the machine completely and go back to handicraft in order to abolish a good deal of useless machinery and burdensome routine: we merely have to use imagination and intelligence and social discipline in our traffic with the machine itself. In the last century or two of social disruption, we were tempted by an excess of faith in the machine to do everything by means of it. We were like a child left alone with a paint brush who applies it impartially to unpainted wood, to varnished furniture, to the tablecloth, to his toys, and to his own face. When, with increased knowledge and judgment, we discover that some of these uses are inappropriate, that others are redundant, that others are inefficient substitutes for a more vital adjustment, we will contract the machine to those areas in which it serves directly as an instrument of human purpose. The last, it is plain, is a large area: but it is probably smaller than that now occupied by the machine. One of the uses of this period of indiscriminate mechanical experiment was to disclose unsuspected points of weakness in society itself. Like an old-fashioned menial, the arrogance of the machine grew in proportion to its master's feebleness and folly. With a change in ideals from material conquest, wealth, and power to life, culture, and expression, the machine like the menial with a new and more confident master, will fall back into its proper place: our servant, not our tyrant.

Quantitatively, then, we shall probably be less concerned with production in future than we were forced to be during the period of rapid expansion that lies behind us. So, too, we shall probably use fewer mechanical instruments than we do at present, although we shall have a far greater range to select from, and shall have more skillfully designed, more finely calibrated, more economical and reliable contrivances than we now possess. The machines of the future, if our present technics continues, will surpass those in use at present as the Parthenon surpassed a neolithic wood-hut: the transformation will be both toward durability and to refinement of forms. The dissociation of production from the acquisitive life will favor technical conservatism on a high level rather than a flashy experimentalism on a low level.

But this change will be accompanied by a qualitative change in interest, too: in general a change from mechanical interest to vital and psychal and social interests. This potential change in interest is generally ignored in predictions about the future of the machine. Yet once its importance is grasped it plainly alters every purely quantitative prediction that is based upon the assumption that the interests which for three centuries have operated chiefly within a mechanical framework will continue to remain forever within that framework. On the contrary, proceeding under the surface in the work of poets and painters and biological scientists, in a Goethe, a Whitman, a von Mueller, a Darwin, a Bernard, there has been a steady shift in attention from the mechanical to the vital and the social: more and more, adventure and exhilarating effort will lie here, rather than within the already partly exhausted field of the machine.

Such a shift will change the incidence of the machine and profoundly alter its relative position in the whole complex of human thought and activity. Shaw, in his Back to Methuselah, put such a change in a remote future; and risky though prophecy of this nature be, it seems to me that it is probably already insidiously at work. That such a movement could not take place, certainly not in science and its technical applications, without a long preparation in the inorganic realm is now fairly obvious: it was the relative simplicity of the original mechanical abstractions that enabled us to develop the technique and the confidence to approach more complicated phenomena. But while this movement toward the organic owes a heavy debt to the machine, it will not leave its parent in undisputed possession of the field. In the very act of enlarging its dominion over human thought and practice, the machine has proved to a great degree self-eliminating: its perfection involves in some degree its disappearance —as a communal water-system, once built, involves less daily attention and less expense on annual replacements than would a hundred thousand domestic wells and pumps. This fact is fortunate for the race. It will do away with the necessity, which Samuel Butler satirically pictured in Erewhon, for forcefully extirpating the dangerous troglodytes of the earlier mechanical age. The old machines will in part die out, as the great saurians died out, to be replaced by smaller, faster, brainer, and more adaptable organisms, adapted not to the mine, the battlefield and the factory, but to the positive environment of life.

12: Toward a Dynamic Equilibrium

The chief justification of the gigantic changes that took place during the nineteenth century was the fact of change itself. No matter what happened to human lives and social relations, people looked upon each new invention as a happy step forward toward further inventions, and society went on blindly like a caterpillar tractor, laying down its new road in the very act of lifting up the old one. The machine was supposed to abolish the limits of movement and of growth: machines were to become bigger: engines were to become more powerful: speeds were to become faster: mass production was to multiply more vastly: the population itself was to keep on increasing indefinitely until it finally outran the food supply or exhausted the soil of nitrogen. So went the nineteenth century myth.

Today, the notion of progress in a single line without goal or limit seems perhaps the most parochial notion of a very parochial century. Limits in thought and action, norms of growth and development, are now as present in our consciousness as they were absent to the contemporaries of Herbert Spencer. In our technics, countless improvements of course remain to be made, and there are doubtless numerous fresh fields still to be opened: but even in the realm of pure mechanical achievement we are already within sight of natural limits, not imposed by human timidity or lack of resources or immature technics, but by the very nature of the elements with which we work. The period of exploration and unsystematic, sporadic advance, which seemed to the nineteenth century to embody the essential characteristics of the new economy, is rapidly coming to an end. We are now faced with the period of consolidation and systematic assimilation. Western Civilization as a whole, in other words, is in the condition that new pioneering countries like the United States found themselves in, once all their free lands had been taken up and their main lines of transportation and communication laid out: it must now begin to settle down and make the most of what it has. Our machine system is beginning to approach a state of internal equilibrium. Dynamic equilibrium, not indefinite progress, is the mark of the opening age: balance, not rapid one-sided advance: conservation, not reckless pillage. The parallel between neolithic and neotechnic times holds even here: for the main advances which were consolidated in neolithic times remained stable, with minor variations within the pattern, for between 2500 and 3500 years. Once we have generally reached a new technical plateau we may remain on that level with very minor ups and down for thousands of years. What are the implications of this approaching equilibrium?

First: equilibrium in the environment. This means first the restoration of the balance between man and nature. The conservation and restoration of soils, the re-growth wherever this is expedient and possible, of the forest cover to provide shelter for wild life and to maintain man's primitive background as a source of recreation, whose importance increases in proportion to the refinement of his cultural heritage. The use of tree crops where possible as substitutes for annuals, and the reliance upon kinetic energy—sun, falling water, wind—instead of upon limited capital supplies. The conservation of minerals and metals: the larger use of scrap metals. The conservation of the environment itself as a resource, and the fitting of human needs into the pattern formed by the region as a whole: hence the progressive restoration out of such unbalanced regions as the over-urbanized metropolitan areas of London and New York. Is it necessary to point out that all this marks the approaching end of the miner's economy? Not mine and move, but stay and cultivate are the watchwords of the new order. Is it also necessary to emphasize that with respect to our use of metals, the conservative use of the existing supply will lower the importance of the mine in relation to other parts of the natural environment?

Second: equilibrium in industry and agriculture. This has rapidly been taking place during the last two generations in the migration of modern technics from England to America and to the rest of Europe, and from all these countries in turn to Africa and Asia. No one center is any longer the home of modern industry or its sole focal point: the finest work in rapid motion picture photography has been done in Japan, and the most astounding instrument of cheap mass production is the Bata Shoe Factories of Czechoslovakia. The more or less uniform distribution of mechanical industry over every portion of the planet tends to produce a balanced industrial life in every region: ultimately a state of balance over the earth itself. A similar advance remains to be worked out more largely for agriculture. With the decentralization of population into new centers, encouraged by motor and aerial transportation and by giant power, and with the application of scientific methods to the culture of soils and the processes of agriculture, as so admirably practiced today in Belgium and Holland, there is a tendency to equalize advantage between agricultural regions. With economic regionalism the area of market gardening and mixed farming—already favored by the scientific transformation of our diet—will widen again, and specialized farming for world export will tend to diminish except where, as in industry, some region produces specialties that cannot easily be duplicated.

Once the regional balance between industry and agriculture is worked out in detail, production in both departments will be on a more stable basis. This stability is the technical side of the normalization of consumption with which I have already dealt. Since at bottom the profit-motive arose out of and was furthered by uncertainty and speculation, whatever stability specialized capitalism had in the past rested on its capacity for promoting change, and taking advantage of it. Its safety rested upon its progressive tendency to revolutionize the means of production, promote new shifts in population, and take advantage of the speculative disorder. The equilibrium of capitalism, in other words, was the equilibrium of chaos. Per contra, the forces that work toward a normalization of consumption, toward a planned and rationed production, toward a conservation of resources, toward a planned distribution of population are in sharp opposition by reason of their essential technics to the methods of the past: hence an inherent conflict between this technology and the dominant capitalist methods of exploitation. As we approach an industrial and agricultural equilibrium part of the *raison d'être* of capitalism itself will vanish.

Third: equilibrium in population. There are parts of the Western World in which there is a practical balance between the number of births and deaths: most of these countries, France, Great Britain, the

United States, the Scandinavian countries, are in a relatively high state of technical and cultural development. The blind animal pressure of births, responsible for so many of the worst features of nineteenth century development, is now characteristic in the main of backward countries, countries in a state of political or technical inferiority. If equilibrium takes place here during the next century one may look forward to a rational re-settlement of the entire planet into the regions most favorable to human habitation: an era of deliberate recolonization will take the place of those obstreperous and futile conquests which began with the explorations of the Spaniards and the Portuguese in the sixteenth century and which have continued without any essential change down to the most recent raids of the Japanese. Such an internal re-settlement is already taking place in many countries: the movement of industries into Southern England, the development of the French Alps, the settlement of new farmers in Palestine and Siberia, are first steps toward achieving a state of equilibrium. The balancing off of the birth-rate and death-rate, and the balancing off of rural and urban environments—with the wholesale wiping out of the blighted industrial areas inherited from the past—are all part of a single integration.

This state of balance and equilibrium—regional, industrial, agricultural, communal—will work a further change within the domain of the machine itself: a change of tempo. The temporary fact of increasing acceleration, which seemed so notable to Henry Adams when he surveyed the progress from twelfth century unity to twentieth century multiplicity, the fact which was later accompanied by a belief in change and speed for their own sake—will no longer characterize our society. It is not the absolute speed assumed by any part of the machine system that indicates efficiency: what is important is the relative speed of the various parts with a view to the ends to be accomplished: namely, the maintenance and development of human life. Efficiency, even on the technical level alone, means a gearing together of the various parts so that they may deliver the correct and the predictable amounts of power, goods, services, utilities. To achieve this efficiency, it may be necessary to lower the tempo rather than to increase it in this or that department; and as larger portions of our days go to leisure and smaller portions to work, as our think-ing becomes synthetic and related, instead of abstract and pragmatic, as we turn to the cultivation of the whole personality instead of cen-tering upon the power elements alone—as all these things come about we may look forward to a slowing of the tempo throughout our lives, even as we may look forward to a lessening of the number of unnecessary external stimuli. Mr. H. G. Wells has characterized the approaching period as the Era of Rebuilding. No part of our life, our thought, or our environment can escape that necessity and that obligation.

The problem of tempo: the problem of equilibrium: the problem of organic balance: in back of them all the problem of human satis-faction and cultural achievement—these have now become the critical and all-important problems of modern civilization. To face these problems, to evolve appropriate social goals and to invent appropri-ate social and political instruments for an active attack upon them, and finally to carry them into action: here are new outlets for social intelligence, social energy, social good will.

13: Summary and Prospect

We have studied the origins, the advances, the triumphs, the lapses, and the further promises of modern technics. We have observed the limitations the Western European imposed upon himself in order to create the machine and project it as a body outside his personal will: we have noted the limitations that the machine has imposed upon men through the historic accidents that accompanied its development. We have seen the machine arise out of the denial of the organic and the living, and we have in turn marked the reaction of the organic and the living upon the machine. This reaction has two forms. One of them, the use of mechanical means to return to the primitive, means a throwback to lower levels of thought and emotion which will ultimately lead to the destruction of the machine itself and the higher types of life that have gone into its conception. The other in-volves the rebuilding of the individual personality and the collective group, and the re-orientation of all forms of thought and social ac-tivity toward life: this second reaction promises to transform the nature and function of our mechanical environment and to lay wider and firmer and safer foundations for human society at large. The issue is not decided: the results are not certain: and where in the present chapter I have used the prophetic form I have not been blind to the fact that while all the tendencies and movements I have pointed to are real, they are still far from being supreme: so when I have said "it will" I have meant "we must."

In discussing the modern technics, we have advanced as far as seems possible in considering mechanical civilization as an isolated system: the next step toward re-orienting our technics consists in bringing it more completely into harmony with the new cultural and regional and societal and personal patterns we have co-ordinately begun to develop. It would be a gross mistake to seek wholly within the field of technics for an answer to all the problems that have been raised by technics. For the instrument only in part determines the character of the symphony or the response of the audience: the composer and the musicians and the audience have also to be considered.

What shall we say of the music that has so far been produced? Looking backward on the history of modern technics, one notes that from the tenth century onward the instruments have been scraping and tuning. One by one, before the lights were up, new members had joined the orchestra, and were straining to read the score. By the seventeenth century the fiddles and the wood-wind had assembled, and they played in their shrill high notes the prelude to the great opera of mechanical science and invention. In the eighteenth century the brasses joined the orchestra, and the opening chorus, with the metals predominating over the wood, rang through every hall and gallery of the Western World. Finally, in the nineteenth century, the human voice itself, hitherto subdued and silent, was timidly sounded through the systematic dissonances of the score, at the very moment that imposing instruments of percussion were being introduced. Have we heard the complete work? Far from it. All that has happened up to now has been little more than a rehearsal, and at last, having recognized the importance of the singers and the chorus, we will have to score the music differently, subduing the insistent brasses and the kettle-drums and giving more prominence to the violins and the voices. But if this turns out to be so, our task is even more difficult: for we will have to re-write the music in the act of playing it, and change the leader and re-group the orchestra at the very moment that we are re-casting the most important passages. Impossible? No: for however far modern science and technics have fallen short of their inherent possibilities, they have taught mankind at least one lesson: Nothing is impossible.

INVENTIONS

1: Introduction

This list of inventions makes no pretence to being exhaustive. It is meant merely to provide an historical framework of technical facts for the social interpretations of the preceding pages. While I have attempted to choose the more important inventions and processes, I have doubtless left out many that have equal claim to appear. The most comprehensive guide to this subject are the compilations by Darmstaedter and by Feldhaus; but I have drawn from a variety of sources. The dates and attributions of many inventions, as every technician knows, must remain somewhat arbitrary. Unlike a human baby, one often cannot say at what date an invention is born: frequently, indeed, what was apparently a still birth may be resuscitated a few years after its first unhappy appearance. And again, with inventions the family lineage often is hard to establish; for, as W. F. Ogburn and Dorothy S. Thomas have demonstrated, inventions are often practically simultaneous: the result of a common heritage and a common need. While I have endeavored to be both accurate and impartial in giving the date of the invention and the name of the putative inventor, the reader should keep in mind that these data are offered only for his convenience in looking further. Instead of a single date one finds usually a series of dates which mark progress from the state of pure fantasy to concrete realization in the form that has been most acceptable to the capitalist *mores*—that of a commercial success. As a result of these *mores* far too much stress has usually been laid upon the individual who put the title of private ownership upon this social process by taking out patent rights on "his" invention. But observe: inventions are often patented long before they can be practicably used, and, on the other hand, they are often ready for use long before industrial enterprisers are willing to take advantage of them. Since modern science and technology are part of the common stock of Western Civilization, I have refused to attribute inventions to one country or another and I have done my best to avoid an unconscious bias in weighting the list in behalf of my own country—trusting by my good example to shame the scholars who permit their most childish impulses to flaunt themselves in this field. If any bias or misinformation still exists, I will welcome corrections.

2: List of Inventions

Summary of the existing technics before the tenth century. Fire: its application in furnaces, ovens, kilns. The simple machines: inclined plane, screw, etc. Thread, cord, rope. Spinning and weaving. Advanced agriculture, including irrigation, terrace-cultivation, and soil regeneration (lapsed in Northern Europe). Cattle breeding and the use of the horse for transport. Glass-making, pottery-making, basket-making. Mining, metallurgy and smithing, including the working of iron. Power machines: water-mills, boats with sails, probably windmills. Machine-tools: bow-drills and lathes. Handicraft tools with tempered metal cutting edges. Paper. Water-clocks. Astronomy, mathematics, physics, and the tradition of science. In Northern Europe a scattered and somewhat decayed technological tradition based on Rome; but South and East, from Spain to China, an advanced and still active technology, whose ideas were filtering into the West and North through traders, scholars, and soldiers.

TENTH CENTURY

Use of water-clocks and water-mills. The iron horse-shoe and an effective harness for horses. Multiple yoke for oxen. Possible invention of the mechanical clock.

999: Painted glass windows in England

ELEVENTH CENTURY

1041-49: Movable type (Pi Sheng)
1050: First real lenses (Alhazen)
1065: Oliver of Malmesbury attempts flight
1080: Decimal system (Azachel)

TWELFTH CENTURY

Military use of gunpowder in China. The magnetic compass, known to the Chinese 1160 B.C., comes into Europe, via the Arabs.

1105: First recorded windmill in Europe (France)
1100: Bologna University
1118: Cannon used by Moors
1144: Paper (Spain)
1147: Use of wood cuts for Capital letters. (Benedictine monastery at Engelberg)
1180: Fixed steering rudder
1188: Bridge at Avignon. 18 stone arches—3,000 ft. long

1190: Paper mill (at Hérault, France)
1195: Magnetic compass in Europe (English Citation)

THIRTEENTH CENTURY

Mechanical clocks invented.
1232: Hot-air balloons (in China)
1247: Cannon used in defence of Seville
1269: Pivoted magnetic compass (Petrus Peregrinus)
1270: Treatise on lenses (Vitellio)
    Compound lenses (Roger Bacon)
1272: Silk reeling machine (Bologna)
1280: Opus Ruralium Commodorum— Compendium of Agricultural Practice (Petrus de Crescentis)
1285-1299: Spectacles
1289: Block printing (Ravenna)
1290: Paper mill (Ravensburg)
1298: Spinning wheel

FOURTEENTH CENTURY

Mechanical clock becomes common. Water-power used to create draft for blast furnace: makes cast iron possible. Treadle loom (inventor unknown). Invention of rudder and beginning of canalization. Improved glass-making.

1300: Wooden type (Turkestan)

1315: Beginnings of Scientific Anatomy through dissection of human body (Raimondo de Luzzi of Bologna)

1320: Water-driven iron works, near Dobrilugk

1322: Sawmill at Augsburg

1324: Cannon [Gunpowder: 846 A.D. (Magnus Graecus)]

1330: Crane at Lüneburg

1345: Division of hours and minutes into sixties

1338: Guns

1350: Wire-pulling machine (Rudolph of Nürnberg)

1370: Perfected mechanical clock (von Wyck)

1382: Giant cannon—4.86 metres long

1390: Metal types (Korea)

1390: Paper mill

FIFTEENTH CENTURY

Use of wind-mill for land drainage. Invention of turret windmill. Introduction of knitting. Iron drill for boring cannon. Trip-hammer. Two-masted and three-masted ship.

1402: Oil painting (Bros. van Eyck)

1405: Diving suit (Konrad Kyeser von Eichstadt)

1405: Infernal machine (Konrad Kyeser von Eichstadt)

1409: First book in movable type (Korea)

1410: Paddle-wheel boat designed

1418: Authentic wood engraving

1420: Observatory at Samarkand

1420: Sawmill at Madeira

1420: Velocipede (Fontana)

1420: War-wagon (Fontana)

1423: First European woodcut

1430: Turret windmill

1436: Scientific cartography (Banco)

1438: Wind-turbine (Mariano)

1440: Laws of perspective (Alberti)

1446: Copperplate engraving

1440-1460: Modern printing (Gutenberg and Schoeffer)

1457: Rediscovery of wagon on springs referred to by Homer

1470: Foundations of trigonometry (J. Müller Regiomontanus)

1471: Iron cannon balls

1472: Observatory at Nürnberg by Bernard Walther

1472-1519: Leonardo da Vinci made the following inventions:

Centrifugal pump
Dredge for canal-building
Polygonal fortress with outworks
Breech-loading cannon
Rifled firearms
Antifriction roller bearing
Universal joint
Conical screw
Rope-and-belt drive
Link chains
Submarine-boat
Bevel gears
Spiral gears
Proportional and paraboloid
Compasses
Silk doubling and winding apparatus
Spindle and flyer
Parachute
Lamp-chimney
Ship's log
Standardized mass-production house

1481: Canal lock (Dionisio and Petro Domenico)

1483: Copper etching (Wenceslaus von Olnutz)

1492: First globe (Martin Behaim)

SIXTEENTH CENTURY

Tinning for preservation of iron. Windmills of 10 H.P. become common. Much technical progress and mechanization in mining industries, spread of blast-furnaces and iron-moulding. Introduction of domestic clock.

1500: First portable watch with iron main-spring (Peter Henlein)

1500: Mechanical farming drill (Cavallina)

1500-1650: Intricate cathedral clocks reach height of development

1508: Multicolored woodcut

1511: Pneumatic beds (Vegetius)
1518: Fire-engine (Platner)
1524: Fodder-cutting machine
1528: Re-invention of taxi meter for coaches
1530: Foot-driven spinning wheel (Jürgens)
1534: Paddle-wheel boat (Blasco de Garay)
1535: Diving bell (Francesco del Marchi)
1539: First astronomical map (Alessandro Piccolomini)
1544: Cosmographia Universalis (Sebastian Münster)
1544: Elaboration of algebraic symbols (Stifel)
1545: Modern surgery (Ambroise Paré)
1546: Railway in German mines
1548: Water supply by pumping works (Augsburg)
1550: First known suspension bridge in Europe (Palladio)
1552: Iron-rolling machine (Brulier)
1558: Military tank
1558: Camera with lens and stop for diaphragm (Daniello Barbaro)
1560: Accademia Secretorum Naturae at Naples (first scientific society)
1565: Lead pencil (Gesner)
1569: Industrial exhibition at Rathaus, Nürnberg
1575: Hero's Opera (translation)
1578: Screw lathe (Jacques Besson)
1579: Automatic ribbon loom at Dantzig
1582: Gregorian calendar revision
1582: Tide-mill pump for London (Morice)
1585: Decimal system (Simon Stevin)
1589: Knitting frame (William Lee)
1589: Man-propelled wagon (Gilles de Bom)
1590: Compound microscope (Jansen)
1594: Use of clock to determine longitude
1595: Design for metal bridges—arch and chain (Veranzio)

1595: Wind-turbine (Veranzio)
1597: Revolving theater stage

SEVENTEENTH CENTURY

Water wheels of 20 H.P. introduced: transmission by means of reciprocating rods over distance of one-quarter mile. Glass hothouse comes into use. Foundations of modern scientific method. Rapid developments in physics.

1600: Dibbling of wheat to increase yield (Plat)
1600: Treatise on terrestrial magnetism and electricity (Gilbert)
1600: Pendulum (Galileo)
1603: Accademia dei Lincei at Rome
1608: Telescope (Lippersheim)
1609: First law of motion (Galileo)
1610: Discovery of gases (Van Helmont)
1613: Gunpowder in mine blasting
1614: Discovery of logarithms by John Napier
1615: Use of triangulation system in surveying by Willebrord Snell van Roijen (1581-1626)
1617: First logarithm table (Henry Briggs)
1618: Machine for plowing, manuring and sowing (Ramsay and Wilgoose)
1619: Use of coke instead of charcoal in blast furnace (Dudley)
1619: Tile-making machine
1620: Adding machine (Napier)
1624: Submarine (Cornelius Drebbel). Went two miles in test between Westminster and Greenwich
1624: First patent law protecting inventions (England)
1628: Steam engine (described 1663 by Worcester)
1630: Patent for steam engine (David Ramsey)
1635: Discovery of minute organisms (Leeuwenhoek)
1636: Infinitesimal calculus (Fermat)
1636: Fountain pen (Schwenter)
1636: Threshing machine (Van Berg)

1637: Periscope (Hevel, Danzig)
1643: Barometer (Torricelli)
1647: Calculation of focusses of all forms of lens
1650: Calculating machine (Pascal)
1650: Magic lantern (Kircher)
1652: Air pump (v. Guericke)
1654: Law of probability (Pascal)
1657: Pendulum clock (Huygens)
1658: Balance spring for clocks (Hooke)
1658: Red corpuscles in blood (Schwammerdam)
1660: Probability law applied to insurance (Jan de Witt)
1665: Steam automobile model (Verbiest, S. J.)
1666: Mirror telescope (Newton)
1667: Cellular structure of plants (Hooke)
1667: Paris Observatory
1669: Seed drill (Worlidge)
1671: Speaking tube (Morland)
1673: New Type fortification (Vauban)
1675: First determination of speed of light (Roemer)
1675: Greenwich Observatory founded
1677: Foundation of Ashmolean Museum
1678: Power loom (De Gennes)
1679-1681: First modern tunnel for transport, 515 feet long, in Languedoc Canal
1680: First power dredge (Cornelius Meyer)
1680: Differential calculus (Leibniz)
1680: Gas engine using gunpowder (Huygens)
1682: Law of gravitation (Newton)
1682: 100 H.P. pumping works at Marly (Ranneguin)
1683: Industrial Exhibition at Paris
1684: Fodder-chopper run by water-power (Delabadie)
1685: Foundation of scientific obstetrics (Van Deventer)
1687: Newton's *Principia*
1688: Distillation of gas from coal (Clayton)

1695: Atmospheric steam engine (Papin)

EIGHTEENTH CENTURY

Rapid improvements in mining and textile machinery. Foundation of modern chemistry.

1700: Water power for mass-production (Polhem)
1705: Atmospheric steam engine (Newcomen)
1707: Physician's pulse watch with second hand (John Floger)
1708: Wet sand iron casting (Darby)
1709: Coke used in blast furnace (Darby)
1710: First stereotype (Van der Mey and Müller)
1711: Sewing machine (De Camus)
1714: Mercury thermometer (Fahrenheit)
1714: Typewriter (Henry Mill)
1716: Wooden railways covered with iron
1719: Three color printing from copper plate (Le Blond)
1727: First exact measurement of blood pressure (Stephen Hales)
1727: Invention of stereotype (Ged)
1727: Light-images with silver nitrate (Schulze: see 1839)
1730: Stereotyping process (Goldsmith)
1733: Flying shuttle (Kay)
1733: Roller spinning (Wyatt and Paul)
1736: Accurate chronometer (Harrison)
1736: Commercial manufacture of sulphuric acid (Ward)
1738: Cast-iron rail tramway (at Whitehaven, England)
1740: Cast steel (Huntsman)
1745: First technical school divided from army engineering at Braunschweig
1749: Scientific calculation of water resistance to ship (Euler)
1755: Iron wheels for coal cars

1756: Cement manufacture (Smeaton)

1763: Modern type chronometer (Le Roy)

1761: Air cylinders; piston worked by water wheel. More than tripled production of blast furnace (Smeaton)

1763: First exhibition of the industrial arts. Paris.

1763: Slide rest (French encycl.)

1765-1769: Improved steam pumping engine with separate condenser (Watt)

1767: Cast iron rails at Coalbrookdale

1767: Spinning jenny (Hargreaves)

1769: Steam carriage (Cugnot)

1770: Caterpillar tread (R. L. Edgeworth: see 1902)

1772: Description of ball-bearing (Narlo)

1774: Boring machine (Wilkinson)

1775: Reciprocative engine with wheel

1776: Reverberatory furnace (Brothers Cranege)

1778: Modern water closet (Bramah)

1778: Talking automaton (von Kempelen)

1779: Bridge cast-iron sections (Darby and Wilkinson)

1781-1786: Steam engine as prime mover (Watt)

1781: Steamboat (Joufroy)

1781: Drill plow (Proude: also used by Babylonians: 1700-1200 B.C.)

1782: Balloon (J. M. and J. E. Montgolfier). Original invention Chinese

1784: Puddling process—reverberatory furnace (Cort)

1784: Spinning mule (Crompton)

1785: Interchangeable parts for muskets (Le Blanc)

1785: First steam spinning mill at Papplewick

1785: Power loom (Cartwright)

1785: Chlorine as bleaching agent (Berthollet)

1785: Screw propeller (Bramah)

1787: Iron boat (Wilkinson)

1787: Screw propeller steamboat (Fitch)

1788: Threshing machine (Meikle)

1790: Manufacture of soda from NaCl (Le Blanc)

1790: Sewing machine first patented (M. Saint—England)

1791: Gas engine (Barker)

1792: Gas for domestic lighting (Murdock)

1793: Cotton gin (Whitney)

1793: Signal telegraph (Claude Chappe)

1794: Ecole Polytechnique founded

1795-1809: Food-canning (Appert)

1796: Lithography (Senefelder)

1796: Natural cement (J. Parker)

1796: Toy helicopter (Cayley)

1796: Hydraulic press (Bramah)

1797: Screw-cutting lathe (Maudslay). Improved slide-rest metal lathe (Maudslay)

1799: Humphry Davy demonstrates anesthetic properties of nitrous oxide

1799: Conservatoire Nationale des Arts et Métiers (Paris)

1799: Manufactured bleaching powder (Tennant)

NINETEENTH CENTURY

Enormous gains in power conversion. Mass-production of textiles, iron, steel, machinery. Railway building era. Foundations of modern biology and sociology.

1800: Galvanic cell (Volta)

1801: Public railroad with horsepower —Wandsworth to Croydon, England

1801: Steamboat *Charlotte Dundas* (Symington)

1801-1802: Steam carriage (Trevithick)

1802: Machine dresser for cotton warps (necessary for power weaving)

1802: Planing machine (Bramah)

1803: Side-paddle steamboat (Fulton)

1804: Jacquard loom for figured fabrics

1804: Oliver Evans amphibian steam carriage
1805: Twin screw propeller (Stevens)
1807: First patent for gas-driven automobile (Isaac de Rivaz)
1807: Kymograph—moving cylinder for recording continuous movement (Young)
1813: Power loom (Horrocks)
1814: Grass tedder (Salmon)
1814: Steam printing press (Koenig)
1817: Push-cycle (Drais)
1818: Milling machine (Whitney)
1818: Stethoscope (Laennec)
1820: Bentwood (Sargent)
1820: Incandescent lamp (De la Rue)
1820: Modern planes (George Rennie)
1821: Iron steamboat (A. Manby)
1822: First Scientific Congress at Leipzig
1822: Steel alloys (Faraday)
1823: Principle of motor (Faraday)
1823-1843: Calculating machines (Babbage)
1824: Portland cement (Aspdin)
1825: Electro-magnet (William Sturgeon)
1825: Stockton and Darlington Railway
1825-1843: Thames tunnel (Marc I. Brunel)
1826: Reaping machine (Bell). First used in Rome and described by Pliny
1827: Steam automobile (Hancock)
1827: High pressure steam boiler—1,400 lbs. (Jacob Perkins)
1827: Chromo-lithography (Zahn)
1828: Hot blast in iron production (J. B. Nielson)
1828: Machine-made steel pen (Gillot)
1829: Blind print (Braille)
1829: Filtration plant for water (Chelsea Water Works, London)
1829: Liverpool and Manchester Railway
1829: Sewing machine (Thimonnier)
1829: Paper matrix stereotype (Genoux)
1830: Compressed air for sinking shafts and tunnels under water (Thomas Cochrane)

1830: Elevators (used in factories)
1831: Reaping machine (McCormick)
1831: Dynamo (Faraday)
1831: Chloroform
1832: Water turbine (Fourneyron)
1833: Magnetic telegraph (Gauss and Weber)
1833: Laws of Electrolysis (Faraday)
1834: Electric battery in power boat (M. H. Jacobi)
1834: Anilin dye in coal tar (Runge)
1834: Workable liquid refrigerating machine (Jacob Perkins)
1835: Application of statistical method to social phenomena (Quetelet)
1835: Commutator for dynamo
1835: Electric telegraph
1835: Electric automobile (Davenport)
1836: First application of electric telegraph to railroads (Robert Stephenson)
1837: Electric motor (Davenport)
1837: Needle telegraph (Wheatstone)
1838: Electro-magnetic telegraph (Morse)
1838: Single wire circuit with ground (Steinheil)
1838: Steam drop hammer (Nasmyth)
1838: Two-cycle double-acting gas engine (Barnett)
1838: Propeller steamship (Ericsson: see 1805)
1838: Boat driven by electric motor (Jacobi)
1839: Manganese steel (Heath)
1839: Electrotype (Jacobi)
1839: Callotype (Talbot)
1839: Daguerreotype (Niépce and Daguerre)
1839: Hot vulcanization of rubber (Goodyear)
1840: Grove's incandescent lamp
1840: Corrugated iron roof—East Counties Railroad Station
1840: Micro-photography (Donne)
1840: First steel cable suspension bridge, Pittsburgh (Roebling)
1841: Paper positives in photography (Talbot)

1841: Conservation of energy (von Mayer)
1842: Electric engine (Davidson)
1842: Conservation of energy (J. R. von Mayer)
1843: Aerostat (Henson)
1843: Typewriter (Thurber)
1843: Spectrum analysis (Miller)
1843: Gutta percha (Montgomery)
1844: Carbon arc lamp (Poucault)
1844: Nitrous oxide application (Dr. Horace Wells): see 1799
1844: Practical wood-pulp paper (Keller)
1844: Cork-and-rubber linoleum (Galloway)
1845: Electric arc patented (Wright)
1845: Modern high speed sewing machine (Elias Howe)
1845: Pneumatic tire (Thomson)
1845: Mechanical boiler-stoker
1846: Rotating cylinder press (Hoe)
1846: Ether (Warren and Morton)
1846: Nitroglycerine (Sobrero)
1846: Gun-cotton (C. F. Schönbein)
1847: Chloroform-anaesthetics (J. Y. Simpson)
1847: Electric locomotive (M. G. Farmer)
1847: Iron building (Bogardus)
1848: Modern safety match (R. C. Bottger)
1848: Rotary fan (Lloyd)
1849: Electric locomotive (Page)
1850: Rotary ventilator (Fabry)
1850: Ophthalmoscope
1851: Crystal Palace. First International Exhibition of Machines and the Industrial Arts (Joseph Paxton)
1851: Electric motor car (Page)
1851: Electro-magnetic clock (Shepherd)
1851: Reaper (McCormick)
1853: Science Museum (London)
1853: Great Eastern steamship—680 feet long—watertight compartments
1853: Mechanical ship's log (William Semens)

1853: Mass-production watches (Denison, Howard and Curtis)
1853: Multiple telegraph on single wire (Gintl)
1854: Automatic telegraph message recorder (Hughes)
1855: Commercial production of aluminum (Deville)
1855: 800 H.P. water turbine at Paris
1855: Television (Caselle)
1855: Iron-plated gunboats
1855: Safety lock (Yale)
1856: Open hearth furnace (Siemens)
1856: Bessemer converter (Bessemer)
1856: Color photography (Zenker)
1858: Phonautograph. Voice vibrations recorded on revolving cylinder (Scott)
1859: Oil mining by digging and drilling (Drake)
1859: Storage cell (Planté)
1860: Ammonia refrigeration (Carre)
1860: Asphalt paving
1860-1863: London "Underground"
1861-1864: Dynamo motor (Pacinnoti)
1861: Machine gun (Gatling)
1862: *Monitor* (Ericsson)
1863: Gas engine (Lenoir)
1863: Ammonia soda process (Solvay)
1864: Theory of light and electricity (Clerk-Maxwell)
1864: Motion picture (Ducos)
1864 and 1875: Gasoline engine motor car (S. Marcus)
1865: Pasteurization of wine (L. Pasteur)
1866: Practical dynamo (Siemens)
1867: Dynamite (Nobel)
1867: Re-enforced concrete (Monier)
1867: Typewriter (Scholes)
1867: Gas engine (Otto and Langen)
1867: Two-wheeled bicycle (Michaux)
1868: Tungsten steel (Mushet)
1869: Periodic table (Mendelejev and Lothar Meyer)
1870: Electric steel furnace (Siemens)
1870: Celluloid (J. W. and I. S. Hyatt)
1870: Application of hypnotism in psychopathology (Charcot)
1870: Artificial madder dye (Perkin)

1871: Aniline dye for bacteria staining (Weigert)
1872: Model airplane (A. Penaud)
1872: Automatic airbrake (Westinghouse)
1873: Ammonia compression refrigerator—Carle Linde (München)
1874: Stream-lined locomotive
1875: Electric car (Siemens)
1875: Standard time (American railroads)
1876: Bon Marché at Paris (Boileau and G. Eiffel)
1876: Discovery of toxins
1876: Four-cycle gas engine (Otto)
1876: Electric telephone (Bell)
1877: Microphone (Edison)
1877: Bactericidal properties of light established (Downes & Blunt)
1877: Compressed air refrigerator (J. J. Coleman)
1877: Phonograph (Edison)
1877: Model flying machine (Kress)
1878: Centrifugal cream separator (De Laval)
1879: Carbon glow lamp (Edison)
1879: Electric railroad
1880: Cup and cone ball-bearing in bicycle
1880: Electric elevator (Siemens)
1882: First central power station (Edison)
1882: Motion picture camera (Marly)
1882: Steam turbine (De Laval)
1883: Dirigible balloon (Brothers Tissandier)
1883: High speed gasoline engine (Daimler)
1884: Steel-frame skyscraper (Chicago)
1884: Cocaine (Singer)
1884: Linotype (Mergenthaler)
1884: Turbine for High Falls (Pelton)
1884: Smokeless powder (Duttenhofer)
1884: Steam turbine (Parsons)
1885: International standard time
1886: Aluminum by electrolytic process (Hall)
1886: Hand camera (Eastman)
1886: Aseptic surgery (Bergmann)
1886: Glass-blowing machine
1887: Polyphase alternator (Tesla)

1887: Automatic telephone
1887: Electro-magnetic waves (Hertz)
1887: Monotype (Leviston)
1888: Recording adding machine (Burroughs)
1889: Artificial silk of cotton refuse (Chardonnet)
1889: Hard rubber phonograph records
1889: Eiffel Tower
1889: Modern motion picture camera (Edison)
1890: Detector (Branly)
1890: Pneumatic tires on bicycles
1892: Calcium carbide (Willson and Moissan)
1893-1898: Diesel motor
1892: Artificial silk of wood pulp (Cross, Bevan and Beadle)
1893: Moving picture (Edison)
1893: By-product coke oven (Hoffman)
1894: Jenkin's "Phantoscope"—first moving picture of modern type
1895: Motion picture projector (Edison)
1895: X-ray (Roentgen)
1896: Steam-driven aerodrome flight—one half mile without passenger (Langley)
1896: Radio-telegraph (Marconi)
1896: Radio activity (Becquerel)
1898: Osmium lamp (Welsbach)
1898: Radium (Curie)
1898: Garden City (Howard)
1899: Loading coil for long distance telegraphy and telephony (Pupin)

TWENTIETH CENTURY

General introduction of scientific and technical research laboratories.

1900: High speed tool steel (Taylor & White)
1900: Nernst lamp
1900: Quantum theory (Planck)
1901: National Bureau of Standards—United States
1902: Caterpillar tread improved. [See 1770]
1902: Radial type airplane engine (Charles Manly)

1903: First man-lifting airplane (Orville and Wilbur Wright)
1903: Electric fixation of nitrogen
1903: Arc process nitrogen fixation (Birkeland and Eyde)
1903: Radio-telephone
1903: Deutsches Museum (München)
1903: Oil-burning steamer
1903: Tantalum lamp (von Bolton)
1904: Fleury tube
1904: Moore tube light
1905: Rotary mercury pump (Gaede)
1905: Cyanamide process for nitrogen fixation (Rothe)
1906: Synthetic resins (Baekeland)
1906: Audion (De Forest)

1907: Automatic bottle machine (Owen)
1907: Tungsten lamp
1907: Television-photograph (Korn)
1908: Technisches Museum für Industrie und Gewerbe (Wien)
1909: Duralumin (Wilm)
1910: Gyro-compass (Sperry)
1910: Synthetic ammonia process for nitrogen fixation (Haber)
1912: Vitamins (Hopkins)
1913: Tungsten filament light (Coolidge)
1920: Radio broadcasting
1922: Perfected color-organ (Wilfred)
1927: Radio television
1933: Aerodynamic motor car (Fuller)

BIBLIOGRAPHY

1: General Introduction

Books cannot take the place of first-hand exploration: hence any study of technics should begin with a survey of a region, working through from the actual life of a concrete group to the detailed or generalized study of the machine. This approach is all the more necessary for the reason that our intellectual interests are already so specialized that we habitually begin our thinking with abstractions and fragments which are as difficult to unify by the methods of specialism as were the broken pieces of Humpty-Dumpty after he had fallen off the wall. Open-air observation in the field, and experience as a worker, taking an active part in the processes around us, are the two fundamental means for overcoming the paralysis of specialism. As a secondary means for going deeper into technical operations and equipment, particularly for laymen whose training and scope of experience are limited, the Industrial Museum is helpful. The earliest of these is the Conservatoire des Arts et Métiers in Paris: educationally however it is a mere storehouse. The most exhaustive is the Deutsches Museum in München; but its collections have a little over-reached themselves in bigness and one loses sight of the forest for the trees. Perhaps the best sections in it are the dramatic reconstructions of mines; this feature has been copied at the Rosenwald Museum in Chicago. The Museums in Wien and in London both have educational value, without being overwhelming. One of the best of the small museums is the Museum of Science and Industry in New York. The new museum of the Franklin Institute in Philadelphia, and that of the Smithsonian Institution in Washington are respectively the latest and the oldest in the United States. The Museum of the Bucks County Historical Society at Doylestown, Pennsylvania, is full of interesting eotechnic relics.

Up to the present the only general introductions in English of any value have been Stuart Chase's *Men and Machines* and Harold Rugg's *The Great Technology*. Each has the limitation of historical foreshortening; but Chase is good in his description of modern technical improvements and Rugg is

447

valuable for his various educational suggestions. There is no single, comprehensive and adequate history of technics in English. Usher's *A History of Mechanical Inventions* is the nearest approach to it. While it does not cover every aspect of technics, it treats critically and exhaustively whatever it does touch, and the earlier chapters on the equipment of antiquity and the development of the clock are particularly excellent summaries. It is perhaps the most convenient and accurate work in English. In German the series of books done by Franz Marie Feldhaus, particularly his *Ruhmesblätter der Technik*, would be valuable for their illustrations alone; they form the core of any historical library. Both Usher and Feldhaus are useful for their comments on sources and books. Topping all these books is that monument of twentieth century scholarship, *Der Moderne Kapitalismus*, by Werner Sombart. There is scarcely any aspect of Western European life from the tenth century on that has escaped Sombart's eagle-like vision and mole-like industry; and his annotated bibliographies would almost repay publication by themselves. *The Evolution of Modern Capitalism*, by J. A. Hobson, parallels Sombart's work; and while the original edition drew specially on English sources his latest edition openly acknowledges a debt to Sombart. In America Thorstein Veblen's works, taken as a whole, including his less-appreciated books like *Imperial Germany* and *The Nature of Peace*, form a unique contribution to the subject. For the resources of modern technics Erich Zimmerman's recent survey of *World Resources and World Industries* fills what up to recently had been a serious gap; this is complemented, in a degree, by H. G. Wells's somewhat diffuse study of the physical processes of modern life in his *The Work, Wealth and Happiness of Mankind*.

For further comment on some of the more important books see the following list. The Roman numerals in brackets refer to the relevant chapter or chapters.

2: List of Books

Ackerman, A. P., and Dana, R. T.: *The Human Machine in Industry.* New York: 1927.

Adams, Henry: *The Degradation of the Democratic Dogma.* New York: 1919.
Adams's attempt to adapt the Phase Rule to social phenomena, though unsound, resulted in a very interesting prediction for the final phase, which corresponds, in effect, to our neotechnic one. [v]

Agricola, Georgius: *De Re Metallica.* First Edition: 1546. Translated from edition of 1556 by H. C. Hoover and Lou Henry Hoover, 1912.
One of the great classics in technics. Gives a cross section of advanced technical practices in the heavy industries in the early sixteenth century. Important for any just estimate of eotechnic achievement. [II, III, IV]

Albion, R. G.: *Introduction to Military History.* New York: 1929. [II]

Allport, Floyd A.: *Institutional Behavior*. Chapel Hill: 1933.
A critical and on the whole fair analysis of the defects in the current gospel of labor-saving and enforced leisure: much better than Borsodi though afflicted with a little of the same middle class suburban romanticism. [VI, VIII]

Andrade, E. N.: *The Mechanism of Nature*. London: 1930.

Annals of the American Academy of Political and Social Science: *National and World Planning*. Philadelphia: July 1932.

Appier, Jean, and Thybourel, F.: *Recueil de Plusieurs Machines Militaires et Feux Artificiels Pour la Guerre et Recreation*. Pont-a-Mousson: 1620. [II]

Ashton, Thomas S.: *Iron and Steel in the Industrial Revolution*. New York: 1924.
Useful introduction to the subject, perhaps the best in English. But see Ludwig Beck. [II, IV, V]

Babbage, Charles: *On the Economy of Machinery and Manufactures*. Second Edition. London: 1832. [IV]
One of the landmarks in paleotechnic thought, by a distinguished British mathematician.
Exposition of 1851; or, Views of the Industry, the Science and the Government of England. Second Edition. London: 1851.

Bacon, Francis: *Of the Advancement of Learning*. First Edition. London: 1605.
A synoptic survey of the gaps and achievements of eotechnic knowledge: pre-Galilean in its conception of scientific method but nevertheless highly suggestive. [I, III]
Novum Organum. First Edition. London: 1620.
The New Atlantis. First Edition. London: 1660.
An incomplete utopia, useful only as an historical document. For a more intimate view of current technics and a new industrial order, see J. V. Andreae's *Christianopolis*.

Bacon, Roger: *Opus Majus*. Translated by Robert B. Burke. Two vols. Philadelphia: 1928. [I, III]
To be read in connection with Thorndike, who perhaps is a little too depreciative of Bacon, in reaction against the praise of those who know no other example of medieval science.

Baker, Elizabeth: *Displacement of Men by Machines; Effects of Technological Change in Commercial Printing*. New York: 1933. [V, VIII]
Good factual study of the changes within a single industry that combines tradition and steady technical progress.

Banfield, T. C.: *Organization of Industry*. London: 1848.

Barclay, A.: *Handbook of the Collections Illustrating Industrial Chemistry*. Science Museum, South Kensington. London: 1929. [IV, V]
Like the other handbooks put out by the Science Museum it is admirable in scope and method and lucidity: more than mere handbooks, these essays should not be absent from a working library on modern technics.

Barnett, George: *Chapters on Machinery and Labor*. Cambridge: 1926.
Factual discussion of the displacement of labor by automatic machines. [v, viii]

Bartels, Adolph: *Der Bauer in der Deutschen Vergangenheit*. Second Edition. Jena: 1924.
Like the other books in this series, richly illustrated.

Bavink, Bernhard: *The Anatomy of Modern Science*. Translated from German. Fourth Edition. New York: 1932.
A useful survey whether or not one accepts Bavink's metaphysics [i]

Bayley, R. C.: *The Complete Photographer*. Ninth Edition. London: 1926.
The best general book in English on the history and technique of modern photography. [v, vii]

Beard, Charles A. (Editor): *Whither Mankind*. New York: 1928.
Toward Civilization. New York: 1930 [vii, viii]
The first book attempts to answer how far and in what manner various aspects of life have been affected by science and the machine. The second is a confident and somewhat muddled apology for modern technics, which however is prefaced by an excellent critical essay by the editor.

Bechtel, Heinrich: *Wirtschaftsstil des Deutschen Spätmittelalters*. München: 1930. [iii]
Follows in detail the trail blazed by Sombart: treats art and architecture along with industry and commerce. Good section on mining.

Beck, Ludwig: *Die Geschichte des Eisens in Technischer und Kulturgeschichtlicher Beziehung*. Five vols. Braunschweig: 1891-1903. [ii, iii, iv, v]
A monumental work of the first order.

Beck, Theodor: *Beiträge zur Geschichte des Machinenbaues*. Second Revised Edition. Berlin: 1900. [i, iii, iv]
Because it summarizes the achievements and the technical books of the early Italian and German engineers, it has special value for the historical student.

Beckmann, J.: *Beiträge zur Geschichte der Erfindungen*. Five vols. Leipzig: 1783-1788. Translated: *A History of Inventions, Discoveries and Origins*. London: 1846.
The first treatise on the history of modern technics; not to be lightly passed over even today. Particularly interesting because, like Adam Smith's classic, it shows the bent of eotechnic thought before the paleotechnic revolution.

Bellamy, Edward: *Looking Backward*. First Edition. Boston: 1888. New Edition. Boston: 1931. [viii]
A somewhat dehumanized utopia which has nevertheless gained rather than lost ground during the last generation. It is in the tradition of Cabet rather than Morris.

Bellet, Daniel: *La Machine et la Main-d'Œuvre Humaine*. Paris: 1912. *L'Evolution de l'Industrie*. Paris: 1914.

Bennet and Elton: *History of Commercial Milling*. [iii]
Useful work. But see Usher's criticism.

Bennett, C. N.: *The Handbook of Kinematography*. Second Edition. London: 1913.

Bent, Silas: *Machine Made Man*. New York: 1930.

Berdrow, Wilhelm: *Alfred Krupp*. Two vols. Berlin: 1927. [IV]
Exhaustive picture of one of the great paleotects: but curiously incomplete in its
lack of reference to his pioneer work in housing.

Berle, Adolf A., Jr.: *The Modern Corporation and Private Property*. New
York: 1933. [VIII]
Excellent factual study of the concentration of modern finance in the United States
and the difficulty of applying our usual legal concepts to the situation. But cautious
to the point of downright timidity in its recommendations.

Besson, Jacques: *Theatre des Instruments Mathématiques et Méchaniques*.
Genève: 1626. [III]
The work of a sixteenth century mathematician who was also a brilliant technician.

Biringucci, Vannuccio: *De la Pirotechnia*. Venice: 1540. Translated into
German. Braunschweig: 1925. [III]

Blake, George G.: *History of Radiotelegraphy and Telephony*. London: 1926.
[V]

Bodin, Charles: *Economie Dirigée, Economie Scientifique*. Paris: 1932.
Conservative opposition.

Boissonade, Prosper: *Life and Work in Mediaeval Europe: Fifth to Fifteenth
Centuries*. New York: 1927. [III]
A good contribution to a well-conceived and well-edited series.

Booth, Charles: *Life and Labor in London*. Seventeen vols. Begun 1889.
London: 1902. [IV]
Factual picture, massive and complete, of the level of life in a great imperial metrop-
olis. See also the later and more compact survey.

Borsodi, Ralph: *This Ugly Civilization*. New York: 1929. [VI]
An attempt to show that with the aid of the electric motor and modern machines
household industry may compete with mass production methods. See Kropotkin for
a far sounder statement of this thesis.

Böttcher, Alfred: *Das Scheinglück der Technik*. Weimar: 1932. [VI]

Bourdeau, Louis: *Les Forces de l'Industrie: Progrés de la Puissance Humaine*.
Paris: 1884.

Bouthoul, Gaston: *L'Invention*. Paris: 1930. [I]

Bowden, Witt: *Industrial Society in England Toward the End of the Eighteenth Century*. New York: 1925. [IV]
Should be supplemented with Mantoux and Halévy.

Boyle, Robert: *The Sceptical Chymist*. London: 1661.

Bragg, William: *Creative Knowledge: Old Trades and New Science*. New
York: 1927.

Brandt, Paul: *Schaffende Arbeit und Bildende Kunst*. Vol. I: "Im Altertum
und Mittelalter." [I, II, III] Vol. II: "Vom Mittelalter bis zur Gegenwart."
Leipzig: 1927. [III, IV]
Draws on the important illustrations of Stradanus, Ammann, Van Vliet and Luyken
for presentation of eotechnic industry. But fails to utilize French sources sufficiently

Branford, Benchara: *A New Chapter in the Science of Government.* London: 1919. [VIII]

Branford, Victor (Editor): *The Coal Crisis and the Future: A Study of Social Disorders and Their Treatment.* London: 1926. [V]
Coal—Ways to Reconstruction. London: 1926.

Branford, Victor, and Geddes, P.: *The Coming Polity.* London: 1917. [V]
An application of Le Play and Comte to the contemporary situation.
Our Social Inheritance. London: 1919. [VIII]

Branford, Victor: *Interpretations and Forecasts: A Study of Survivals and Tendencies in Contemporary Society.* New York: 1914.
Science and Sanctity. London: 1923. [I, VI, VIII]
The most comprehensive statement of Branford's philosophy: at times obscure, at times wilful, it is nevertheless full of profound and penetrating ideas.

Brearley, Harry C.: *Time Telling Through the Ages.* New York: 1919. [I]

Brocklehurst, H. J., and Fleming, A. P. M.: *A History of Engineering.* London: 1925.

Browder, E. R.: *Is Planning Possible Under Capitalism?* New York: 1933.

Buch der Erfindungen, Gewerbe und Industrien. Ten vols. Ninth Edition. Leipzig: 1895-1901.

Bücher, Karl: *Arbeit und Rhythmus.* Leipzig: 1924. [I, II, VII]
A unique contribution to the subject which has been expended and modified in the course of numerous editions. A fundamental discussion of esthetics and industry.

Buckingham, James Silk: *National Evils and Practical Remedies.* London: 1849. [IV]
The quintessence of paleotechnic reformism: a utopia whose defects like that of Richardson's *Hygeia*, bring out the characteristics of the period.

Budgen, Norman F.: *Aluminium and Its Alloys.* London: 1933. [V]

Burr, William H.: *Ancient and Modern Engineering.* New York: 1907.

Butler, Samuel: *Erewhon, or Over the Range.* First Edition. London: 1872.
Describes an imaginary country where people have given up machines and carrying a watch is a crime. While looked upon as pure sport and satire in Victorian times, it points to an unconscious fear of the machine that still survives, not without some reason.

Butt, I. N., and Harris, I. S.: *Scientific Research and Human Welfare.* New York: 1924.
Popular.

Buxton, L. H. D.: *Primitive Labor.* London: 1924. [II]

Byrn, Edward W.: *Progress of Invention in the Nineteenth Century.* New York: 1900. [IV]
Useful synopsis of inventions and processes.

Campbell, Argyll, and Hill, Leonard: *Health and Environment.* London: 1925. [IV, V]
Full of valuable data on the defects of the paleotechnic environment.

Capek, Karel: *R.U.R.* New York: 1923. [V]
A play that antedated Mr. Televox, the modern automaton. Its drama, dealing with the revolt of the mechanized robot upon becoming slightly human, is spoiled by a sloppy ending. A signpost in the revolt against excessive mechanization: like Rice's *The Adding Machine* and O'Neill's *The Hairy Ape.*

Carter, Thomas F.: *The Invention of Printing in China and Its Spread Westward.* New York: 1931. [III]
A brilliant book which adds an important supplement to Usher's chapter on printing. All but establishes the last link in the chain that binds the appearance of printing in Europe to its earlier development—including cast metal types—in China and Korea.

Casson, H. N.: *Kelvin: His Amazing Life and Worldwide Influence.* London: 1930. [V]
History of the Telephone. Chicago: 1910.

Chase, Stuart: *Men and Machines.* New York: 1929. [IV, V, VIII]
Superficial but suggestive.
The Nemesis of American Business. New York: 1931. [V]
See study of A. O. Smith plant.
The Promise of Power. New York: 1933. [V]
Technocracy; an Interpretation. New York: 1933.
The Tragedy of Waste. New York: 1925. [V, VIII]
The best of Chase's books to date, probably: full of useful material on the perversions of modern commerce and industry.

Chittenden, N. W.: *Life of Sir Isaac Newton.* New York, 1848.

Clark, Victor S.: *History of Manufactures in the United States.* (1607-1928.) Three vols. New York: 1929. [III, IV]
Since the eotechnic period lingered, even in advanced parts of the country, till the third quarter of the nineteenth century this work is a valuable study of late eotechnic practices—including surface mining.

Clay, Reginald S., and Court, Thomas H.: *The History of the Microscope.* London: 1932. [III]

Clegg, Samuel: *Architecture of Machinery: An Essay on Propriety of Form and Proportion.* London: 1852. [VII]

Cole, G. D. H.: *Life of Robert Owen.* London: 1930.
Good study of an important industrialist and utopian whose pioneer ideas on industrial management and city building are still bearing fruit.
Modern Theories and Forms of Industrial Organisation. London: 1932. [VIII]

Cooke, R. W. Taylor: *Introduction to History of Factory System.* London: 1886.
Good historic perspective; but must now be supplemented by Sombart's data. [III, IV]

Coudenhove-Kalergi, R. N.: *Revolution Durch Technik.* Wien: 1932.

454   TECHNICS AND CIVILIZATION

Coulton, G. G.: *Art and the Reformation.* New York: 1928. [I, III]

Court, Thomas H., and Clay, Reginald S.: *The History of the Microscope.* London: 1932. [III]

Crawford, M. D. C.: *The Heritage of Cotton.* New York: 1924. [IV]

Cressy, Edward: *Discoveries and Inventions of the Twentieth Century.* Third Edition. New York: 1930. [V]
For the layman.

Dahlberg, Arthur: *Jobs, Machines and Capitalism.* New York: 1932. [V, VIII]
An attempt to solve the problem of labor displacement under technical improvement.

Dampier, Sir William: *A History of Science and Its Relations with Philosophy and Religion.* New York: 1932. [I]

Dana, R. T., and Ackerman, A. P.: *The Human Machine in Industry.* New York: 1927.

Daniels, Emil: *Geschichte des Kriegswesens.* Six vols. (Sammlung Goschen) Leipzig: 1910-1913. [II, III, IV]
Perhaps the best small general introduction to the development of warfare.

Darmstaedter, Ludwig, and others: *Handbuch zur Geschichte der Naturwissenschaften und der Technik: In Chronologischer Darstellung.* Second Revised and Enlarged Edition. Berlin: 1908. [I-VIII]
An exhaustive compendium of dates, but better for science than technics.

Demmin, Auguste Frédéric: *Weapons of War: Being a History of Arms and Armour from the Earliest Period to the Present Time.* London: 1870. [II]

Descartes, René: *A Discourse on Method.* First Edition. Leyden: 1637.
One of the foundation stones of seventeenth century metaphysics: not seriously challenged in science—except among physiologists like Claude Bernard—till Mach.

Dessauer, Friedrich: *Philosophie der Technik.* Bonn: 1927.
A book with a high reputation in Germany; but a little given to laboring the obvious.

Deutsches Museum: *Amtlicher Führer durch die Sammlungen.* München: 1928.

Diamond, Moses: *Evolutionary Development of Reconstructive Dentistry.* Reprinted from the *New York Medical Journal and Medical Record.* New York: August, 1923. [V]

Diels, Hermann: *Antike Technik.* First Edition. Berlin: 1914. Second Edition. 1919.

Dixon, Roland B.: *The Building of Cultures.* New York: 1928.

Dominian, L.: *The Frontiers of Language and Nationality in Europe.* New York: 1917. [VI]

Douglas, Clifford H.: *Social Credit.* Third Edition. London: 1933.

Dulac, A., and Renard, G.: *L'Evolution Industrielle et Agricole depuis Cent Cinquante Ans.* [IV, V]
Good picture of the last century and a half's development.

Dyer, Frank L., and Martin, T. C.: *Edison: His Life and Inventions.* New York: 1910.

Eckel, E. C.: *Coal, Iron and War: A Study in Industrialism, Past and Future.* New York: 1920.
Interesting study arising in part out of the stresses of the World War.

Economic Significance of Technological Progress: A Report to the Society of Industrial Engineers. New York: 1933. [V, VIII]
A summary by a committee of which Polakov was chairman: see Polakov.

Eddington, A. S.: *The Nature of the Physical World.* New York: 1929. [VIII]

Egloff, Gustav: *Earth Oil.* New York: 1933. [V]

Ehrenberg, Richard: *Das Zeitalter der Fugger.* Jena: 1896. Translated. *Capital and Finance in the Age of the Renaissance.* New York: 1928. [I, II, III]

Elton, John, and Bennett, Richard: *History of Corn Milling.* Four vols. London: 1898-1904.

Encyclopédie (en folio) *des Sciences, des Arts et des Métiers. Recueil de Planches.* Paris: 1763. [III]
A cross section of European technics in the middle of the eighteenth century, with special reference to France, which by then had taken the lead from Holland. The detailed explanation and illustration of processes give it special importance. The engravings I have used are typical of the whole work. The *Encyclopédie* has been slighted by German historians of technics. In its illustration of the division of labor it is a graphic commentary on Adam Smith.

Engelhart, Viktor: *Weltanschauung und Technik.* Leipzig: 1922.

Engels, Friedrich: *The Condition of the Working Class in England in 1844.* Translated. London: 1892. [IV]
Firsthand picture of the horrors of paleotechnic industrialism during one of its greatest crises: further documentation has enriched, but not lightened, Engels' description. See the Hammonds.

Engels, Friedrich, and Marx, Karl: *Manifesto of the Communist Party.* New York: 1930. [IV]

Enock, C. R.: *Can We Set the World in Order? The Need for a Constructive World Culture; An Appeal for the Development and Practice of a Science of Corporate Life . . . a New Science of Geography and Industry Planning.* London: 1916. [V, VIII]
A book whose pertinent criticisms and originality atones for the streak of crotchetiness in it.

Erhard, L.: *Der Weg des Geistes in der Technik.* Berlin: 1929.

Espinas, Alfred: *Les Origines de la Technologie.* Paris: 1899.

Ewing, J. Alfred: *An Engineer's Outlook.* London: 1933. [V, VIII]

Drastic criticism of the failure of morals and politics to keep pace with the machine: suggestion for reducing the tempo of invention till we have mastered our difficulties. Noteworthy because of Ewing's professional eminence.

Eyth, Max: *Lebendige Krafte; Sieben Vortrage aus dem Gebiete der Technik.* First Edition. Berlin: 1904. Third Edition. Berlin: 1919.

Farnham, Dwight T., and others: *Profitable Science in Industry.* New York: 1925.

Feldhaus, Franz Maria: *Leonardo; der Techniker und Erfinder.* Jena: 1913. [III]
Die Technik der Vorzeit; der Geschichtlichen Zeit und der Naturvölker. Leipzig: 1914.
Ruhmesblätter der Technik von der Urerfindungen bis zur Gegenwart. Two vols. Second Edition. Leipzig: 1926. [I-VIII]
An invaluable work.
Kulturgeschichte der Technik. Two vols. Berlin: 1928. [I-VIII]
Lexikon der Erfindungen und Entdeckungen auf den Gebieten der Naturwissenschaften und Technik. Heidelberg: 1904.
Technik der Antike und des Mittelalters. Potsdam: 1931. [III]
Although not always exhaustive in his treatment of sources outside Germany or the German literature of the subject, Feldhaus has placed the student of the historical development of technics under a constant debt.

Ferrero, Gina Lombroso: *The Tragedies of Progress.* New York: 1931.
A weak book which exaggerates the virtues of the past and does not succeed in presenting a drastic enough criticism of the present, despite the obvious bias against it. [VI]

Field, J. A.: *Essays on Population.* Chicago: 1931. [V]

Flanders, Ralph: *Taming Our Machines: The Attainment of Human Values in a Mechanized Society.* New York: 1931. [V, VIII]
Essays by an engineer who realizes that the machine age is not a pure utopia.

Fleming, A. P. M., and Brocklehurst, H. J.: *A History of Engineering.* London: 1925.

Fleming, A. P. M., and Pearce, J. G.: *Research in Industry.* London: 1917.

Föppl, Otto: *Die Weiterentwicklung der Menschheit mit Hilfe der Technik.* Berlin: 1932.

Ford, Henry: *Today and Tomorrow.* New York: 1926.
Moving Forward. New York: 1930.
My Life and Work. New York: 1926. [V, VIII]
Important because of Ford's industrial power and his almost instinctive recognition of the necessities for neotechnic reorganization of industry: but vitiated by the cant that is so often associated with an American's good intentions, particularly when he must justify his arbitrary financial power.

Form, Die. Fortnightly organ of the Deutscher Werkbund.
Between 1925 and January 1933 the most important periodical dealing with all the arts of form, both in the hand-crafts and the machine-crafts. While the leadership here has now passed back again to France, Belgium, Holland, and the Scandinavian countries

Die Form remains an indispensable record of Germany's short but genuinely creative outburst. [VII]

Fournier, Edouard: *Curiosités des Inventions et Decouvertes.* Paris: 1855.

Fox, R. M.: *The Triumphant Machine.* London: 1928.

Frank, Waldo: *The Rediscovery of America.* New York: 1929. [VI]
Some valuable comments on the subjective effects of mechanization.

Freeman, Richard A.: *Social Decay and Regeneration.* London: 1921. [VI]
An upper class criticism of the machine from the standpoint of human deterioration resulting. See Allport for a more intelligent statement.

Frémont, Charles: *Origines et Evolution des Outils.* Paris: 1913.

Frey, Dagobert: *Gotik und Renaissance als Grundlagen der Modernen Welanschauung.* Augsburg: 1929. [I, VII]
A brilliant and well-illustrated study of a difficult, delicate and fascinating subject.

Friedell, Egon: *A Cultural History of the Modern Age.* Three vols. New York: 1930-1932.
Usually witty, sometimes inaccurate, occasionally obscurantist: not to be trusted about matters of fact, but, like Spengler, occasionally valuable for oblique revelations not achieved by more academically competent minds.

Frost, Dr. Julius: *Die Hollandische Landwirtschaft; Ein Muster Moderner Rationalisierung.* Berlin: 1930.

Gage, S. H.: *The Microscope.* Revised Edition. Ithaca: 1932. [III]

Galilei, Galileo: *Dialogues Concerning Two New Sciences.* New York: 1914. [I, III]
A classic.

Gantner, Joseph: *Revision der Kunstgeschichte.* Wien: 1932. [VII]
Suggests the necessity of revision in historical judgments upon the basis of new interests and values. The author was editor of the brilliant if short-lived *Die Neue Stadt.*

Gantt, H. L.: *Work, Wages and Profits.* New York: 1910.
One of the landmarks of the efficiency movement by a contemporary of Taylor's who had advanced beyond the master's original narrow position.

Garrett, Garret: *Ouroboros, or the Future of the Machine.* New York: 1926.

Gaskell, P.: *Artisans and Machinery; The Moral and Physical Condition of the Manufacturing Population Considered with Reference to Mechanical Substitutes for Human Labour.* London: 1836. [IV]
Gaskell, writing with a belief in the established order, presents a pretty damning view of early paleotechnic industry, whose defects revolted him.

Gast, Paul: *Unsere Neue Lebensform.* München: 1932.

Geddes, Norman Bel: *Horizons.* Boston: 1932. [V, VII]
Suggestions of new forms for machines and utilities, with a full utilization of aerodynamic principles and modern materials. While it owes more to publicity than scholarship, it is useful because of its illustrations.

Geddes, Patrick: *An Analysis of the Principles of Economics.* Edinburgh: 1885. [VIII]

Geddes, Patrick: *The Classification of Statistics*. Edinburgh: 1881.
Early papers by Geddes still suggestive to those capable of carrying Geddes's clues to their conclusion. The first sociological application of the modern concept of energy.
An Indian Pioneer of Science; the Life and Work of Sir Jagadis Bose. London: 1920.
Cities in Evolution. London: 1915.
Geddes's earlier essays distinguishing the paleotechnic from the neotechnic period appear here.

Geddes, Patrick, and Thomson, J. A.: *Life; Outlines of General Biology.* Two vols. New York: 1931.
Biology. New York: 1925.
The smaller book gives the skeleton of the larger work in dwarf form. The later chapters in Volume II of *Life* are perhaps the best epitome of Geddes's thought as yet available. He projected a similar work in *Sociology* but did not live to complete it.

Geddes, Patrick, and Slater, G.: *Ideas at War.* London: 1917. [II, IV]
A brilliant enlarged sketch of Geddes's smaller article on Wardom and Peacedom that appeared in the *Sociological Review.*

Geer, William C.: *The Reign of Rubber.* New York: 1922. [V]
One of the few available books on a subject that calls for more extended and scholarly treatment than it has yet enjoyed.

Geitel, Max (Editor): *Der Siegeslauf der Technik.* Three vols. Berlin: 1909.

George, Henry: *Progress and Poverty.* New York: 1879.
While George's overemphasis of the rôle of the private appropriation of the rent of land caused him to give a highly one-sided account of modern industrialism, his work, like Marx's, is a landmark in criticism.

Giese, Fritz: *Bildungsideale im Maschinenzeitalter.* Halle, a.S.: 1931.

Glanvill, Joseph: *Scepsis Scientifica; or Confessed Ignorance the Way to Science.* London: 1665. [I]

Glauner, Karl, Th.: *Industrial Engineering.* Des Moines: 1931.

Gloag, John: *Artifex, or The Future of Craftsmanship.* New York: 1927.

Glockmeier, Georg: *Von Naturalwirtschaft zum Millardentribut: Ein Langschnitt durch Technik, Wissenschaft und Wirtschaft zweier Jahrtausende.* Zurich: 1931.

Goodyear, Charles: *Gum Elastic and Its Varieties.* 1853. [V]

Gordon, G. F. C.: *Clockmaking, Past and Present; with which Is Incorporated the More Important Portions of "Clocks, Watches and Bells" by the late Lord Grimthorpe.* London: 1925. [I, III]

Graham, J. J.: *Elementary History of the Progress of the Art of War.* London: 1858. [II]

Gras, N. S. B.: *Industrial Evolution.* Cambridge: 1930. [I-V]
A useful series of concrete studies of the development of industry.
An Introduction to Economic History. New York: 1922.

Green, A. H., and others: *Coal; Its History and Uses.* London: 1878. [IV]

Grossmann, Robert: *Die Technische Entwicklungen der Glasindustrie in ihrer Wirtschaftlichen Bedeutung.* Leipzig: 1908. [III]

Guerard, A. L.: *A Short History of the International Language Movement.* London: 1922. [VI]
An excellent summary of the case for an international language and the status of the movement a dozen years ago. Ogden's work on Basic English, while valuable for its suggestions in logic and grammar, has never presented an adequate defense for the use of a living language for international intercourse.

Hale, W. J.: *Chemistry Triumphant.* Baltimore: 1933. [V]

Halévy, Elie: *The Growth of Philosophic Radicalism.* London: 1928. [IV]
The best history of the ideology of the utilitarians.

Hammond, John Lawrence and Barbara: *The Rise of Modern Industry.* New York: 1926. [III, IV]
The Town Labourer. (1760-1832).
The Skilled Labourer (1760-1832). New York: 1919. [IV]
The Village Labourer. London: 1911. [III, IV]
This series of books, even the more general one on the rise of modern industry, is based almost exclusively on British documentation. Within these limits it constitutes the most vivid, massive, and unchallengable picture of the beginnings of the paleotechnic régime and its proud progress that has been done. Cf. Engels, Mantoux, and for contrast Ure. The pattern described by the Hammonds was followed, with minor variations, in every other country.

Hamor, William A., and Weidlein, E. R.: *Science in Action.* New York: 1931.

Harris, L. S., and Butt, I. N.: *Scientific Research and Human Welfare.* New York: 1924. [V]

Harrison, H. S.: *Pots and Pans.* London: 1923. [II]
The Evolution of the Domestic Arts. Second Edition. London: 1925.
Travel and Transport. London: 1925. [II]
War and Chase. London: 1929. [II]
An excellent series of introductions: but note particularly that on war and the chase.

Hatfield, H. Stafford: *The Inventor and His World.* New York: 1933.

Hauser, Henri: *La Modernité du XVIe Siècle.* Paris: 1930. [I]

Hausleiter, L.: *The Machine Unchained.* New York: 1933.
Worthless.

Hart, Ivor B.: *The Mechanical Investigations of Leonardo da Vinci.* London: 1925. [III]
With Feldhaus's work on Leonardo an excellent summary of Leonardo's achievements. See also the chapter in Usher.
The Great Engineers. London: 1928.

Havemeyer, Loomis: *Conservation of Our Natural Resources* (based on Van Hise). New York: 1930. [V]
Recognition by the engineer of the facts on the waste and destruction of the environment first clearly put by George Perkins Marsh in the sixties.

Henderson, Fred: *Economic Consequences of Power Production.* London: 1931. [v, viii]
Able and well-reasoned study of the tendencies to automatism and remote control in neotechnic production.

Henderson, Lawrence J.: *The Order of Nature.* Cambridge: 1925. [i]
The Fitness of the Environment; An Inquiry into the Biological Significance of the Properties of Matter. New York: 1927. [i, viii]
A brilliant and original contribution which reverses the usual treatment of adaptation.

Hendrick, B. J.: *The Life of Andrew Carnegie.* New York: 1932. [iv]

Hill, Leonard, and Campbell, Argyll: *Health and Environment.* London: 1925. [iv, v]
Valuable.

Hine, Lewis: *Men at Work.* New York: 1932. [v]
Photographs of modern workers on the job. The kind of study that should be done systematically if Geddes's Encyclopedia Graphica is ever to be done.

Hobson, John A.: *The Evolution of Modern Capitalism; a Study of Machine Production.* New Edition (Revised). London: 1926. [i-v]
Incentives in the New Industrial Order. London: 1922. [viii]
Wealth and Life; a Study in Values. London: 1929. [viii]
One of the most intelligent, clear-thinking and humane of the modern economists. These books are a useful corrective to uncritical dreams of the "new capitalism" so fashionable in America between 1925 and 1930.

Hocart, A. M.: *The Progress of Man.* London: 1933.
Brief critical survey of the various fields of anthropology, including technics.

Hoe, R.: *A Short History of the Printing Press.* New York: 1902.

Holland, Maurice, and Pringle, H. F.: *Industrial Explorers.* New York: 1928.

Hollandsche Molen: *Eerste Jaarboekje.* Amsterdam: 1927. [iii]
Report of the society for preserving the old mills of Holland.

Holsti, R.: *Relation of War to the Origin of the State.* Helsingfors: 1913. [ii]
A book that challenges the complacent old-fashioned notion which made war a peculiar property of savage peoples. Demonstrates the ritualistic nature of much primitive warfare.

Holzer, Martin: *Technik und Kapitalismus.* Jena: 1932. [viii]
A keen criticism of technicism and pseudo-efficiency fostered by modern large scale finance.

Hooke, Robert: *Micrographia.* London: 1665. [i]
Posthumous Works. London: 1705.

Hopkins, W. M.: *The Outlook for Research and Invention.* New York: 1919. [v]

Hough, Walter: *Fire as an Agent in Human Culture.* Smithsonian Institution, Bulletin 139. Washington: 1926. [ii]

Howard, Ebenezer: *Tomorrow; A Peaceful Path to Reform.* London: 1898. Second Edition entitled: *Garden Cities of Tomorrow.* London: 1902. [v]

A book which describes one of the most important neotechnic inventions, the garden-city. See also Kropotkin and Geddes's *Cities in Evolution.*

Iles, George: *Inventors at Work.* New York: 1906.
Leading American Inventions. New York: 1912.

Jameson, Alexander (Editor): *A Dictionary of Mechanical Science, Arts, Manufactures and Miscellaneous Knowledge.* London: 1827. [III, IV]

Jeffrey, E. C.: *Coal and Civilization.* New York: 1925. [IV, V]

Jevons, H. Stanley: *Economic Equality in the Cooperative Commonwealth.* London: 1933. [VIII]
Detailed suggestions for a typically English and orderly passage to communism.

Jevons, W. Stanley: *The Coal Question.* London: 1866. [IV]
A book which called attention to the fundamentally insecure basis of the paleotechnic economy.

Johannsen, Otto: *Louis de Geer.* Berlin: 1933. [III]
Short account of a Belgian capitalist who waxed fat in the munitions industry in seventeenth century Sweden. See also the account of Christopher Polhem in Usher.

Johnson, Philip: *Machine Art.* New York: 1934.
A study of the basic esthetic elements in machine forms.

Jones, Bassett: *Debt and Production.* New York: 1933. [VIII]
An attempt to prove that the rate of industrial production is decreasing while the structure of debt rises. An important thesis.

Kaempffert, Waldemar: *A Popular History of American Invention.* New York: 1924. [IV, V]

Kapp, Ernst: *Grundlinien einer Philosophie der Technik.* Braunschweig: 1877.

Keir, R. M.: *The Epic of Industry.* New York: 1926. [IV, V]
Deals with the development of American industry. Well illustrated.

Kessler, Count Harry: *Walter Rathenau: His Life and Work.* New York: 1930. [V]
Sympathetic account of perhaps the leading neotechnic financier and industrialist: a biographic appendix to Veblen's theory of business enterprise showing the conflict between pecuniary and technical standards in a single personality.

Kirby, Richard S., and Laurson, P. G.: *The Early Years of Modern Civil Engineering.* New Haven: 1932. [IV]
Some interesting American material.

Klatt, Fritz: *Die Geistige Wendung des Maschinenzeitalters.* Potsdam: 1930.

Knight, Edward H.: *Knight's American Mechanical Dictionary.* New York: 1875. [V]
A very creditable compilation, considering the time and place, which gives a useful cross section of paleotechnic industry.

Koffka, Kurt: *The Growth of the Mind.* New York: 1925.

Kollmann, Franz: *Schönheit der Technik.* München: 1928. [VII]

Good study with numerous photographs which already needs a supplement dealing
with later forms.

Kraft, Max: *Das System der Technischen Arbeit.* Four vols. Leipzig: 1902.

Krannhals, Paul: *Das Organische Weltbild.* Two vols. München: 1928.
Der Weltsinn der Technik. München: 1932. [I]
Der Weltsinn is an attempt to form a critical philosophy of technics and relate it to
other aspects of life.

Kropotkin, P.: *Fields, Factories and Workshops; or Industry Combined with
Agriculture and Brainwork with Manual Work.* First Edition, 1898. Re-
vised Edition. London: 1919. [V, VIII]
An early attempt to trace out the implications of the neotechnic economy, greatly re-
enforced by later developments in electricity and factory production. See Howard.
Mutual Aid. London: 1904.

Kulischer, A. M., and Y. M.: *Kriegs und Wanderzüge; Weltgeschichte als
Völkerbewegung.* Berlin: 1932. [II, IV]
Able analysis of the relation between war and the migrations of peoples.

Labarte: *Histoire des Arts Industrielles au Moyen Age et à L'Epoque de la
Renaissance.* Three vols. Paris: 1872-1875.
Does not live up to the promise of its title. See Boissonade and Renard.

Lacroix, Paul: *Military and Religious Life in the Middle Ages and . . .
the Renaissance.* London: 1874. [II]

Landauer, Carl: *Planwirtschaft und Verkehrswirtschaft.* München: 1931.

Langley, S. P.: *Langley Memoir on Mechanical Flight.* Part I. 1887-1896.
Washington: 1911. [V]

Launay, Louis de: *La Technique Industrielle.* Paris: 1930.

Laurson, P. G., and Kirby, R. S.: *The Early Years of Modern Civil Engineering.* New Haven: 1932. [IV]

Le Corbusier: *L'Art Decoratif d'Aujourdui.* Paris: 1925.
Vers Une Architecture. Paris: 1922. Translated. London: 1927. [VII]
Following the work of Sullivan and Wright and Loos more than a generation later,
Le Corbusier re-discovered the machine for himself and is perhaps the chief polemical
advocate of machine forms.

Lee, Gerald Stanley: *The Voice of the Machines; An Introduction to the
Twentieth Century.* Northampton: 1906.
A sentimental book.

Leith, C. K.: *World Minerals and World Politics.* New York: 1931. [V]

Lenard, Philipp: *Great Men of Science; A History of Human Progress.*
London: 1933.

Leonard, J. N.: *Loki; The Life of Charles P. Steinmetz.* New York: 1929. [V]

Le Play, Frederic: *Les Ouvriers Européens.* Six vols. Second Edition. Tours:
1879. [II]

One of the great landmarks of modern sociology: the failure to follow it up reveals the limitations of the major schools of economists and anthropologists. The lack of such concrete studies of work and worker and working environment is a serious handicap in writing a history of technics or appraising current forces.

Leplay House: *Coal: Ways to Reconstruction.* London: 1926. [v]
Application of neotechnic thought to a backward industry.

Levy, H.: *The Universe of Science.* London: 1932.
Good introduction. [I, v]

Lewis, Gilbert Newton: *The Anatomy of Science.* New Haven: 1926. [I, v]
Excellent exposition of the contemporary approach to science: see also Poincaré, Henderson, Levy, and Bavink.

Lewis, Wyndham: *Time and Western Man.* New York: 1928. [I]
Critical tirade against time-keeping and all the timed-arts by an eye-minded advocate of the spatial arts. One-sided but not altogether negligible.

Liehburg, Max Eduard: *Das Deue Weltbild.* Zurich: 1932.

Lilje, Hanns: *Das Technische Zeitalter.* Berlin: 1932.

Lindner, Werner, and Steinmetz, G.: *Die Ingenieurbauten in Ihrer Guten Gestaltung.* Berlin: 1923. [VII]
Particularly good in its relation of older forms of industrial construction to modern works: plenty of illustrations. See Le Corbusier and Kollmann.

Lombroso, Ferrero Gina: *The Tragedies of Progress.* New York: 1931.
(See Ferrero.)

Lucke, Charles E.: *Power.* New York: 1911.

Lux, J. A.: *Ingenieur-Aesthetik.* München: 1910. [VII]
One of the early studies. See Lindner.

MacCurdy, G. G.: *Human Origins.* London: 1923. New York: 1924. [I, II]
Good factual account of tools and weapons in prehistoric cultures.

MacIver, R. M.: *Society: Its Structure and Changes.* New York: 1932.
Well-balanced and penetrating introduction.

Mackaye, Benton: *The New Exploration.* New York: 1928. [v, VIII]
Pioneer treatise on geotechnics and regional planning to be put alongside Marsh and Howard.

Mackenzie, Catherine: *Alexander Graham Bell.* New York: 1928. [v]

Mâle, Emile: *Religious Art in France, XIII Century.* Translated from Third Edition. New York: 1913. [I]

Malthus, T. R.: *An Essay on Population.* Two vols. London: 1914. [IV]

Man, Henri de: *Joy in Work.* London: 1929. [VI]
A factual study of the psychological rewards of work, based however upon very limited observation and an insufficient number of cases. Any useful observations on the subject await studies in the fashion of Terpenning's work on the Village. See Le Play.

Manley, Charles M.: *Langley Memoir on Mechanical Flight.* Part II. Washington: 1911. [v]

Mannheim, Karl: *Ideologie und Utopie*. Bonn: 1929.
A very suggestive if difficult work.

Mantoux, Paul: *La Revolution Industrielle du XVIIIe Siècle*. Paris: 1906.
Translated.
Industrial Revolution. First Edition. Paris: 1905. Translated. New York:
1928. [IV]
Deals with the technical and industrial changes in eighteenth century England, and
is perhaps the best single book on the subject that has so far been produced.

Marey, Etienne Jules: *Animal Mechanism; A Treatise on Terrestrial and
Aërial Locomotion*. New York: 1874. [v]
Movement. New York: 1895.
Important physiological studies which were destined to stimulate a renewed interest
in flight. See Pettigrew.

Marot: Helen: *The Creative Impulse in Industry*. New York: 1918. [VIII]
Appraisal of potential educational values in modern industrial organizations. Still
full of pertinent criticism and suggestion.

Martin, T. C., and Dyer, F. L.: *Edison: His Life and Inventions*. New York:
1910. [v]

Marx, Karl, and Engels, Friedrich: *Manifesto of the Communist Party*. New
York.
Capital. Translated by Eden and Cedar Paul. Two vols. London: 1930.
A classic work whose historic documentation, sociological insight, and honest human
passion outweigh the defects of its abstract economic analysis. The first adequate
interpretation of modern society in terms of its technics.

Mason, Otis T.: *The Origins of Invention; A Study of Industry Among Primitive Peoples*. New York: 1895. [I, II]
A good book in its time that now cries for a worthy successor.

Mataré, Franz: *Die Arbeitsmittel, Maschine, Apparat, Werkzeug*. Leipzig:
1913. [I, v]
Important. Emphasizes the rôle of the apparatus and the utility and demonstrates the
neotechnic tendencies of the advanced chemical industries as regards scientific organization, the proportionately higher number of technicians, and the increasing
automatism of the work.

Matschoss, Conrad (Editor): *Männer der Technik*. Berlin: 1925.
Series of biographies, criticized by Feldhaus for various omissions and errors.

Matschoss: Conrad: *Die Entwicklung der Dampfmaschine; eine Geschichte
der Ortsfesten Dampfmaschine und der Lokomobile, der Schiffsmaschine
und Lokomotive*. Two vols. Berlin: 1908. [IV]
An exhaustive study of the steam engine. For a shorter account see Thurston.
Technische Kulturdenkmäler. Berlin: 1927.

Mayhew, Charles: *London Labor and the London Poor*. Four vols. London:
1861.

Mayo, Elton: *The Human Problems of an Industrial Civilization*. New York:
1933. [v]
Useful study of the relation of efficiency to rest-periods and interest in work. See
Henri de Man.

McCartney, Eugene S.: *Warfare by Land and Sea*. (Our Debt to Greece and Rome Series.) Boston: 1923. [II]

McCurdy, Edward: *Leonardo da Vinci's Notebooks*. New York: 1923. [I, III]
The Mind of Leonardo da Vinci. New York: 1928. [I, III]

Meisner, Erich: *Weltanschauung Eines Technikers*. Berlin: 1927.

Meyer, Alfred Gotthold: *Eisenbauten—Ihre Geschichte und Esthetik*. Esslingen a.N.: 1907. [IV, V, VII]
Very important: an able critical and historical work.

Middle West Utilities Company: *America's New Frontier*. Chicago: 1929. [V]
Despite its origin, a very useful study of the relation of electricity to industrial and urban decentralization.

Milham, Willis I.: *Time and Time-Keepers*. New York: 1923. [I, III, IV]

Moholy-Nagy, L.: *The New Vision* (translated by Daphne Hoffman). New York. (Undated.) [VII]
Malerei Fotografie Film. München: 1927. [VII]
While it does not live up to the promise of its early chapters, *The New Vision* is still one of the best presentations of modern experiments in form initiated at the Bauhaus in Dessau under Gropius and Moholy-Nagy. Even the failures and blind alleys in these experiments do not lack interest—if only because those who are new to the subject tend to repeat them.

Morgan, C. Lloyd: *Emergent Evolution*. New York: 1923.

Mory, L. V. H., and Redman, L. V.: *The Romance of Research*. Baltimore: 1933.

Mumford, Lewis: *The Story of Utopias*. New York: 1922. [VI, VIII]
Summary of the classic utopias which, while often superficial, sometimes open up a neglected trail.

Neuburger, Albert: *The Technical Arts and Sciences of the Ancients*. New York: 1930.
Voluminous. But see Feldhaus.

Neudeck, G.: *Geschichte der Technik*. Stuttgart: 1923.
Sometimes useful for historical facts. Comprehensive but not first-rate.

Nummenhoff, Ernst: *Der Handwerker in der Deutschen Vergangenheit*. Jena: 1924.
Profusely illustrated.

Nussbaum, Frederick L.: *A History of the Economic Institutions of Modern Europe*. New York: 1933.
A condensation of Sombart.

Obermeyer, Henry: *Stop That Smoke!* New York: 1933. [IV, V]
Popular account of the cost and extent of the paleotechnic smoke-pall which hangs heavy over our manufacturing centers even today.

Ogburn, W. F.: *Living with Machines*. New York: 1933. [IV, V]
Social Change. New York: 1922.

Ortega y Gasset, José: *The Revolt of the Masses*. New York: 1933 [VI]

Ostwald, Wilhelm: *Energetische Grundlagen der Kulturwissenschaften.* Leipzig: 1909.
See Geddes's *The Classification of Statistics,* written a generation earlier.

Ozenfant, Amédée: *Foundations of Modern Art.* New York: 1931. [VII]
Uneven, but sometimes penetrating.

Pacoret, Etienne: *Le Machinisme Universel; Ancien, Moderne et Contemporain.* Paris: 1925.
One of the most useful introductions in French.

Parrish, Wayne William: *An Outline of Technocracy.* New York: 1933.

Pasdermadjian, H.: *L'Organisation Scientifique du Travail.* Geneva: 1932.

Pasquet, D.: *Londres et Les Ouvriers de Londres.* Paris: 1914.

Passmore, J. B., and Spencer, A. J.: *Agricultural Implements and Machinery.* [III, IV]. *A Handbook of the Collections in the Science Museum, London.* London: 1930.
Useful.

Paulhan, Frédéric: *Psychologie de l'Invention.* Paris: 1901.
Wisely deals with mechanical invention, not as a special gift of nature, but as a particular variety of a more general human trait common to all the arts.

Peake, Harold J. E.: *Early Steps in Human Progress.* London: 1933. [I, II]
Good; but see Renard.

Peake, Harold, and Fleure, H. J.: *The Corridors of Time.* Eight vols. Oxford: 1927.

Péligot, Eugène M.: *Le Verre; Son Histoire, sa Fabrication.* Paris: 1877. [III]

Penty, Arthur: *Post-Industrialism.* London: 1922. [VI]
Criticism of modern finance and the machine and prediction of the downfall of the system at a time when this position was far less popular than at present.

Petrie, W. F.: *The Arts and Crafts of Ancient Egypt.* Second Edition. London: 1910. [I, II]
The Revolutions of Civilization. London: 1911. [III]

Pettigrew, J. Bell: *Animal Locomotion; or Walking, Swimming and Flying; with a Dissertation on Aeronautics.* New York: 1874. [V]
Important contribution. See Marey.

Poincaré, Henri: *Science and Method.* London: 1914.
A classic in the philosophy of science.

Polakov, Walter N.: *The Power Age; Its Quest and Challenge.* New York: 1933. [V, VIII]
Excellent presentation of the implication of the new forms of utilizing electric power and organizing modern industry. Unfortunate in its assumption that the use of power is the distinguishing feature of neotechnic industry.

Popp, Josef: *Die Technik Als Kultur Problem.* München: 1929.

Poppe, Johann H. M. von: *Geschichte Aller Erfindungen und Entdeckungen im Bereiche der Gewerbe, Künste und Wissenschaften.* Stuttgart: 1837. [III]
Beckmann's nearest successor: containing some facts that have been dropt by the roadside since.

Porta, Giovanni Battista della: *Natural Magick.* London: 1658. [III]
English translation of a sixteenth century classic.

Porter, George R.: *Progress of the Nation.* Three vols. in one. London: 1836-1843. [IV]
Useful as documentation.

Pound, A.: *Iron Man in Industry.* Boston: 1922. [V]
Discussion of automatism in industry and the need to compensate for it.

Pupin, Michael J.: *Romance of the Machine.* New York: 1930.
Trivial.

Rathenau, Walter: *The New Society.* New York: 1921. [V, VIII]
In Days to Come. London: 1921. [VIII]
Die Neue Wirtschaft. Berlin: 1919. [VIII]
Aware of the dangers of iron-bound mechanization, Rathenau, though sometimes a little shrill and almost hysterical, wrote a series of sound criticisms of the existing order; and in *In Days to Come* and *The New Society* he outlined a new industrial society. He differed from many social democrats and communists in recognizing the critical importance of the moral and educational problems involved in the new orientation.

Read, T. T.: *Our Mineral Civilization.* New York: 1932.

Recent Social Trends in the United States. Two vols. New York: 1933.

Recent Economic Changes in the United States. Two vols. New York: 1929. [IV, V]
An inquiry, still useful for its data, which would have been even more important had its facts been mustered in such a fashion as to point more clearly to its properly dubious and pessimistic conclusion.

Recueil de Planches, sur les Science, les Art Liberaux, et les Art Mechanique. (Supplement to Diderot's *Encyclopedia*). Paris: 1763. [III]
See *Encyclopédie.*

Redman, L. V., and Mory, L. V. H.: *The Romance of Research.* Baltimore: 1933.

Redzich, Constantin: *Das Grosse Buch der Erfindungen und deren Erfinder.* Two vols. Leipzig: 1928.

Renard, George F.: *Guilds in the Middle Ages.* London: 1919. [III]
Life and Work in Primitive Times. New York: 1929. [II]
Penetrating and suggestive study of a subject whose scant materials require an active yet prudent imagination.

Renard, George F., and Dulac, A.: *L'Evolution Industrielle et Agricole depuis Cent Cinquante Ans.* Paris: 1912. [IV, V]
A standard work.

Renard, George F., and Weulersse, G.: *Life and Work in Modern Europe; Fifteenth to Eighteenth Centuries.* London: 1926. [III]
Excellent.

Reuleaux, Franz: *The Kinematics of Machinery; Outlines of a Theory of Machines.* London: 1876.
The most important systematic morphology of machines: a book so good that it has discouraged rivals.

Richards, Charles R.: *The Industrial Museum.* New York: 1925.
Critical survey of existing types of industrial museum.

Rickard, Thomas A.: *Man and Metals; A History of Mining in Relation to the Development of Civilization.* Two vols. New York: 1932. [II-V]
Compendious and fairly exhaustive.

Riedler, A.: *Das Maschinen-Zeichnen.* Second Edition. Berlin: 1913.
An influential treatise in Germany.

Robertson, J. Drummond: *The Evolution of Clockwork; with a Special Section on the Clocks of Japan.* London: 1931.
Recent data on a subject whose early history has many pitfalls. See Usher.

Roe, Joseph W.: *English and American Tool Builders.* New Haven: 1916. [IV]
Valuable. See Smiles.

Rossman, Joseph: *The Psychology of the Inventor.* New York: 1932.

Routledge, Robert: *Discoveries and Inventions of the Nineteenth Century.* London: 1899. [IV]

Rugg, Harold O.: *The Great Technology; Social Chaos and the Public Mind.* New York: 1933. [V, VIII]
Concerned with the educational problem of realizing the values of modern industry and of controlling the machine.

Russell, George W.: *The National Being.* New York: 1916.

Salter, Arthur: *Modern Mechanization.* New York: 1933.

Sarton, George: *Introduction to the History of Science.* Three Vols. Baltimore: 1927-1931. [I]
The life-work of a devoted scholar.

Sayce, R. U.: *Primitive Arts and Crafts; An Introduction to the Study of Material Culture.* New York: 1933. [II]
Suggestive.

Schmidt, Robert: *Das Glas.* Berlin: 1922. [III]

Schmitthenner, Paul: *Krieg und Kriegführung im Wandel der Weltgeschichte.* Potsdam: 1930. [II, III, IV]
Well illustrated with an excellent bibliography.

Schneider, Hermann: *The History of World Civilization from Prehistoric Times to the Middle Ages.* Volume I. New York: 1931.

Schregardus, J., Visser, Door C., and Ten Bruggencate, A.: *Onze Hollandsche Molen.* Amsterdam: 1926.
Well illustrated.

Schulz, Hans: *Die Geschichte der Glaserzeugung.* Leipzig: 1928. [III]
Das Glas. München: 1923. [III]

Schumacher, Fritz: *Schöpferwille und Mechanisierung.* Hamburg: 1933.
Der Fluch der Technik. Hamburg: 1932.
Says more in a few pages than many more pretentious treatises succeed in doing in a tome. Schumacher's humane and rational mind compares with Spengler's as his admirable schools and communities in Hamburg compare with the decayed esthetic obscurantism of the Böttcherstrasse in Bremen. It is important to recognize that both strains are characteristic of German thought, although at the moment that represented by Schumacher is in eclipse.

Schuyler, Hamilton: *The Roeblings; A Century of Engineers, Bridge-Builders and Industrialists.* Princeton: 1931. [IV]
More important for its subject than for what the author has added to it.

Schwarz, Heinrich: *David Octavius Hill; Master of Photography.* New York: 1931. [V, VII]
Good.

Schwarz, Rudolph: *Wegweisung der Technik.* Potsdam. (No date.) [VII]
Some interesting comparisons between the strong north gothic of Lubeck and modern machine-forms. Note also that this holds with the bastides of Southern France.

Science at the Crossroads. Papers presented to the International Congress of the History of Science and Technology by the delegates of the U.S.S.R. London: 1931.
Suggestive, if often teasingly obscure papers, on communism and Marxism and modern science.

Scott, Howard: *Introduction to Technocracy.* New York: 1933.
A book whose political callowness, historical ignorance and factual carelessness did much to discredit the legitimate conclusions of the so-called technocrats.

Soule, George: *A Planned Society.* New York: 1932. [VIII]

Sheard, Charles: *Life-giving Light.* New York: 1933. [V]
One of the better books in the very uneven Century of Progress Series.

Singer, Charles: *From Magic to Science.* New York: 1928. [I]
A Short History of Medicine. New York: 1928.

Slosson, E. E.: *Creative Chemistry.* New York: 1920. [V]

Smiles, Samuel: *Industrial Biography; Iron Workers and Toolmakers.* London: 1863. [IV]
Lives of the Engineers. Four vols. London: 1862-1866. Five vols. London: 1874. New vols. London: 1895. [IV]
Men of Invention and Industry. 1885. [IV]
Smiles, perhaps better known for his complacently Victorian moralizings on self-help and success, was a pioneer in the field of industrial biography; and his studies, which were often close to their sources, are important contributions to the history of tech

nics. His accounts of Maudslay, Bramah and their followers make one wish that people of his particular bent and industry had appeared more often.

Smith, Adam: *An Inquiry into the Nature and Causes of the Wealth of Nations.* Two vols. London: 1776. [III]
A cross-section of the late eotechnic economy, as the division of the process was reducing the worker to a mere cog in the mechanism. See the *Encyclopédie* for pictures.

Smith, Preserved: *A History of Modern Culture.* Vol. I. New York: 1930. [III]
Excellent discussion of every subject but technics.

Soddy, Frederick: *Wealth, Virtual Wealth and Debt.* London: 1926. Second Edition, Revised. New York: 1933. [VIII]
The application of energetics to finance.

Sombart, Werner: *Gewerbewesen.* Two vols. Berlin: 1929.
The Quintessence of Capitalism. New York: 1915.
Krieg und Kapitalismus. München: 1913. [II, III, IV]
Invaluable study of the social, technical and financial relations between war and capitalism, with particular emphasis on the important changes that took place in the sixteenth and seventeenth centuries.
Luxus und Kapitalismus. München: 1913. [II, III]
Penetrating social and economic account of the rôle of the court and the courtesan and the cult of luxury developed during the Renascence.
Der Moderne Kapitalismus. Four vols. München: 1927. [I-V]
A work conceived and carried out on a colossal scale. It parallels the present history of technics, as the Mississippi might be said to parallel the railway train that occasionally approaches its banks. While sometimes Sombart's generalizations seem to me too neat and confident—as in the change from the organic to the inorganic as the increasing mark of modern technics—I have differed from his weighty scholarship only when no other course was open.

Spencer, A. J., and Passmore, J. B.: *Agricultural Implements and Machinery. A Handbook of the Collections in the Science Museum, London.* London: 1930.

Spengler, Oswald: *The Decline of the West.* Two vols. New York: 1928.
While Spengler makes many generalizations about technics this is one department where this sometimes penetrating and original (but crotchety) thinker is particularly unreliable. In typical nineteenth century fashion he dismisses the technical achievements of other cultures and gives a fake air of uniqueness to the *early* Faustian inventions, which borrowed heavily from the more advanced Arabs and Chinese. Partly his errors derive from his theory of the absolute isolation of cultures: a counterpart curiously to the unconscious imperialism of the British theory of absolute diffusion from a single source.
Man and Technics. New York: 1932.
A book heavily burdened by a rancid mysticism, tracing back to the weaker sides of Wagner and Nietzsche.

Stenger, Erich: *Geschichte der Photographie.* Berlin: 1929. [V]
Useful summary.

Stevers, Martin: *Steel Trails; The Epic of the Railroads.* New York: 1933. [IV]
Popular, but not without technical interest.

Strada, Jacobus de: *Kunstlicher Abriss Allerhand Wasser, Wind, Ross und Handmühlen*. Frankfurt: 1617. [III]

Survey Graphic: Regional Planning Number. May, 1925. [v]
Predicted the breakdown of the present metropolitan economy and sketched outlines of a neotechnic regionalism.

Sutherland, George: *Twentieth Century Inventions; A Forecast*. New York: 1901.

Taussig, F. E.: *Inventors and Moneymakers*. New York: 1915.
Over-rated.

Tawney, R. H.: *Equality*. New York: 1931.
Religion and the Rise of Capitalism. New York: 1927. [I]
The Acquisitive Society. New York: 1920.
The work of an able economist and a humane mind.

Taylor, Frederic W.: *The Principles of Scientific Management*. New York: 1911. [v]
One of those classics whose reputation is incomprehensible without a direct acquaintance with the personality behind it.

Taylor Society (Person, H. S., Editor): *Scientific Management in American Industry*. New York: 1929. [v]
Survey of more recent applications of Taylor's and Gantt's principles.

Thompson, Holland: *The Age of Invention*. New Haven: 1921. [IV, v]
The story of technics in America. Readable but not exhaustive. See Kaemffert.

Thomson, J. A., and Geddes, Patrick: *Life; Outlines of General Biology*. New York: 1931.
Biology. New York: 1925.
See Geddes.

Thorndike, Lynn: *A History of Magic and Experimental Science During the First Thirteen Centuries of Our Era*. Two vols. New York: 1923. [I, III]
Science and Thought in the Fifteenth Century. New York: 1929. [I, III]
Both invaluable.

Thorpe, T. E. (Editor), Green, Miall and others: *Coal; Its History and Uses*. London: 1878. [IV]

Thurston, R. H.: *A History of the Growth of the Steam Engine*. First Edition. 1878. Fourth Edition. 1903. [IV]
Very good.

Tilden, W. A.: *Chemical Discovery and Invention in the Twentieth Century*. London: 1916. [v]

Tilgher, Adriano: *Work; What Is Has Meant to Men Through the Middle Ages*. New York: 1930.
A disappointing work.

Tomlinson's Encyclopedia of the Useful Arts. Two vols. London: 1854.

Traill, Henry D.: *Social England*. Six vols. London: 1909.
Well-illustrated background.

Tryon, F. G., and Eckel, E. C.: *Mineral Economics.* New York: 1932. [v]
Useful.

Tugwell, Rexford Guy: *Industry's Coming of Age.* New York: 1927.
A little glib and over-sanguine about the prospects of a transformation of industry under existing leadership.

Unwin, George: *Industrial Organization in the Sixteenth and Seventeenth Centuries.* Oxford: 1904.

Updike, D. B.: *Printing Types; Their History, Forms and Use.* Two vols. Cambridge: 1922. [III]
Important.

Ure, Andrew: *The Philosophy of Manufactures; or An Exposition of the Scientific, Moral and Commercial Economy of the Factory System of Great Britain.* First Edition. London: 1835. [IV]. Third Edition. London: 1861.
Perhaps the chief example of paleotechnic apologetics in which the author unconsciously hangs himself by his own rope.
Dictionary of Arts, Manufactures and Mines. Seventh Edition. Edited by Robert Hunt and F. W. Hudler. London: 1875.

Usher, Abbott Payson: *A History of Mechanical Inventions.* New York: 1929. [I-v]
See Introduction.

Van Loon, Hendrick: *Man the Miracle Maker.* New York: 1928.
The Fall of the Dutch Republic. New York: 1913. [III]
Some useful data on trade and transportation in Holland.

Veblen, Thorstein: *The Instinct of Workmanship and the State of the Industrial Arts.* New York: 1914.
Imperial Germany and the Industrial Revolution. New York: 1915.
The Theory of Business Enterprise. New York: 1905.
The Theory of the Leisure Class. New York: 1899.
The Place of Science in Modern Civilization. New York: 1919.
The Engineers and the Price System. New York: 1921. [v, VIII]
An Inquiry into the Nature of Peace and the Terms of Its Perpetuation. New York: 1917.
After Marx, Veblen shares with Sombart the distinction of being perhaps the foremost sociological economist. His various works, taken together, form a unique contribution to the theory of modern technics. Perhaps the most important from the standpoint of technics are *The Theory of Business Enterprise* and *Imperial Germany and the Industrial Revolution:* but there are valuable sections in *The Theory of the Leisure Class* and in *The Instinct of Workmanship.* While a believer in rationalized industry, Veblen did not regard adaptation as the passive adjustment of an organism to an inflexible physical and mechanical environment.

Vegetius, Renatus Flavius: *Military Institutions.* London: 1767. [II]
Eighteenth century translation of a fifteenth century classic.

Verantius, Faustus: *Machinae Novae.* Venice: 1595. [III]

Vierendeel, A.: *Esquisse d'une Histoire de la Technique.* Brussels: 1921.

Von Dyck, W.: *Wege und Ziele des Deutschen Museums*. Berlin: 1929.

Voskuil, Walter H.: *Minerals in Modern Industry*. New York: 1930. [v]
The Economics of Water Power Development. New York: 1928. [v]
Good summary.

Vowles, Hugh P., and Margaret W.: *The Quest for Power; from Prehistoric Times to the Present Day*. London: 1931. [I-v]
A valuable study of the various forms of prime-mover.

Warshaw, H. T.: *Representative Industries in the United States*. New York: 1928.

Wasmuth, Ewald: *Kritik des Mechanisierten Weltbildes*. Hellerau: 1929.

Webb, Sidney, and Beatrice: *A History of Trades Unionism*. First Edition. London: 1894.
Industrial Democracy. Two vols. London: 1897.
Classic accounts with special reference to England.

Weber, Max: *General Economic History*. New York: 1927.
The Protestant Ethic and the Spirit of Capitalism. London: 1930. [I]

Weinreich, Hermann: *Bildungswerte der Technik*. Berlin: 1928.
Useful mainly for bibliography.

Wells, David L.: *Recent Economic Changes*. New York: 1886.
Compare with the similar volume of 1929.

Wells, H. G.: *Anticipation of the Reaction of Mechanical and Scientific Progress*. London: 1902.
The Work, Wealth and Happiness of Mankind. Two vols. New York: 1931. [v]

Wendt, Ulrich: *Die Technik als Kulturmacht*. Berlin: 1906.
One of the best historical commentaries on technics.

Westcott, G. F.: *Pumping Machinery. A Handbook of the Science Museum*. London: 1932. [III, IV]

Whitehead, Alfred North: *Science and the Modern World*. New York: 1925.
The Concept of Nature. Cambridge: 1926.
Adventures of Ideas. New York: 1933.

Whitney, Charles S.: *Bridges: A Study in Their Art, Science and Evolution*. New York: 1929.

World Economic Planning; The Necessity for Planned Adjustment of Productive Capacity and Standards of Living. The Hague: 1932. [v, VIII]
Exhaustive introduction to the subject, from almost every possible angle.

Worringer, Wilhelm: *Form in Gothic*. London: 1927.
Interesting, if not always substantiated: has a bearing on form in general.

Zimmer, George F.: *The Engineering of Antiquity*. London: 1913.

Zimmerman, Erich W.: *World Resources and Industries; An Appraisal of Agricultural and Industrial Resources*. New York: 1933. [IV, v]
Very useful; with an adequate bibliography.

Zimmern, Alfred: *The Greek Commonwealth*. Oxford: 1911. [II]
　Nationality and Government. London: 1918. [VI]

Zonca, Vittorio: *Novo Teatro di Machine et Edifici*. Padua: 1607. [III]

Zschimmer, Eberhard: *Philosophie der Technik*. Jena: 1919.

ACKNOWLEDGMENTS

My principal debt, throughout this study, has been to my master, the late Patrick Geddes. His published writings do but faint justice to the magnitude and range and originality of his mind; for he was one of the outstanding thinkers of his generation, not alone in Great Britain, but in the world. From Geddes's earliest papers on *The Classification of Statistics* to his latest chapters in the two volume study of *Life,* written with J. Arthur Thomson, he was steadily interested in technics and economics as elements in that synthesis of thought and that doctrine of life and action for which he laid the foundations. Geddes's unpublished papers are now being collected and edited at the Outlook Tower in Edinburgh. Only second to the profound debt I owe Geddes is that which I must acknowledge to two other men: Victor Branford and Thorstein Veblen. With all three I had the privilege of personal contact; and for those who can no longer have that opportunity I have included in the bibliography a fairly full list of their works, including some which do not bear directly upon the subject in hand.

In the preparation of *Technics and Civilization* I am indebted to the helpful interest and aid of the following men: Mr. Thomas Beer, Dr.-Ing. Walter Curt Behrendt, Mr. M. D. C. Crawford, Dr. Oskar von Miller, Professor R. M. MacIver, Dr. Henry A. Murray, Jr., Professor Charles R. Richards, and Dr. H. W. Van Loon. For the criticism of certain chapters of the manuscript I must give my warm thanks to Mr. J. G. Fletcher, Mr. J. E. Spingarn and Mr. C. L. Weis. For vigilant and searching criticism of the book in one draft or another, by Miss Catherine K. Bauer, Professor Geroid Tanquary Robinson, Mr. James L. Henderson, and Mr. John Tucker, Jr., I am under an obligation that would be almost unbearable were friendship not willing to underwrite it. For aid in gathering historical illustrations I am particularly obliged to Mr. William M. Ivins and his assistants at the Metropolitan Museum of Art. Finally, I must give my cordial thanks to the John Simon Guggenheim Foundation for the partial fellowship in 1932 that enabled me to spend four months in research and meditation in Europe—not less because those fruitful months altered the scope and scale of the entire work. L. M.

INDEX

477